Self-Similar Groups

Mathematical
Surveys
and
Monographs

Volume 117

Self-Similar Groups

Volodymyr Nekrashevych

American Mathematical Society

EDITORIAL COMMITTEE

Jerry L. Bona Peter S. Landweber
Michael G. Eastwood Michael P. Loss

J. T. Stafford, Chair

2000 *Mathematics Subject Classification.* Primary 20F65, 37B10;
Secondary 37F20, 37F15, 20E08, 22A22.

For additional information and updates on this book, visit
www.ams.org/bookpages/surv-117

Library of Congress Cataloging-in-Publication Data

Nekrashevych, Volodymyr, 1975–
    Self-similar groups / Volodymyr Nekrashevych.
        p. cm. — (Mathematical surveys and monographs, ISSN 0076-5376 ; v. 117)
    Includes bibliographical references and index.
    ISBN 0-8218-3831-8 (alk. paper)
    1. Geometric group theory. 2. Symbolic dynamics. 3. Self-similar processes. I. Title. II. Mathematical surveys and monographs ; no. 117.

QA183.N45   2005
512′.2—dc22   2005048021

Copying and reprinting. Individual readers of this publication, and nonprofit libraries acting for them, are permitted to make fair use of the material, such as to copy a chapter for use in teaching or research. Permission is granted to quote brief passages from this publication in reviews, provided the customary acknowledgment of the source is given.

Republication, systematic copying, or multiple reproduction of any material in this publication is permitted only under license from the American Mathematical Society. Requests for such permission should be addressed to the Acquisitions Department, American Mathematical Society, 201 Charles Street, Providence, Rhode Island 02904-2294, USA. Requests can also be made by e-mail to reprint-permission@ams.org.

© 2005 by the American Mathematical Society. All rights reserved.
The American Mathematical Society retains all rights
except those granted to the United States Government.
Printed in the United States of America.

∞ The paper used in this book is acid-free and falls within the guidelines
established to ensure permanence and durability.
Visit the AMS home page at http://www.ams.org/

10 9 8 7 6 5 4 3 2 1   10 09 08 07 06 05

Contents

Preface vii

Chapter 1. Basic Definitions and Examples 1
  1.1. Rooted tree X^* and its boundary X^ω 1
  1.2. Groups acting on rooted trees 2
  1.3. Automata 3
  1.4. Wreath products 9
  1.5. Self-similar actions 10
  1.6. The Grigorchuk group 12
  1.7. The adding machine and self-similar actions of $\mathbb{Z}^n$ 16
  1.8. Branch groups 17
  1.9. Other examples 21
  1.10. Bi-reversible automata and free groups 23

Chapter 2. Algebraic Theory 31
  2.1. Permutational bimodules 31
  2.2. Bases of a covering bimodule and wreath recursions 32
  2.3. Tensor products and self-similar actions 33
  2.4. The left G-space $\mathfrak{M}^{\otimes \omega}$ 36
  2.5. Virtual endomorphisms 37
  2.6. The linear recursion 40
  2.7. Invariant subgroups and the kernel of a self-similar action 41
  2.8. Recurrent actions 44
  2.9. Example: free abelian groups 46
  2.10. Rigidity 50
  2.11. Contracting actions 57
  2.12. Finite-state actions of $\mathbb{Z}^n$ 62
  2.13. Defining relations and word problem 64

Chapter 3. Limit Spaces 71
  3.1. Introduction 71
  3.2. The limit G-space $\mathcal{X}_G$ 73
  3.3. Digit tiles 78
  3.4. Axiomatic description of $\mathcal{X}_G$ 82
  3.5. Connectedness of $\mathcal{X}_G$ 91
  3.6. The limit space $\mathcal{J}_G$ 92
  3.7. Limit spaces of self-similar subgroups 96
  3.8. The limit space $\mathcal{J}_G$ as a hyperbolic boundary 97
  3.9. Groups of bounded automata 102
  3.10. One-dimensional subdivision rules 110

3.11.	Uniqueness of the limit space	113

Chapter 4. Orbispaces — 117
- 4.1. Pseudogroups and étale groupoids — 117
- 4.2. Orbispaces — 119
- 4.3. Open sub-orbispaces and coverings — 122
- 4.4. Coverings and skew-products — 124
- 4.5. Partial self-coverings — 127
- 4.6. The limit orbispace $\mathcal{J}_G$ — 128
- 4.7. Paths in an orbispace — 131

Chapter 5. Iterated Monodromy Groups — 137
- 5.1. Definition of iterated monodromy groups — 137
- 5.2. Standard self-similar actions of $\mathrm{IMG}(p)$ on X^* — 142
- 5.3. Iterated monodromy groups of limit dynamical systems — 146
- 5.4. Length structures and expanding maps — 148
- 5.5. Limit spaces of iterated monodromy groups — 150
- 5.6. Iterated monodromy group of a pull-back — 154
- 5.7. The limit solenoid and inverse limits of self-coverings — 156

Chapter 6. Examples and Applications — 161
- 6.1. Expanding self-coverings of orbifolds — 161
- 6.2. Limit spaces of free Abelian groups — 165
- 6.3. Examples of self-coverings of orbifolds — 169
- 6.4. Rational functions — 174
- 6.5. Combinatorial equivalence and Thurston's Theorem — 176
- 6.6. "Twisted rabbit" question of J. Hubbard — 179
- 6.7. Abstract kneading automata — 185
- 6.8. Topological polynomials and critical portraits — 190
- 6.9. Iterated monodromy groups of complex polynomials — 193
- 6.10. Polynomials from kneading automata — 196
- 6.11. Quadratic polynomials — 203
- 6.12. Examples of iterated monodromy groups of polynomials — 208
- 6.13. Matings — 215

Bibliography — 223

Index — 229

Preface

Self-similar groups (groups generated by automata) appeared in the early eighties as interesting examples. It was discovered that very simple automata generate groups with complicated structure and exotic properties which are hard to find among groups defined by more "classical" methods.

For example, the Grigorchuk group [**Gri80**] can be defined as a group generated by an automaton with five states over an alphabet of two letters. This group is a particularly simple example of an infinite finitely generated torsion group and is the first example of a group whose growth is intermediate between polynomial and exponential. Another interesting example is a group discovered in [**GŻ02a**], which is generated by a three-state automaton over the alphabet of two letters. This group can be defined as the *iterated monodromy group* of the polynomial $z^2 - 1$ (see Chapter 5 of this book). It is the first example of an amenable group (see [**BV**]), which cannot be constructed from groups of sub-exponential growth by the group-theoretical operations preserving amenability.

Many other interesting groups were constructed using self-similar actions and automata. This includes, for instance, groups of finite width, groups of non-uniform exponential growth, new just-infinite groups, etc.

The definition of a self-similar group action is as follows. Let X be a finite alphabet and let X^* denote the set of all finite words over X. A faithful action of a group G on X^* is said to be *self-similar* if for every $g \in G$ and $x \in X$ there exist $h \in G$ and $y \in X$ such that

$$g(xw) = yh(w)$$

for all words $w \in X^*$. Thus, self-similar actions agree with the self-similarity of the set X^* given by the shift map $xw \mapsto w$.

One of the aims of these notes is to show that self-similar groups are not just isolated examples, but that they have close connections with dynamics and fractal geometry.

We will show, for instance, that self-similar groups appear naturally as *iterated monodromy groups* of self-coverings of topological spaces (or *orbispaces*) and encode combinatorial information about the dynamics of such self-coverings. Especially interesting is the case of a post-critically finite rational function $f(z)$. We will see that iterated monodromy groups give a convenient algebraic way of characterizing combinatorial (Thurston) equivalence of rational functions and that the Julia set of f can be reconstructed from its iterated monodromy group.

In the other direction, we will associate a *limit dynamical system* to every *contracting* self-similar action. The limit dynamical system consists of the *limit (orbi)space* $\mathcal{J}_G$ and of a continuous finite-to-one surjective map $\mathsf{s} : \mathcal{J}_G \longrightarrow \mathcal{J}_G$, which becomes a partial self-covering if we endow $\mathcal{J}_G$ with a natural orbispace structure.

Since the main topics of these notes are geometry and dynamics of self-similar groups and algebraic interpretation of self-similarity, we do not go deep into the rich and various algebraic aspects of groups generated by automata such as just-infiniteness, branch groups, computation of spectra, Lie methods, etc. A reader interested in these topics may read the surveys [**BGŠ03, Gri00, BGN03**].

The first chapter, "Basic definitions and examples", serves as an introduction. We define the basic terminology used in the study of self-similar groups: automorphisms of rooted trees, automata and wreath products. We define the notion of a self-similar action by giving several equivalent definitions and conclude with a sequence of examples illustrating different aspects of the subject.

The second chapter, "Algebraic theory", studies self-similarity of groups from the algebraic point of view. We show that self-similarity can be interpreted as a *permutational bimodule*, i.e., a set with two commuting (left and right) actions of the group. The bimodule associated to a self-similar action is defined as the set $\mathfrak{M}$ of transformations $v \mapsto xg(v)$ of the set of words X^*, where $x \in \mathsf{X}$ is a letter and $g \in G$ is an element of the self-similar group. It follows from the definition of a self-similar action that for every $m \in \mathfrak{M}$ and $h \in G$ the compositions $m \cdot h$ and $h \cdot m$ are again elements of $\mathfrak{M}$. We get in this way two commuting (left and right) actions of the self-similar group G on $\mathfrak{M}$. The bimodule $\mathfrak{M}$ is called the *self-similarity bimodule*. The self-similarity bimodules can be abstractly described as bimodules for which the right action is free and has a finite number of orbits. A self-similarity bimodule together with a choice of a *basis* (an orbit transversal) of the right action uniquely determines the self-similar action. Change of a basis of the bimodule changes the action to a conjugate one.

Virtual endomorphisms are another convenient tool used to construct permutational bimodules and hence self-similar actions. A virtual endomorphism ϕ of a group G is a homomorphism from a subgroup of finite index $\mathrm{Dom}\,\phi \leq G$ to G. We show that the set of formal expressions of the form $\phi(g)h$ (with natural identifications) is a permutational bimodule and that one gets a self-similar action in this way. If we start from a self-similar action, then the *associated virtual endomorphism* ϕ is defined on the stabilizer G_x of a letter $x \in \mathsf{X}$ in G by the condition that

$$g(xw) = x\phi(g)(w)$$

for every $w \in \mathsf{X}^*$ and $g \in \mathrm{Dom}\,\phi = G_x$.

For example, the *adding machine* action, i.e., the natural action of $\mathbb{Z}$ on the ring of diadic integers $\mathbb{Z}_2 \geq \mathbb{Z}$, where $\mathbb{Z}_2$ is encoded in the usual way by infinite binary sequences, is the self-similar action defined by the virtual endomorphism $\phi : n \mapsto n/2$. In this sense self-similar actions may be viewed as generalizations of *numeration systems*. In Section 2.9 of Chapter 2, we apply the developed technique to describe self-similar actions of the free abelian groups $\mathbb{Z}^n$, making the relation between self-similar actions and numeration systems more explicit.

Section 2.11 introduces the main class of self-similar actions for these notes. It is the class of the so-called *contracting actions*. An action is called contracting if the associated virtual endomorphism ϕ asymptotically shortens the length of the elements of the group. Contraction of a self-similar action corresponds to the condition of expansion of a dynamical system. We show in the next chapters that if a self-covering of a Riemannian manifold (or orbifold) is expanding, then its iterated monodromy group is contracting with respect to a *standard* self-similar action.

The *limit spaces* and the *limit dynamical systems* of contracting self-similar actions are constructed and studied in Chapter 3. If $\mathfrak{M}$ is the permutational bimodule associated to a self-similar action of a group G, then its tensor power $\mathfrak{M}^{\otimes n}$ is defined in a natural way. It describes the action of G on the set of words of length n and is interpreted as the nth iteration of the self-similarity of the group. Passing to the (appropriately defined) limits as n goes to infinity, we get the left G-module (G-space) $\mathfrak{M}^{\otimes \omega} = \mathfrak{M} \otimes \mathfrak{M} \otimes \ldots$ and the right G-module $\mathfrak{M}^{\otimes -\omega} = \ldots \otimes \mathfrak{M} \otimes \mathfrak{M}$. The left G-space $\mathfrak{M}^{\otimes \omega}$ is naturally interpreted as the action of G on the space of infinite words $\mathsf{X}^\omega = \{x_1 x_2 \ldots \ : \ x_i \in \mathsf{X}\}$.

The right G-space $\mathcal{X}_G = \mathfrak{M}^{\otimes -\omega}$ (if the action is contracting) is a finite-dimensional metrizable locally compact topological space with a proper co-compact right action of G on it. The limit space $\mathcal{X}_G$ can also be described axiomatically as the unique proper co-compact G-space with a *contracting self-similarity* (Theorem 3.4.13). A right G-space $\mathcal{X}$ is called *self-similar* if the actions $(\mathcal{X}, G)$ and $(\mathcal{X} \otimes_G \mathfrak{M}, G)$ are topologically conjugate. For the notion of a contracting self-similarity see Definition 3.4.11.

Another construction is the quotient (orbispace) $\mathcal{J}_G$ of $\mathcal{X}_G$ by the action of G (Section 3.6). The limit space $\mathcal{J}_G$ can be alternatively defined as the quotient of the space of the left-infinite sequences $\mathsf{X}^{-\omega} = \{\ldots x_2 x_1 \ : \ x_i \in \mathsf{X}\}$ by the equivalence relation, which identifies two sequences $\ldots x_2 x_1$ and $\ldots y_2 y_1$ if there exists a bounded sequence $g_k \in G$ such that $g_k (x_k \ldots x_1) = y_k \ldots y_1$ for all k. Here a sequence is called bounded if it takes a finite set of values. One can prove that this equivalence is described by a finite graph labeled by pairs of letters and that equivalence classes are finite. This gives us a nice symbolic presentation of the space $\mathcal{J}_G$.

The limit space $\mathcal{J}_G$ comes together with a natural *shift map* $\mathsf{s} : \mathcal{J}_G \longrightarrow \mathcal{J}_G$ and with a Markov partition of the dynamical system $(\mathcal{J}_G, \mathsf{s})$. The shift is induced by the usual shift $\ldots x_2 x_1 \mapsto \ldots x_3 x_2$, and the elements of the Markov partition are the images of the cylindrical sets of the described symbolic presentation of $\mathcal{J}_G$. The elements of the Markov partition are called *(digit) tiles*. Digit tiles can also be defined for the limit G-space $\mathcal{X}_G$, and they are convenient tools for the study of the topology of $\mathcal{X}_G$.

The most well-studied contracting groups are the self-similar groups generated by *bounded automata*. They can be defined as the groups whose digit tiles have finite boundary. We show that this condition is equivalent to a condition studied by S. Sidki in [**Sid00**] and show an iterative algorithm which constructs approximations of the limit spaces $\mathcal{J}_G$ of such groups. Groups generated by bounded automata are defined and studied in Section 3.9; their limit spaces are considered in Section 3.10 and Section 3.11, where we prove that in some cases the limit spaces depend only on the algebraic structure of the group and thus can be used to distinguish the groups up to isomorphisms.

Chapter 4, "Orbispaces", is a technical chapter in which we collect the basic definitions related to the theory of orbispaces. Orbispaces are structures represented locally as quotients of topological spaces by finite homeomorphism groups. They are generalizations of a more classical notion of an *orbifold* introduced by W. Thurston (see [**Thu90**] and [**Sco83**]). A similar notion of a V-manifold was introduced earlier by I. Satake [**Sat56**]. We use in our approach pseudogroups and étale groupoids, following [**BH99**]. Most constructions in this chapter are well known, though we present some new (and we hope natural) definitions, like the definition of an open map between orbispaces and the notion of an open sub-orbispace. We also define the *limit orbispace* $\mathcal{J}_G$ of a contracting self-similar action and show that the shift map $\mathsf{s} : \mathcal{J}_G \longrightarrow \mathcal{J}_G$ is a covering of the limit orbispace by an open sub-orbispace (is a *partial self-covering*).

The orbispace structure on $\mathcal{J}_G$ comes from the fact that the limit space $\mathcal{J}_G$ is the quotient of the limit space $\mathcal{X}_G = \mathfrak{M}^{\otimes -\omega}$ by the action of the group G. Introduction of this additional structure on $\mathcal{J}_G$ makes it possible to reconstruct the group G itself from the partial self-covering s of $\mathcal{J}_G$ as the iterated monodromy group IMG (s) (see Theorem 5.3.1). Hence, if we want to be able to go back and forth between self-similar groups and dynamical systems, then we need to define iterated monodromy groups in the general setting of orbispace mappings.

One cannot avoid using orbispaces even in more classical situations like iterations of rational functions. W. Thurston associated with every post-critically finite rational function its *canonical orbispace*, playing an important role in the study of dynamics (see [**DH93, Mil99**]).

Chapter 5 defines and studies iterated monodromy groups. If $p : \mathcal{M}_1 \longrightarrow \mathcal{M}$ is a covering of a topological space (or an orbispace) $\mathcal{M}$ by an open subset (an open sub-orbispace) $\mathcal{M}_1$, then the fundamental group $\pi_1(\mathcal{M}, t)$ acts naturally by the monodromy action on the set of preimages $p^{-n}(t)$ of the basepoint under the nth iteration of p. Let us denote by K_n the kernel of the action. Then the *iterated monodromy group* of p (denoted IMG (p)) is the quotient $\pi_1(\mathcal{M}, t) \big/ \bigcap_{n \geq 0} K_n$.

The disjoint union $T = \bigsqcup_{n \geq 0} p^{-n}(t)$ of the sets of preimages has a natural structure of a rooted tree. It is the tree with the root t, where a vertex $z \in p^{-n}(t)$ is connected by an edge with the vertex $p(z) \in p^{-(n-1)}(t)$. The iterated monodromy group acts faithfully on this tree in a natural way.

We define a special class of isomorphisms of the tree of preimages T with the tree of words X^* using preimages of paths in $\mathcal{M}$. After conjugation of the natural action of IMG (p) on T by such an isomorphism, we get a *standard* faithful self-similar action of IMG (p) on X^*. The standard action depends on a choice of paths connecting the basepoint to its preimages, but a different choice of paths corresponds to a different choice of a basis of the associated self-similarity bimodule. In particular, two different standard actions of IMG (p) are conjugate, and if the actions are contracting, then the limit spaces $\mathcal{X}_{\mathrm{IMG}(p)}$ and $\mathcal{J}_{\mathrm{IMG}(p)}$ (and the limit dynamical system) depend only on the partial self-covering p.

The main result of the chapter is Theorem 5.5.3, which shows that the limit space $\mathcal{J}_{\mathrm{IMG}(p)}$ of the iterated monodromy group of an expanding partial self-covering $p : \mathcal{M}_1 \longrightarrow \mathcal{M}$ is homeomorphic to the Julia set of p (to the attractor of the backward orbits) and, moreover, that the limit dynamical system $\mathsf{s} : \mathcal{J}_{\mathrm{IMG}(p)} \longrightarrow \mathcal{J}_{\mathrm{IMG}(p)}$ is topologically conjugate to the restriction of p onto the Julia set. The respective orbispace structures of the Julia set and the limit space also agree.

The last chapter shows different examples of iterated monodromy groups and their applications. We start with the case when a self-covering $p : \mathcal{M} \longrightarrow \mathcal{M}$ is defined on the whole (orbi)space $\mathcal{M}$. The case when $\mathcal{M}$ is a Riemannian manifold and p is expanding was studied by M. Shub, J. Franks and M. Gromov. They showed that $\mathcal{M}$ is in this case an *infra-nil* manifold and that p is induced by an expanding automorphism of a nilpotent Lie group (the universal cover of $\mathcal{M}$). We show how results of M. Shub and J. Franks follow from Theorem 5.5.3, also proving them in a slightly more general setting. A particular case, when $\mathcal{M}$ is a torus $\mathbb{R}^n/\mathbb{Z}^n$, corresponds to numeration systems on $\mathbb{R}^n$ and is related to self-affine *digit tilings* of the Euclidean space, which were studied by many mathematicians.

Another interesting class of examples are the iterated monodromy groups of post-critically finite rational functions. A rational function $f(z) \in \mathbb{C}(z)$ is called *post-critically finite* if the orbit of every critical point under the iterations of f is finite. If P is the union of the orbits of the critical points, then f is a partial self-covering of the punctured sphere $\widehat{\mathbb{C}} \setminus P$. Then the iterated monodromy group of f is, by definition, the iterated monodromy group of this partial self-covering.

The closure of the iterated monodromy group of a rational function f in the automorphism group of the rooted tree is isomorphic to the Galois group of an extension of the field of functions $\mathbb{C}(t)$. This is the extension obtained by adjoining the solutions of the equation $f^{\circ n}(x) = t$ to $\mathbb{C}(t)$ for all n. These Galois groups were considered by Richard Pink, who was the first to define the profinite iterated monodromy groups.

Every post-critically finite rational function is an expanding self-covering of the associated *Thurston orbifold* by an open sub-orbifold, so Theorem 5.5.3 can be applied, and we get a symbolic presentation of the action of the rational function on the Julia set.

Iterated monodromy groups are rather exotic from the point of view of group theory. The only known finitely presented examples are the iterated monodromy groups of functions with "smooth" Julia sets: z^d, Chebyshev polynomials and Lattè examples. Some iterated monodromy groups of rational functions are groups of intermediate growth (for instance IMG $(z^2 + i)$), while some are essentially new examples of amenable groups (like IMG $(z^2 - 1)$).

Chapter 6 concludes with a complete description of automata generating iterated monodromy groups of polynomials and with an example showing how iterated monodromy groups can be used to construct and to understand plane-filling curves originating from matings of polynomials.

Acknowledgments. I was introduced to the fascinating subject of groups generated by automata by Rostislav Grigorchuk and Vitalij Sushchansky, to whom I am very grateful for their tremendous support of my research.

I also want to use this opportunity to thank Laurent Bartholdi, Yevgen Bondarenko, Anna Erschler, Yaroslav Lavreniuk, Kevin Pilgrim, Richard Pink, Dierk Schleicher, Said Sidki and the referees for helpful suggestions and collaboration.

A great part of this work was done during visits to Geneva University, sponsored by the Swiss National Science Foundation, during the stay in Heinrich Heine University of Düsseldorf as a fellow of the Alexander von Humboldt Foundation and during the stay at International University Bremen. I gratefully acknowledge the support of these foundations and institutions and want especially to thank Pierre de la Harpe for invitations and hospitality during my visits to Geneva and Fritz Grunewald for hosting my visit to Düsseldorf.

CHAPTER 1

Basic Definitions and Examples

1.1. Rooted tree X^* and its boundary X^ω

We recall here the basic notions and facts about rooted trees and their automorphism groups. For more details see the papers [**GNS00, BGŠ03, Sid98**].

Let X be a finite set, which we call *alphabet*. By X^* we denote the set

$$\{x_1 x_2 \ldots x_n \; : \; x_i \in X\}$$

of all finite words over the alphabet X, including the empty word $\varnothing$. In other terms, X^* is the free monoid generated by X. The length of a word $v = x_1 x_2 \ldots x_n$ (the number of letters in it) is denoted $|v|$.

The set X^* is naturally a vertex set of a rooted tree in which two words are connected by an edge if and only if they are of the form v and vx, where $v \in X^*, x \in X$. The empty word $\varnothing$ is the root of the tree X^*. See Figure 1.1 for the case $X = \{0, 1\}$.

The set $X^n \subset X^*$ is called the *nth level* of the tree X^*. A map $f : X^* \longrightarrow X^*$ is an *endomorphism* of the tree X^* if it preserves the root and adjacency of the vertices; i.e., if for any two adjacent vertices $v, vx \in X^*$ the vertices $f(v)$ and $f(vx)$ are also adjacent, so that there exist $u \in X^*$ and $y \in X$ such that $f(v) = u$ and $f(vx) = uy$. It is easy to prove by induction on n that if f is an endomorphism, then $f(X^n) \subseteq X^n$. An *automorphism* is a bijective endomorphism.

An interesting object is the *boundary* of the tree X^*. The boundary of a tree is the set of the *ends*, i.e., the infinite simple paths starting at some fixed vertex (e.g. at the root). The boundary of the tree X^* is naturally identified with the set X^ω of all infinite sequences (words) $x_1 x_2 \ldots$, where $x_i \in X$. Here a sequence $x_1 x_2 \ldots \in X^\omega$

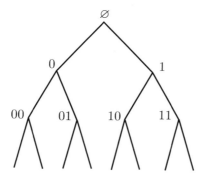

FIGURE 1.1. Binary tree

is identified with the end
$$\varnothing, \quad x_1, \quad x_1 x_2, \quad x_1 x_2 x_3, \ldots$$
of the tree X^*.

The set X^ω is a countable direct power $\mathsf{X}^\mathbb{N}$ of the set X. We introduce the topology of the direct product of the discrete sets X on X^ω or, in other words, the topology of the coordinate-wise convergence. The space X^ω is totally disconnected, metrizable, compact and without isolated points; thus it is homeomorphic to the Cantor set.

The disjoint union $\mathsf{X}^* \sqcup \mathsf{X}^\omega$ also has a natural topology, in which the sequence $x_1, x_1 x_2, x_1 x_2 x_3, \ldots$ converges to the infinite word $x_1 x_2 x_3 \ldots$. It is the topology defined by the basis of open sets
$$\{v\mathsf{X}^* \sqcup v\mathsf{X}^\omega \ : \ v \in \mathsf{X}^*\},$$
where $v\mathsf{X}^*$ and $v\mathsf{X}^\omega$ are the sets of (resp. finite and infinite) words starting with the word v.

The space $\mathsf{X}^* \sqcup \mathsf{X}^\omega$ is compact and X^* is a countable dense subset of isolated points. This space is the natural compactification of the tree X^* by its boundary X^ω (see [**Gro87**] for this compactification in a more general setting of hyperbolic spaces).

Every endomorphism $f : \mathsf{X}^* \longrightarrow \mathsf{X}^*$ can be extended uniquely to a continuous map $f : \mathsf{X}^* \sqcup \mathsf{X}^\omega \longrightarrow \mathsf{X}^* \sqcup \mathsf{X}^\omega$. It is also easy to see that $f : \mathsf{X}^* \longrightarrow \mathsf{X}^*$ is uniquely determined by the induced map on X^ω, since $f(v)$ is the beginning of length $|v|$ of $f(vx_1 x_2 \ldots)$ for any infinite word $x_1 x_2 \ldots$.

1.2. Groups acting on rooted trees

We are using left actions in most cases. So, the image of a point x under the action of an element g of a group is denoted $g(x)$, and in the product $g_1 g_2$ the element g_2 acts first.

Let us denote by $\operatorname{Aut}\mathsf{X}^*$ the group of all automorphisms of the rooted tree X^*.

DEFINITION 1.2.1. An action of a group G by automorphisms of the tree X^* is said to be *level-transitive* if it is transitive on every level X^n of the tree X^*.

An action is level-transitive if and only if the induced action on the boundary X^ω is *minimal* (an action is said to be minimal if all its orbits are dense).

We have the following standard subgroups of a group acting on a rooted tree.

DEFINITION 1.2.2. Let $G \le \operatorname{Aut}\mathsf{X}^*$ be an automorphism group of the rooted tree X^*.

(1) The *vertex stabilizer* is the subgroup $G_v = \{g \in G \ : \ g(v) = v\}$, where $v \in \mathsf{X}^*$ is a vertex.
(2) The *nth level stabilizer* is the subgroup $\operatorname{St}_G(n) = \bigcap_{v \in \mathsf{X}^n} G_v$.
(3) The *rigid stabilizer* of a vertex $v \in \mathsf{X}^*$ is the group $G[v]$ of all automorphisms acting non-trivially only on the vertices of the form $vu, u \in \mathsf{X}^*$; i.e., $G[v] = \{g \in G \ : \ g(w) = w \text{ for all } w \notin v\mathsf{X}^*\}$.
(4) The *nth level rigid stabilizer* $\operatorname{RiSt}_G(n)$ is the subgroup $\langle G[v] \ : \ v \in \mathsf{X}^n \rangle$ generated by the union of the rigid stabilizers of the vertices of the nth level.

We will write just $\mathsf{St}(n)$ or $\mathsf{RiSt}(n)$ if it is clear which group G is under consideration. We have the following easy properties of these subgroups.

PROPOSITION 1.2.3. *Let G be a level-transitive automorphism group of the rooted tree X^*. Then*

(1) *The vertex stabilizer G_v for $v \in \mathsf{X}^n$ is a subgroup of index $|\mathsf{X}|^n$ in G.*
(2) *For every $v \in \mathsf{X}^*$ and $g \in G$ the equalities $g \cdot G_v \cdot g^{-1} = G_{g(v)}$ and $g \cdot G[v] \cdot g^{-1} = G[g(v)]$ take place.*
(3) *The level stabilizers $\mathsf{St}_G(n)$ are normal finite index subgroups of G and $\bigcap_{n \geq 1} \mathsf{St}_G(n) = \{1\}$.*
(4) *If a word v is a beginning of a word $u \in \mathsf{X}^*$, then $G_u \leq G_v$ and $G[u] \leq G[v]$.*
(5) *If words $v, u \in \mathsf{X}^*$ are such that neither word is a beginning of the other, then $G[v] \cap G[u] = [G[v], G[u]] = \{1\}$.*
(6) *The level rigid stabilizer $\mathsf{RiSt}_G(n)$ is a normal subgroup, which is equal to the direct product $\prod_{v \in \mathsf{X}^n} G[v]$.*

It follows that for a level-transitive group $G \leq \operatorname{Aut} \mathsf{X}^*$ only one of the following two cases is possible.
  a) All but a finite number of the level rigid stabilizers $\mathsf{RiSt}_G(n)$ are trivial.
  b) All the rigid stabilizers $G[v]$ and $\mathsf{RiSt}_G(n)$ are infinite.

DEFINITION 1.2.4. Let $G \leq \operatorname{Aut} \mathsf{X}^*$ be a level-transitive group. If all the rigid stabilizers $\mathsf{RiSt}_G(n)$ are infinite (equivalently, non-trivial), then we say that G is *weakly branch*.

The group G is said to be *branch* if $\mathsf{RiSt}(n)$ has finite index in G for every n.

We will discuss branch groups in Section 1.8 in more detail. We have the following classical fact.

PROPOSITION 1.2.5. *The equality $\mathsf{St}_{\operatorname{Aut} \mathsf{X}^*}(n) = \mathsf{RiSt}_{\operatorname{Aut} \mathsf{X}^*}(n)$ takes place. The subgroups $\mathsf{St}_{\operatorname{Aut} \mathsf{X}^*}(n)$ form a fundamental system of neighborhoods of the identity of a profinite topology on $\operatorname{Aut} \mathsf{X}^*$ coinciding with the topology of the pointwise convergence on X^*.* □

1.3. Automata

1.3.1. Restrictions. Our main objects of investigation are groups acting on the rooted tree X^*. We need some nice way to define automorphisms of rooted trees and to be able to perform computations with them. Different languages were developed for this purpose: automata, wreath products and *tableaux* (due to Leo Kaloujnine).

Let $g : \mathsf{X}^* \longrightarrow \mathsf{X}^*$ be an endomorphism of the rooted tree X^*. Consider a vertex $v \in \mathsf{X}^*$ and the subtrees $v\mathsf{X}^*$ and $g(v)\mathsf{X}^*$. Here $v\mathsf{X}^*$ is the subtree with the root v and with the set of vertices equal to the set of words starting with v. Then we get a map $g : v\mathsf{X}^* \longrightarrow g(v)\mathsf{X}^*$, which is a morphism of the rooted trees (see Figure 1.2).

The subtree $v\mathsf{X}^*$ is naturally isomorphic to the whole tree X^*. The isomorphism is the map $v\mathsf{X}^* \longrightarrow \mathsf{X}^* : vw \mapsto w$. The same is true for $g(v)\mathsf{X}^*$. Identifying $v\mathsf{X}^*$ and $g(v)\mathsf{X}^*$ with X^* we get an endomorphism $g|_v : \mathsf{X}^* \longrightarrow \mathsf{X}^*$. It is uniquely determined by the condition

(1.1) $$g(vw) = g(v)g|_v(w).$$

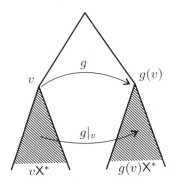

FIGURE 1.2. Restriction

We call the endomorphism $g|_v$ the *restriction* of g in v. We have the following obvious properties of the restrictions:

(1.2) $$g|_{v_1 v_2} = g|_{v_1}|_{v_2},$$
(1.3) $$(g_1 \cdot g_2)|_v = g_1|_{g_2(v)} \cdot g_2|_v.$$

1.3.2. Portraits of automorphisms. Let g be an automorphism of the rooted tree X^*. Then its *portrait* is the tree X^* in which every vertex $v \in X^*$ is labeled by the permutation $\alpha_v \in \mathfrak{S}(X)$ equal to the action of $g|_v$ on X. Here and in the sequel, $\mathfrak{S}(X)$ denotes the symmetric group of permutations of the set X.

For example, if $|X| = 2$, then we just have to distinguish the *active* vertices, i.e., the vertices for which $g|_v$ is non-trivial.

The portrait determines the automorphism g uniquely, since

$$g(x_1 x_2 \ldots x_n) = g(x_1) g|_{x_1}(x_2) g|_{x_1 x_2}(x_3) \ldots g|_{x_1 \ldots x_{n-1}}(x_n).$$

1.3.3. Definition of automata. Let $Q(g) = \{g|_v : v \in X^*\}$ be the set of restrictions of an endomorphism g of the tree X^*. Then $Q(g)$ can be interpreted as a set of internal states of an automaton, which being in a state $g|_v$ and reading on the input tape a letter x, types on the output tape the letter $g|_v(x)$ and goes to the state $g|_v|_x = g|_{vx}$.

Let us define this sort of automata formally. For a general theory of automata see [**Eil74**]. For more facts on (groups of) automatic transformations, see [**GNS00, Sid98, Sus98, Sus99**].

DEFINITION 1.3.1. An *automaton* A over the alphabet X is given by
(1) the *set of the states*, usually also denoted by A;
(2) a map $\tau : \mathsf{A} \times \mathsf{X} \longrightarrow \mathsf{X} \times \mathsf{A}$.

If $\tau(q, x) = (y, p)$, then y and p as functions of (q, x) are called the *output* and the *transition functions*, respectively.

An automaton is said to be finite if it has a finite set of states.

If we want to emphasize that A is an automaton over the alphabet X, then we denote it (A, X).

We introduce the following notation. If $\tau(q, x) = (y, p)$, then we write

(1.4) $$q \cdot x = y \cdot p$$

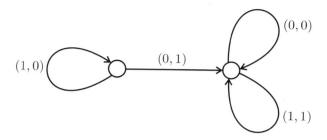

FIGURE 1.3. A Moore diagram

and
$$y = q(x), \qquad p = q|_x.$$
The last two notations agree with the interpretation of the endomorphisms of trees as automata.

It is convenient to define automata using their *Moore diagrams*. It is a directed labeled graph with the vertices identified with the states of the automaton. If $q \cdot x = y \cdot p$, then we have an arrow starting in q, ending in p and labeled by (x, y). So, the arrows show the state transitions, and their labels show the input and the output. See Figure 1.3 for an example.

1.3.4. Automaton (A, X^n). If q is the current state of the automaton A and it gets on the input a finite word $v \in X^*$, then A processes it letter by letter: it reads the first letter x of v, gives the letter $q(x)$ on the output, goes to the state $q|_x$ and is ready to process the word v further. At the end it will give as output some word of the same length as v and will stop at some state of A.

It is hence natural to consider the automaton (A, X) also as an automaton over the alphabet X^n. The structure of the automaton (A, X^n) is defined by the following recurrent rules:

(1.5) $\qquad q|_\varnothing = q \qquad q|_{xv} = q|_x|_v,$

(1.6) $\qquad q(\varnothing) = \varnothing \qquad q(xv) = q(x)q|_x(v).$

Note that rules (1.5), (1.6) are interpreted as associativity if we use notation (1.4):

$q(xv) \cdot q|_{xv} = q \cdot xv = (q \cdot x) \cdot v$
$\qquad = (q(x) \cdot q|_x) \cdot v = q(x) \cdot (q|_x \cdot v) = q(x) \cdot (q|_x(v) \cdot q|_x|_v)$
$\qquad\qquad = q(x)q|_x(v) \cdot q|_x|_v.$

If we identify a word $v \in X^*$ with the transformation $w \mapsto vw$ of the set X^*, then the equality $q \cdot v = q(v) \cdot q|_v$ becomes a correct equality of compositions of transformations of the set X^*.

The image $q(x_1 \ldots x_n)$ of a word $x_1 \ldots x_n \in X^*$ and the state $q|_{x_1 \ldots x_n}$ can be easily computed using the Moore diagram. There exists a unique directed path starting in q with the consecutive arrows labeled by $(x_1, y_1), \ldots, (x_n, y_n)$ for some $y_1, \ldots, y_n \in X$. Then $q(x_1 \ldots x_n) = y_1 \ldots y_n$ and $q|_{x_1 \ldots x_n}$ is the end of the path.

In the same way the action of $q \in A$ on the space X^ω can be defined and computed. For every q and $w = x_1 x_2 \ldots \in X^\omega$ there exists a unique path in the

Moore diagram starting in q and labeled by $(x_1, y_1), (x_2, y_2), \ldots$ for some $y_1 y_2 \ldots \in \mathsf{X}^\omega$. Then $y_1 y_2 \ldots = q(x_1 x_2 \ldots)$. There is no $q|_{x_1 x_2 \ldots}$ for obvious reasons.

As an example consider the automaton with the Moore diagram shown in Figure 1.3. Its right-hand state defines the trivial transformation of the set X^*. The left-hand state a acts on the infinite sequences by the rule

$$a(\underbrace{11\ldots1}_{k \text{ times}} 0 x_1 x_2 \ldots) = \underbrace{00\ldots0}_{k \text{ times}} 1 x_1 x_2 \ldots.$$

This action coincides with the rule of adding 1 to a dyadic integer. The transformation a is called the *adding machine* or the *odometer*. For more details and generalizations see Section 1.7.

Note that $q(x_1 x_2 \ldots)$ is the limit of $q(x_1 \ldots x_n)$ as n goes to infinity, since $y_1 \ldots y_n = q(x_1 \ldots x_n)$ is a beginning of $q(x_1 x_2 \ldots)$.

1.3.5. Composition of automata. In some sense a dual construction is a *composition*, or *multiplication* of automata.

If (A, X) and (B, X) are two automata over an alphabet X, then their *product* is the automaton, denoted $(\mathsf{A} \cdot \mathsf{B}, \mathsf{X})$, whose set of states is the direct product of the sets of states of A and B and whose transition and output functions are given by

(1.7) $\quad\quad\quad\quad (q_1 q_2)(x) = q_1(q_2(x))$

(1.8) $\quad\quad\quad\quad (q_1 q_2)|_x = q_1|_{q_2(x)} q_2|_x,$

where $q_1 \in \mathsf{A}$, $q_2 \in \mathsf{B}$ and the pair $(q_1 q_2)$ is thus a state of $\mathsf{A} \cdot \mathsf{B}$.

These rules may also be interpreted as associativity if we use notation 1.4:

$q_1 q_2(x) \cdot (q_1 q_2)|_x = q_1 q_2 \cdot x$
$= q_1 \cdot (q_2 \cdot x) = q_1 \cdot (q_2(x) \cdot q_2|_x) = (q_1 \cdot q_2(x)) \cdot q_2|_x$
$= \left(q_1 \left(q_2(x) \right) \cdot q_1|_{q_2(x)} \right) \cdot q_2|_x = q_1 \left(q_2(x) \right) \cdot \left(q_1|_{q_2(x)} q_2|_x \right).$

Note that (1.8) coincides with (1.3) and it is easy to prove by induction that the action of the state $q_1 q_2$ on X^* is equal to the composition of the actions of q_1 and q_2.

An important conclusion is that the set of all transformations defined by (finite) automata is a semigroup under composition.

1.3.6. Dual automaton. The definition of an automaton is symmetric, in the sense that if we interchange the alphabet with the set of states and the output function with the transition function, then we get again an automaton.

DEFINITION 1.3.2. If (A, X) is an automaton, then its *dual* is the automaton $(\mathsf{A}, \mathsf{X})' = (\mathsf{X}', \mathsf{A}')$, where the set of states X' is in a bijective correspondence $\mathsf{X} \longrightarrow \mathsf{X}' : x \mapsto x'$ with the alphabet X and the alphabet A' is in a bijective correspondence $\mathsf{A} \longrightarrow \mathsf{A}' : q \mapsto q'$ with the set of states A of the original automaton. We have then

$$x' \cdot q' = p' \cdot y'$$

in $(\mathsf{X}', \mathsf{A}')$ if and only if we have

$$q \cdot x = y \cdot p$$

in (A, X).

1.3. AUTOMATA

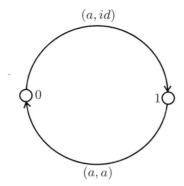

FIGURE 1.4. Dual Moore diagram of the adding machine

If the alphabet X is larger than the set of states of an automaton A, then it may be more convenient to draw not the Moore diagram of (A, X) but the *dual Moore diagram*, i.e., the Moore diagram of the dual automaton (A, X)'.

The arrows of the dual Moore diagram show the action of the states on the alphabet X, and the labels show the state transitions. More precisely, for every pair $(q, x) \in A \times X$ there is an arrow starting in x, ending in $q(x)$ and labeled by $(q, q|_x)$.

Suppose that $q_1 \cdots q_m$ is a product of states of A (i.e., a state of (A^m, X)); then the restrictions $(q_1 \cdots q_m)|_x$ and the image $q_1 \cdots q_m(x)$ can be conveniently computed using the dual Moore diagram. The procedure is the same as that of computing the images of words on Moore diagrams: one has to find the unique path starting in the vertex x and labeled by $(q_m, p_m), (q_{m-1}, p_{m-1}), \ldots (q_1, p_1)$ for some $p_i \in A$. Then $p_1 \cdots p_m = (q_1 \cdots q_m)|_x$ and the end of the path is the letter $q_1 \cdots q_m(x)$.

As an example, consider the dual diagram of the adding machine, shown in Figure 1.4. Note that we usually do not draw the loops corresponding to the trivial state. It is evident from the diagram that

$$a^n|_0 = \begin{cases} a^{n/2} & \text{if } n \text{ is even,} \\ a^{(n-1)/2} & \text{if } n \text{ is odd,} \end{cases} \quad a^n|_1 = \begin{cases} a^{n/2} & \text{if } n \text{ is even,} \\ a^{(n+1)/2} & \text{if } n \text{ is odd.} \end{cases}$$

The dual Moore diagrams of the automata (A, X^n) for a given automaton (A, X) have interesting geometry. They coincide as graphs with the *Schreier graphs* of the action on X^n of the group generated by the states of (A, X). In particular, one of the aims of Chapter 3 will be to show that in the case of so-called "contracting actions" the dual Moore diagrams (A, X^n) converge to some *limit space*.

1.3.7. Automata as endomorphisms of rooted trees. Let (A, X) be an automaton and $q \in A$. Since the beginning of length n of the image $q(w)$ depends only on the beginning of length n of the word $w \in X^*$, the map defined by q is an endomorphism of the rooted tree X^*.

On the other hand, if f is an endomorphism of X^*, then, as was already mentioned, we get an automaton with the set of states $Q(f) = \{f|_v : v \in X^*\}$.

It follows directly from (1.3) that the transformation $g : X^* \longrightarrow X^*$ defined by a state g of this automaton coincides with the original action of g. In particular, this shows that every endomorphism is defined by a state of an automaton (by an *initial automaton*).

FIGURE 1.5. Portrait of the adding machine

For example, Figure 1.5 shows the portrait of the adding machine as an automorphism of the binary tree. We label the active vertices (i.e., the vertices v for which the action of $a|_v$ on the first level X^1 is non-trivial) by arcs. If some vertices are not shown, then we assume that they are not active. The active vertices (the *switches*) go all the way to infinity along the right-most path of the tree.

1.3.8. Inverse automaton. An automaton A is said to be *invertible* if every one of its states defines an invertible transformation of X^*.

It is easy to prove that an automaton is invertible if and only if every one of its states defines an invertible transformation of X.

If (A, X) is an invertible automaton, then its inverse is the automaton (A^{-1}, X), whose states are in a bijective correspondence $A^{-1} \longrightarrow A : g^{-1} \mapsto g$ with the set of states of A, and
$$g^{-1} \cdot x = y \cdot h^{-1}$$
in (A^{-1}, X) is equivalent to
$$g \cdot y = x \cdot h$$
in (A, X).

In particular, if A is finite, then A^{-1} is finite. We get that the set of all automorphisms of the tree X^* defined by states of finite automata is a group. This group is called the *group of finite automata*.

If we have the Moore diagram of an invertible automaton (A, X), then the Moore diagram of the inverse automaton (A^{-1}, X) is obtained by changing every label (x, y) to (y, x). A vertex of the old Moore diagram corresponding to the state $q \in A$ will correspond to the state $q^{-1} \in A^{-1}$ in the new diagram.

If we have the dual Moore diagram of the automaton (A, X), then in order to get the dual Moore diagram of the inverse automaton, we have to change the direction of every arrow and change every label (q, p) to (q^{-1}, p^{-1}).

1.3.9. Reduced automata. An automaton (A, X) is *reduced* if different states of A define different transformations of X^*. If g is an endomorphism of the tree X^*, then the automaton $Q(g) = \{g|_v \; : \; v \in X^*\}$ is reduced. Any automaton can be reduced; i.e., there exists an algorithm which finds a reduced automaton whose states define the same set of transformations as the states of the given automaton. Reduction of automata is described in [**Eil74**].

1.4. Wreath products

Another convenient language and notation for automorphisms of the rooted tree X^* comes from the notion of a *wreath product*.

1.4.1. Permutational wreath products.

DEFINITION 1.4.1. Let H be a group acting (from the left) by permutations on a set X and let G be an arbitrary group. Then the *(permutational) wreath product* $H \wr G$ is the semi-direct product $H \ltimes G^X$, where H acts on the direct power G^X by the respective permutations of the direct factors.

Every element of the wreath product $H \wr G$ can be written in the form $h \cdot g$, where $h \in H$ and $g \in G^X$. If we fix some indexing $\{x_1, \ldots, x_d\}$ of the set X, then g can be written as $(g_1, \ldots, g_d)$ for $g_i \in G$. Here g_i is the coordinate of g, corresponding to x_i. The multiplication rule for the elements $h \cdot (g_1, \ldots, g_d) \in H \wr G$ is given by the formula

$$(1.9) \qquad \alpha(g_1, \ldots, g_d) \cdot \beta(f_1, \ldots, f_d) = \alpha\beta(g_{\beta(1)} f_1, \ldots, g_{\beta(d)} f_d),$$

where $g_i, f_i \in G$, $\alpha, \beta \in H$ and $\beta(i)$ is the image of i under the action of β, i.e., such an index that $\beta(x_i) = x_{\beta(i)}$.

1.4.2. Wreath recursion.
We have the following well known fact.

PROPOSITION 1.4.2. *Denote by* $\operatorname{Aut} X^*$ *the automorphism group of the rooted tree* X^* *and by* $\mathfrak{S}(X)$ *the symmetric group of all permutations of* X. *Fix some indexing* $\{x_1, \ldots, x_d\}$ *of* X. *Then we have an isomorphism*

$$\psi : \operatorname{Aut} X^* \longrightarrow \mathfrak{S}(X) \wr \operatorname{Aut} X^*,$$

given by

$$\psi(g) = \alpha(g|_{x_1}, g|_{x_2}, \ldots, g|_{x_d}),$$

where α *is the permutation equal to the action of* g *on* $X \subset X^*$.

PROOF. It is obvious that the map ψ is a bijection; hence it is sufficient to show that it is a homomorphism. But this follows directly from the definition of wreath products and (1.3):

$$\psi(g)\psi(h) = \alpha\left(g|_{x_1}, g|_{x_2} \ldots, g|_{x_d}\right) \cdot \beta\left(h|_{x_1}, h_{x_2} \ldots, h|_{x_d}\right)$$
$$= \alpha\beta\left(g|_{h(x_1)} h|_{x_1}, g|_{h(x_2)} h_{x_2}, \ldots, g|_{h(x_d)} h|_{x_d}\right)$$
$$= \alpha\beta\left((gh)|_{x_1}, (gh)|_{x_2}, \ldots, (gh)|_{x_d}\right) = \psi(gh).$$

$\square$

We will usually identify $g \in \operatorname{Aut} X^*$ with its image $\psi(g) \in \mathfrak{S}(X) \wr \operatorname{Aut} X^*$, so that we write

$$(1.10) \qquad g = \alpha \cdot (g|_{x_1}, g|_{x_2}, \ldots, g|_{x_d}),$$

where α is the permutation defined by g on the first level X of the tree X^*.

According to this convention, we have $\operatorname{Aut} X^* = \mathfrak{S}(X) \wr \operatorname{Aut} X^*$. The subgroup $(\operatorname{Aut} X^*)^X \leq \mathfrak{S}(X) \wr \operatorname{Aut} X^*$ is the first level stabilizer $\operatorname{St}(1)$. It acts on the tree X^* in the natural way:

$$(g_1, \ldots, g_d)(x_i v) = x_i g_i(v);$$

i.e., the ith coordinate of $(g_1, \ldots, g_d)$ acts on the ith subtree $x_i X^*$.

The subgroup $\mathfrak{S}(\mathsf{X}) \le \mathfrak{S}(\mathsf{X}) \wr \operatorname{Aut} \mathsf{X}^*$ is identified with the group of *rooted automorphisms* $\alpha = \alpha \cdot (1, \ldots, 1)$ acting by the rule

$$\alpha(xv) = \alpha(x)v.$$

Relation (1.10) is called the *wreath recursion*. It is a compact way to define recursively automorphisms of the rooted tree X^*. For example, the relation

$$a = \sigma(1, a),$$

where σ is the transposition $(0, 1)$ of the alphabet $\mathsf{X} = \{0, 1\}$, defines an automorphism of the tree $\{0, 1\}^*$ coinciding with the transformation, defined by the left-hand side state of the automaton, shown in Figure 1.3.

In general, every invertible finite automaton with the set of states $\{g_1, \ldots, g_n\}$ is described by recurrent formulae:

$$\begin{cases} g_1 &= \tau_1 \cdot (h_{11}, h_{12}, \ldots, h_{1d}) \\ g_2 &= \tau_2 \cdot (h_{21}, h_{22}, \ldots, h_{2d}) \\ &\vdots \\ g_n &= \tau_n \cdot (h_{n1}, h_{n2}, \ldots, h_{nd}), \end{cases}$$

where $h_{ij} = g_i|_{x_j}$ and τ_i is the action of g_i on X.

Conversely, any set of formulae of this type, for which τ_i are arbitrary permutations and each h_{ij} belongs to the set $\{g_1, \ldots, g_n\}$, uniquely defines an invertible automaton with the set of states $\{g_1, \ldots, g_n\}$.

1.4.3. Case of a right action. A more classical notation for wreath products uses right actions. Since we use mostly left actions, we keep notation $H \wr G = H \ltimes G^{\mathsf{X}}$ as it was defined above and use notation $G \operatorname{wr} H = G^{\mathsf{X}} \rtimes H$ when H acts on X from the right side.

In this case the elements of $G \operatorname{wr} H$ are written in the form $(g_1, \ldots, g_n)\pi$. The multiplication rule for the elements of the wreath product $G \operatorname{wr} H$ is

(1.11) $\qquad (g_1, \ldots, g_d)\alpha \cdot (h_1, \ldots, h_d)\beta = (g_1 h_{1^\alpha}, \ldots, g_d h_{d^\alpha})\alpha\beta.$

1.5. Self-similar actions

1.5.1. Definitions.

DEFINITION 1.5.1. A faithful action of a group G on X^* (or on X^ω) is said to be *self-similar* if for every $g \in G$ and every $x \in \mathsf{X}$ there exist $h \in G$ and $y \in \mathsf{X}$ such that

(1.12) $\qquad\qquad\qquad g(xw) = yh(w)$

for every $w \in \mathsf{X}^*$ (resp. $w \in \mathsf{X}^\omega$).

We will denote self-similar actions as pairs (G, X), where G is the group and X is the alphabet (that will mean that G acts on X^* or X^ω).

The pair (h, y) is uniquely determined in the conditions of Definition 1.5.1 by the pair (g, x), since the action is faithful. Hence we get an automaton with the set of states G and with the output and transition functions

$$g \cdot x = y \cdot h,$$

i.e., $y = g(x)$ and $h = g|_x$. This automaton is called the *complete automaton of the self-similar action*. It is easy to prove by induction that the action on X^* of the state g is the same as the action of the element g of the group.

The next definition equivalent to 1.5.1 emphasizes this approach:

DEFINITION 1.5.2. A faithful action of a group G on X^* is self-similar if there exists an automaton (G, X) such that the action of $g \in G$ on X^* coincides with the action of the state g of the automaton.

The notation (G, X) for self-similar actions is therefore a partial case of notation (A, X) for automata.

If we have a faithful action of G, then G is isomorphic to a subgroup of $\operatorname{Aut} X^*$, with which it will be identified. So, we will in some cases talk about *self-similar subgroups* of $\operatorname{Aut} X^*$ or *self-similar automorphism groups* of the tree X^*. Definition 1.5.1 is formulated in these terms in the following way.

DEFINITION 1.5.3. An automorphism group G of the rooted tree X^* is *self-similar* (or *state-closed*; see [**Sid98**]) if for every $g \in G$ and $v \in X^*$ we have $g|_v \in G$.

We say that a self-similar action of a group G is *finite-state* if every one of its elements is finite-state as an automorphism of X^*, i.e., if the set $\{g|_v : v \in X^*\}$ is finite for every $g \in G$.

1.5.2. Wreath recursion. Definition 1.5.3 can be written in terms of the wreath recursion in the following way.

DEFINITION 1.5.4. An automorphism group $G \le \operatorname{Aut} X^*$ is *self-similar* if
$$G \le \mathfrak{S}(X) \wr G.$$

Recall that $G \le \mathfrak{S}(X) \wr G$ means that $\psi(G) \le \mathfrak{S}(X) \wr G$, where
$$\psi : \operatorname{Aut} X^* \longrightarrow \mathfrak{S}(X) \wr \operatorname{Aut} X^*$$
is the wreath recursion (see Proposition 1.4.2).

Thus, we get for every self-similar action the homomorphism
$$\psi : G \longrightarrow \mathfrak{S}(X) \wr G,$$
also called *wreath recursion*.

Wreath recursion determines the action of G on X^*, since it determines the complete automaton of the action.

PROPOSITION 1.5.5. *Let G be a group and suppose that we have a homomorphism $\psi : G \longrightarrow \mathfrak{S}(X) \wr G$. Let (A_ψ, X) be the automaton whose output and transition functions are given by*
$$\psi(g) = \sigma \cdot (g|_{x_1}, g|_{x_2}, \ldots, g|_{x_d}),$$
where $\sigma \in \mathfrak{S}(X)$ is such that $g(x) = \sigma(x)$ for all $x \in X$. Then the transformations of X^* defined by the states of A_ψ define an action of the group G on X^*.

PROOF. We must prove that the transformation defined by the state gh is a product of the transformations defined by g and by h. But this follows directly from the definition of a wreath product. □

The action of G on X^* defined in Proposition 1.5.5 is the action *defined by the wreath recursion* ψ.

Note that the action defined by a wreath recursion need not be faithful even if the wreath recursion is an injective homomorphism.

Therefore the next definition is more general than Definition 1.5.1 but equivalent to it in the case of faithful actions.

DEFINITION 1.5.6. A *self-similar action* of a group G over an alphabet X is determined by a homomorphism $\psi : G \longrightarrow \mathfrak{S}(\mathsf{X}) \wr G$. It is the action on X^*, defined in Proposition 1.5.5.

1.5.3. Functionally recursive automorphisms. If G is generated by a finite set $\{g_1, \ldots, g_n\}$, then the wreath recursion $\psi : G \longrightarrow \mathfrak{S}(\mathsf{X}) \wr G$ is uniquely determined by its values on the generators, i.e., by equations of the form

(1.13)
$$\begin{cases} \psi(g_1) &= \tau_1 \cdot (h_{11}, h_{12}, \ldots, h_{1d}) \\ \psi(g_2) &= \tau_2 \cdot (h_{21}, h_{22}, \ldots, h_{2d}) \\ &\vdots \\ \psi(g_n) &= \tau_n \cdot (h_{n1}, h_{n2}, \ldots, h_{nd}), \end{cases}$$

where $\tau_i \in \mathfrak{S}(\mathsf{X})$ and h_{ij} are elements of G. We usually omit ψ in wreath recursion, identifying g with $\psi(g)$.

The elements h_{ij} can be written as group words in g_i. If we know that the action of the group G is faithful, then the wreath recursion (1.13) uniquely determines the group G, since it determines recursively the action of the generators on the tree X^* (determining the complete automaton of the action).

As a corollary we get that the set of all finitely generated self-similar groups is countable. Moreover, for a given alphabet X the union of all finitely generated self-similar groups acting on X^* is a countable group (see [**BS97, Sid98**]) called the *group of functionally recursive automorphisms of* X^*.

1.5.4. Groups generated by automata. If we have in (1.13) that h_{ij} all belong to $\{g_1, \ldots, g_n\}$, then the set $\{g_1, \ldots, g_n\}$ is a finite sub-automaton A of the complete automaton of the action, and we say that the group G *is generated* by the automaton A.

Let us formulate this as a separate definition (see [**Gri88**]).

DEFINITION 1.5.7. Let A be an invertible automaton. The *group generated by the automaton* A is the group $\langle \mathsf{A} \rangle$ generated by the transformations defined by all the states of A.

Groups generated by finite automata are precisely the groups which are defined by wreath recursions (1.13) where $h_{ij} \in \{g_1, \ldots, g_d\}$.

A group generated by a finite automaton is obviously finite-state and finitely generated. Conversely, if the group G is self-similar, finite-state and finitely generated, then it is generated by a finite automaton. One can take all the automata defining the generators of the group G and then take their union.

1.6. The Grigorchuk group

Let us illustrate the introduced notions on one of the most famous examples of a self-similar group. The study of the self-similar actions was stimulated by the discoveries of amazing properties of this and similar groups.

1.6. THE GRIGORCHUK GROUP

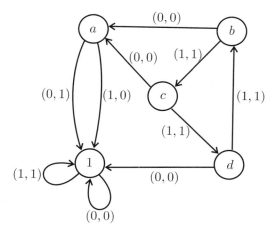

FIGURE 1.6. The automaton generating the Grigorchuk group

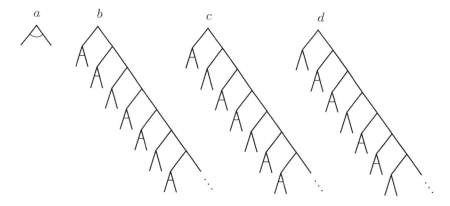

FIGURE 1.7. Portraits of the generators of the Grigorchuk group

The Grigorchuk group acts on X^* for the alphabet $\mathsf{X} = \{0,1\}$ and is generated by four automorphisms a, b, c, d of X^*, defined recursively by the relations

$$\begin{aligned}
a(0w) &= 1w & a(1w) &= 0w \\
b(0w) &= 0a(w) & b(1w) &= 1c(w) \\
c(0w) &= 0a(w) & c(1w) &= 1d(w) \\
d(0w) &= 0w & d(1w) &= 1b(w).
\end{aligned}$$

So, the Grigorchuk group is generated by the automaton shown in Figure 1.6. The portraits of the generators are shown in Figure 1.7. Here also the switches (the active vertices) are marked by arcs. The switches are arranged periodically with period 3 along the right-most paths, as is shown in the figure.

Looking at the transition and the output functions of the automaton, we see that the group G is defined by the wreath recursion

$$\begin{aligned}
a &= \sigma, & b &= (a, c), \\
c &= (a, d), \\
d &= (1, b),
\end{aligned}$$

where σ is the transposition $(0,1) \in \mathfrak{S}(\{0,1\})$. We do not write trivial elements of $G \times G$ and of $\mathfrak{S}(\mathsf{X})$, so that $a = \sigma$ means $a = \sigma \cdot (1,1)$ and $b = (a,c)$ means $b = 1 \cdot (a,c)$.

The Grigorchuk group is a particularly easy example of an infinite finitely generated torsion group (thus it is an answer to one of the Burnside problems). It is also the first example of a group of intermediate growth (answering the Milnor problem). It has many other interesting properties such as just-infiniteness, finite width, etc.

Let us prove that the Grigorchuk group is an infinite torsion group, in order to illustrate the use of self-similarity on a typical example.

The proof essentially coincides with the original proof in [**Gri80**]. Our exposition follows [**Har00**].

Recall that a group G is called a *2-group* if for every element $g \in G$ there exists $n \in \mathbb{N}$ such that
$$g^{2^n} = 1.$$

THEOREM 1.6.1. *The Grigorchuk group G is an infinite 2-group.*

PROOF. **1.** Let us show that $a^2 = b^2 = c^2 = d^2 = 1$ and that $\{1,b,c,d\}$ is a subgroup isomorphic to Klein's Viergruppe $(\mathbb{Z}/2\mathbb{Z}) \times (\mathbb{Z}/2\mathbb{Z})$.

We have $a^2 = \sigma^2 = 1$. We also have
$$b^2 = (a^2, c^2) = (1, c^2), \quad c^2 = (a^2, d^2) = (1, c^2), \quad d^2 = (1, b^2),$$
which implies that $b^2 = c^2 = d^2 = 1$, since we see that the set $\{1, b^2, c^2, d^2\}$ is an automaton in which every state acts trivially on X; hence, by induction on the length of words, every state acts trivially on the whole tree X^*.

We have
$$\begin{aligned} bc &= (a,c)(a,d) &= (1, cd) \\ cd &= (a,d)(1,b) &= (a, db) \\ db &= (1,b)(a,c) &= (a, bc); \end{aligned}$$
hence the triple (bc, cd, db) satisfies the same recurrent relations as the triple (d, b, c). Since the recurrent relations determine the automorphisms uniquely, we get $bc = d$, $cd = b$, $db = c$. Since the elements b, c, d are involutions, we also get $cb = d$, $dc = b$, $bd = c$. This proves the isomorphism $\{1, b, c, d\} \cong (\mathbb{Z}/2\mathbb{Z}) \times (\mathbb{Z}/2\mathbb{Z})$.

Hence, the Grigorchuk group is a quotient of the free product
$$(\mathbb{Z}/2\mathbb{Z}) * ((\mathbb{Z}/2\mathbb{Z}) \times (\mathbb{Z}/2\mathbb{Z})),$$
and every element of the group G can be written in the form
$$s_0 a s_1 a s_2 a \cdots s_{m-1} a s_m,$$
where $s_i \in \{b, c, d\}$ for $i = 1, \ldots, m-1$ and $s_0, s_m \in \{1, b, c, d\}$.

2. Let $G_1 \le G$ be the stabilizer of the first level of the tree X^*, i.e., the subgroup of the elements of G which act trivially on X^1. Every element $g \in G_1$ is written (using the wreath recursion) as $g = (g_0, g_1)$, where $g_0 = g|_0$ and $g_1 = g|_1$. We get from (1.3) that the maps $\phi_i : g \mapsto g_i$ are homomorphisms from G_1 to G. The homomorphisms ϕ_i are called the *virtual endomorphisms* associated to the self-similar action. We will study virtual endomorphisms (homomorphisms from a subgroup of finite index to the group) later in general.

One important observation is that the homomorphisms ϕ_i are onto, since

(1.14) $$b = (a, c), \quad aba = (c, a),$$
(1.15) $$c = (a, d), \quad aca = (d, a),$$
(1.16) $$d = (1, b), \quad ada = (b, 1),$$

so that

$$\phi_0(b) = a, \quad \phi_1(b) = c, \quad \phi_0(aba) = c, \quad \phi_1(aba) = a,$$
$$\phi_0(c) = a, \quad \phi_1(c) = d, \quad \phi_0(aca) = d, \quad \phi_1(aca) = a,$$
$$\phi_0(d) = 1, \quad \phi_1(d) = b, \quad \phi_0(ada) = b, \quad \phi_1(ada) = 1;$$

hence $\phi_0(G_1) = G$ and $\phi_1(G_1) = G$.

We get instantly that G is infinite, since we have a map from a proper subgroup $G_1 < G$ onto G. Another proof follows from the fact that G is level-transitive, which also can be proved using surjectiveness of ϕ_i.

3. Let us prove that for every $g \in G$ there exists $n \in \mathbb{N}$ such that $g^{2^n} = 1$. Recall that g can be written in the form

$$s_0 a s_1 a s_2 a \cdots s_{m-1} a s_m,$$

where $s_i \in \{b, c, d\}$ for $i = 1, \ldots, m-1$ and $s_0, s_m \in \{1, b, c, d\}$. The number of non-unit generators in the shortest representation of g is called the *length* of g.

We will prove the statement by induction on the length of g. It is true for the elements of length 1 (then $g^2 = 1$). It is also easy to see that it is true for the elements of length 2, since

$$(ad)^4 = (\sigma(1,b)\sigma(1,b))^2 = (b,b)^2 = 1$$
$$(ac)^8 = (\sigma(a,d)\sigma(a,d))^4 = (da, ad)^4 = 1$$
$$(ab)^{16} = (\sigma(a,c)\sigma(a,c))^8 = (ca, ac)^8 = 1.$$

(The elements da, ca, ba are conjugate to ad, ac, ab, thus have the same order.)

Suppose that for all $g \in G$ of length less than k there is n such that $g^{2^n} = 1$. If the shortest word $s_0 a s_1 a s_2 a \cdots s_{m-1} a s_m = g$ starts and ends with a (i.e., if $s_0 = s_1 = 1$), then aga is shorter and has the same order as g, since they are conjugate. If the first and the last letter of the word belongs to $\{b, c, d\}$, then we also can find an element $u \in \{b, c, d\}$ so that ugu is shorter than g (one has to take u equal, say to the first letter of the word).

Hence, we may assume (after conjugating g, if necessary, by b, c or d) that g is of the form

$$a s_1 a s_2 \ldots a s_{k/2}$$

for $s_i \in \{b, c, d\}$.

If $k/2$ is even, then

$$g = (as_1 a) \cdot s_2 \cdot (as_3 a) \cdots s_{k/2} = (g_0, g_1).$$

But (1.14)–(1.16) imply that $as_i a \in \{b, c, d\} \times \{a, 1\}$ and $s_i \in \{a, 1\} \times \{b, c, d\}$. Therefore, the lengths of g_0 and g_1 are not greater than $k/2$, and by the induction hypothesis, there exists $n \in \mathbb{N}$ such that $g_0^{2^n} = g_1^{2^n} = 1$. But then also

$$g^{2^n} = \left(g_0^{2^n}, g_1^{2^n}\right) = 1.$$

Suppose now that $k/2$ is odd. Then

(1.17) $\quad g^2 = (as_1a) \cdot s_2 \cdot (as_3a) \cdots (as_{k/2}a) \cdot s_1 \cdot (as_2a) \cdots s_{k/2} = (h_0, h_1),$

where h_i have length at most $2 \cdot k/2 = k$.

We consider the next three cases.

(i) Some s_j is equal to d. Then we will have once $d = (1, b)$ and once $ada = (b, 1)$ in the product (1.17), so that the length of each of h_i is at most $k - 1$. But then there exists $n \in \mathbb{N}$ such that $h_0^{2^n} = h_1^{2^n} = 1$, and we get $g^{2^{n+1}} = 1$.

(ii) Some s_j is equal to c. Then $s_j = (a, d)$ and $as_ja = (d, a)$, so that each of h_i either has length less than k or is equal to a word of length k involving d. In the first case we can apply the induction hypothesis, while the second one was considered in (i).

(iii) If neither (i) nor (ii) holds, then $g = abab \cdots ab$, which is of order at most 16. $\square$

For more on the Grigorchuk group see [**Gri00**] and the last chapter of [**Har00**].

1.7. The adding machine and self-similar actions of $\mathbb{Z}^n$

We describe here a class of self-similar actions of $\mathbb{Z}^n$. The proofs will appear in Section 2.9 after a general theory is developed.

1.7.1. The adding machine. Let us define an automatic transformation a over the alphabet $\{0, 1\}$ by the recursion

$$a = \sigma(1, a),$$

or, in other words, by

$$a(0w) = 1w$$
$$a(1w) = 0a(w).$$

We see that a is defined by a two-state automaton. Its Moore diagram is shown in Figure 1.3 on page 5. This automaton is called the *(binary) adding machine* or the *odometer*.

The action of the infinite cyclic group $\mathbb{Z}$ generated by the transformation a is self-similar and is also called the *adding machine action*.

The recurrent definition of the transformation a coincides with the rule of adding 1 to a dyadic integer. More precisely, one can prove (by induction on $|n|$) that

$$a^n(x_1x_2\ldots x_m) = y_1y_2\ldots y_m$$

if and only if

$$y_1 + y_2 \cdot 2 + y_3 \cdot 2^2 + y_4 \cdot 2^3 + \cdots + y_m \cdot 2^{m-1}$$
$$= (x_1 + x_2 \cdot 2 + x_3 \cdot 2^2 + x_4 \cdot 2^3 + \cdots + x_m \cdot 2^{m-1}) + n \pmod{2^m}.$$

The action of a on the infinite words is interpreted using the bijection

$$\Phi : x_1x_2\ldots \mapsto \sum_{k=1}^{\infty} x_k \cdot 2^{k-1}$$

as the addition of 1 to dyadic integers: $\Phi(a(w)) = \Phi(w) + 1$.

1.7.2. Multi-dimensional adding machines.
The adding machine example can be generalized in a natural way to the free abelian groups $\mathbb{Z}^n$.

Let B be an integral matrix with $|\det B| = d > 1$. Then $B(\mathbb{Z}^n)$ is a subgroup of index d in $\mathbb{Z}^n$. Let $\mathsf{X} = \{r_1, \ldots r_d\}$ be a coset transversal, i.e., a collection of elements of $\mathbb{Z}^n$ such that $\mathbb{Z}^n = \bigsqcup_{i=1}^{d} B(\mathbb{Z}^n) + r_i$.

The matrix B will play the role of the base of a numeration system, and the elements $r_1, \ldots, r_d$ are the "digits". We have $B = 2$ and $\{r_1, r_2\} = \{0, 1\}$ for the case of the binary adding machine and the binary numeration system.

Let $\widehat{\mathbb{Z}}^n$ be the profinite completion of $\mathbb{Z}^n$ with respect to the series of finite index subgroups
$$\mathbb{Z}^n > B(\mathbb{Z}^n) > B^2(\mathbb{Z}^n) > \ldots.$$
Then every element $\gamma \in \widehat{\mathbb{Z}}^n$ can be written uniquely in the form
$$\gamma = r_{i_0} + B(r_{i_1}) + B^2(r_{i_2}) + B^3(r_{i_3}) + \cdots$$
and the map $\Phi : \gamma \mapsto r_{i_0} r_{i_1} \ldots$ is a homeomorphism between $\widehat{\mathbb{Z}}^n$ and X^ω.

If the intersection $\bigcap_{n \in \mathbb{N}} B^n(\mathbb{Z}^n)$ is trivial (we will describe later in Proposition 2.9.2 a simple criterion when this happens), then the group $\mathbb{Z}^n$ is a subgroup of the completion $\widehat{\mathbb{Z}}^n$ and thus acts on it in the natural way. Conjugating by Φ we get an action of $\mathbb{Z}^n$ on X^ω. The respective action on X^* is determined by the condition that
$$g(r_{i_0} r_{i_1} \ldots r_{i_m}) = r_{j_0} r_{j_1} \ldots r_{j_m}$$
is equivalent to
$$g + r_{i_0} + B(r_{i_1}) + \cdots + B^m(r_{i_m}) = r_{j_0} + B(r_{j_1}) + \cdots + B^m(r_{j_m}) \pmod{B^{m+1}(\mathbb{Z}^n)}.$$
It follows that
$$g \cdot r_i = r_j \cdot h$$
in the complete automaton of the action is equivalent to
$$g + r_i = r_j + B(h).$$

For example, if we identify $\mathbb{Z}^2$ with the additive group of the ring of Gaussian integers $\mathbb{Z}[i] \subset \mathbb{C}$, then we can consider the numeration system on $\mathbb{Z}[i]$ with the "base" $(i-1)$ and the digits $\{0, 1\}$: every number $a + bi \in \mathbb{Z}[i]$ can be written in a unique way as a sum
$$a + bi = \sum_{k=0}^{m} c_k \cdot (i-1)^k$$
for some $m \in \mathbb{N}$ and $c_k \in \{0, 1\}$ (see [**Knu69**]). Then we get the corresponding self-similar "adding machine" action of $\mathbb{Z}[i]$ on $\{0, 1\}^*$. Here multiplication by $(i-1)$ corresponds to the matrix $B = \begin{pmatrix} -1 & -1 \\ 1 & 1 \end{pmatrix}$, since the group $\mathbb{Z}[i]$ is identified with $\mathbb{Z}^2$ so that $\{1, i\}$ corresponds to the standard basis of $\mathbb{Z}^2$.

1.8. Branch groups

Recall that a level-transitive group $G \leq \mathsf{Aut}\, \mathsf{X}^*$ is said to be *branch* if $\mathsf{RiSt}_G(n)$ is a subgroup of finite index in G for every n (see Definition 1.2.4).

A group is said to be *just infinite* if it is infinite but all of its proper quotients are finite. According to a theorem of R. Grigorchuk and J. Wilson (see [**Wil71, Gri00, Wil00**]), a just infinite group is either a branch group or contains a subgroup of finite index which is a direct power L^k of a simple or a *hereditary just infinite* group L. A group is said to be hereditary just infinite if it is residually finite and every subgroup of finite index is just infinite.

Many branch groups are defined using their self-similar action on a regular rooted tree (though not all branch groups are self-similar). Actually, self-similarity is one of the most important tools in the study of branch groups, and branch groups were the starting point of the study of self-similar groups in general.

We have already mentioned the Grigorchuk group. Here we present some other examples.

For a detailed account on branch groups see [**Gri00, BGŠ03**].

1.8.1. The Gupta-Sidki group. Let p be an odd prime. The *Gupta-Sidki p-group* is generated by two automorphisms a, t of the tree $\mathsf{X}^* = \{0, 1, \ldots, p-1\}^*$, defined by the recursion

$$a = \sigma, \qquad t = (a, a^{-1}, 1, 1, \ldots, 1, t),$$

where σ is the cyclic permutation $(0, 1, \ldots, p-1) \in \mathfrak{S}(\mathsf{X})$.

It was defined for the first time in [**GS83**]. The Gupta-Sidki group is also an infinite torsion group. (The proof is even a bit shorter than for the Grigorchuk group.) For various interesting properties of this group see the papers [**Sid87b, Sid87a, BG00b, BG02**].

1.8.2. Groups of P. Neumann's type. Let $A \leq \mathfrak{S}(\mathsf{X})$ be a transitive permutation group acting on X. For every $x \in \mathsf{X}$ and $\alpha \in A$ such that $\alpha(x) = x$, define an automorphism $b_{(\alpha,x)}$ of the tree X^* by the recurrent relation

$$\begin{cases} b_{(\alpha,x)} \cdot x = x \cdot b_{(\alpha,x)} \\ b_{(\alpha,x)} \cdot y = \alpha(y) \cdot 1 & \text{if } y \neq x, \end{cases}$$

or, in terms of the wreath recursion

$$b_{(\alpha,x)} = \alpha \left(1, \ldots, 1, b_{(\alpha,x)}, 1, \ldots, 1\right),$$

where $b_{(\alpha,x)}$ on the right-hand side stays in the place corresponding to the letter x. Let $\mathcal{P}(A)$ be the group generated by all such $b_{(\alpha,x)}$.

Recall that a group G is called *perfect* if the commutator subgroup $G' = [G, G]$ coincides with G. A group is perfect if and only if any of its abelian quotient is trivial.

PROPOSITION 1.8.1. *Let A_x denote the stabilizer of a point $x \in \mathsf{X}$ in A. Suppose that $A'_x = A_x$ and that the subgroups $(A_{x_1} \cap A_{x_2})'$, $x_1 \neq x_2$ generate A. Then $\mathcal{P}(A)$ is perfect and $\mathcal{P}(A) = A \wr \mathcal{P}(A)$.*

Here the last equality means that the image of $\mathcal{P}(A)$ under the wreath recursion coincides with $A \wr \mathcal{P}(A)$, since we, as usual, identify $\operatorname{Aut} \mathsf{X}^*$ with $\mathfrak{S}(\mathsf{X}) \wr \operatorname{Aut} \mathsf{X}^*$.

PROOF. The subgroups $G_x = \langle b_{(\alpha,x)} : \alpha \in A, \alpha(x) = x \rangle$ are isomorphic to A_x and thus are perfect. The group $\mathcal{P}(A)$ is generated by the union of the groups G_x; therefore it is also perfect.

If $\alpha_1, \alpha_2 \in A_{x_1} \cap A_{x_2}$ for $x_1 \neq y_1 \in \mathsf{X}$, then

$$\left[b_{(\alpha_1, x_1)}, b_{(\alpha_2, x_2)}\right] = [\alpha_1, \alpha_2] \left(1, \ldots, 1\right);$$

hence $\mathcal{P}(A)$ contains A. Therefore it also contains $(1,\ldots,1,b_{(\alpha,x)},1,\ldots,1) = \alpha^{-1} \cdot b_{(\alpha,x)}$, and, consequently, it contains $\mathcal{P}(A)^{\mathsf{X}}$. Therefore, the image of $\mathcal{P}(A)$ under the wreath product recursion is equal to $A \wr \mathcal{P}(A)$. □

We can take, for example, A equal to the alternating group Alt_6. See [**Neu86**] for more on this example and its properties.

1.8.3. Groups of J. Wilson's type. Let $A \leq \mathfrak{S}(\mathsf{X})$ be a 2-transitive permutation group acting on an alphabet of cardinality ≥ 3. The group $\mathcal{W}(A)$ is generated by two copies of A. One is just A acting at the root of X^*, i.e., the set of the automorphisms $\alpha \cdot (1,\ldots,1)$, for $\alpha \in A$. The other is the set $\overline{A}$ of the automorphisms

$$\overline{\alpha} = (\overline{\alpha}, \alpha, 1, 1, \ldots, 1), \quad \alpha \in A.$$

We fix here two letters $x_1, x_2 \in \mathsf{X}$, corresponding to the first two coordinates in the recursion.

PROPOSITION 1.8.2. *If A is perfect, then $\mathcal{W}(A)$ is also perfect and satisfies the relation $\mathcal{W}(A) = A \wr \mathcal{W}(A)$.*

PROOF. The group $\mathcal{W}(A)$ is generated by two copies of A; therefore is perfect if A is.

Let $\gamma \in A$ be a permutation fixing x_1 and moving x_2 to a different letter. Then, for any two $\alpha, \beta \in A$ we have

$$\left[\overline{\alpha}, \overline{\beta}^\gamma\right] = \left(\left[\overline{\alpha}, \overline{\beta}\right], 1, \ldots, 1\right).$$

Similarly, if $\gamma' \in A$ fixes x_2 and moves x_1 to another letter, then

$$\left[\overline{\alpha}, \overline{\beta}^{\gamma'}\right] = (1, [\alpha, \beta], 1, \ldots, 1).$$

Consequently, $\mathcal{W}(A)$ contains $\mathcal{W}(A)^{\mathsf{X}}$. It also contains A, therefore $\mathcal{W}(A) = A \wr \mathcal{W}(A)$. □

We have the next result from [**Neu86**], which is applicable both to the groups $\mathcal{P}(A)$ and to $\mathcal{W}(A)$ when they satisfy the condition $G = A \wr G$.

THEOREM 1.8.3. *Let A be a non-abelian simple transitive subgroup of $\mathfrak{S}(\mathsf{X})$. If G is a perfect residually finite group such that $G \cong A \wr G$, then*

(1) *All non-trivial normal subgroups of G have finite index; i.e., G is just infinite. (Actually, every non-trivial normal subgroup of G is equal to the stabilizer of a level of X^*.)*
(2) *Every sub-normal subgroup of G is isomorphic to a finite direct power of G, but G does not satisfy the ascending chain condition on subnormal subgroups.*
(3) *G is minimal in the sense of [**Pri80**].*

1.8.4. Non-uniformly exponential growth. The groups $\mathcal{W}(A)$ were used by J. Wilson in [**Wil04b**] to construct the first example of a group of *non-uniform exponential growth*.

A finitely generated group G is said to have a non-uniform exponential growth if for every finite generating set S the number

$$e(S) = \lim_{n \to \infty} \sqrt[n]{|B_S(n)|}$$

is greater than one, but $\inf_S e(S) = 1$. Here $B_S(n) = \{g_1 g_2 \cdots g_n : g_i \in S \cup S^{-1}\}$.

L. Bartholdi has shown in [**Bar03b**] that the group $\mathcal{W}(\mathrm{PSL}(3,2))$, where $\mathrm{PSL}(3,2)$ acts on the projective plane $\mathsf{X} = P^2 \mathbb{F}_2$, is an example of a group of non-uniform exponential growth.

Let us give a sketch of the arguments from the papers [**Wil04b, Bar03b**]. In both cases the group G under consideration is self-similar and satisfies the condition $G = A \wr G$ for some perfect group A (it is Alt_{31} in J. Wilson's example). Then with any generating set $S = \{a_1, \ldots, a_k\}$ of G (with the size k and the order of each of a_i specified), a new set $\hat{S} = \{\hat{a}_1, \ldots, \hat{a}_k\}$ is defined by

$$\hat{a}_i = x_i \cdot (1, 1, \ldots, a_i, \ldots, 1),$$

where $x_1, \ldots, x_k$ is a fixed generating set of A and a_i on the right-hand side of the equality are placed on specially chosen places u_i.

The choice of x_i and u_i is done in such a way that simple manipulations show that if S is a generating set of G, then $\hat{S}$ is also a generating set. Here the decomposition $G = A \wr G$ and perfectness of G are crucial.

Consider now a group word w in the generators $\hat{a}_i$. Application of the wreath recursion gives us $w = \pi \cdot (w_1, \ldots, w_d)$, where w_i are words in the generators a_i. It follows from the definition of $\hat{a}_i$ that the sum of the lengths of the words w_i is not greater than the length of the word w. We may assume that the word w is *reduced*, i.e., is a shortest representation of an element of G. But even then there may happen some cancellations in the words w_i so that the sum of the lengths of the words w_i will be strictly smaller than the length of w. One can find a list Δ of words such that if w contains a subword belonging to this list, then this additional cancellation in the words w_i does happen.

For $\eta \in (0, 1)$ let $W_n^{>\eta}$ be the set of reduced words of length $\leq n$ having at least ηn disjoint subwords belonging to Δ and let $W_n^{<\eta}$ be its complement in the set of all reduced words of length $\leq n$. One finds (by more or less straightforward combinatorial considerations) an estimate $\lim_{n \to \infty} \sqrt[n]{|W_n^{<\eta}|} \leq F(\eta)$. It is $F(\eta) = \frac{2^\eta (1+\eta)^{1+\eta}}{\eta^\eta}$ in the proof of J. Wilson and $F(\eta) = \frac{30^\eta}{\eta^\eta (1-\eta)^{1-\eta}}$ in L. Bartholdi's example.

On the other hand, it is not hard to prove that $\lim_{n \to \infty} \sqrt[n]{|W_n^{>\eta}|} \leq e(S)^{(1-\eta)}$. One has to use the fact that an element $w \in G$ is uniquely determined by its wreath decomposition $\pi(w_1, \ldots w_d)$ and that the sum of the lengths of w_i is not greater than roughly $1 - \eta$ times the length of w.

It follows now that $e\left(\hat{S}\right) \leq \inf_{\eta \in (0,1)} \max\left\{e(S)^{1-\eta}, F(\eta)\right\}$. The last step is to show that iteration of the function

$$E(s) = \inf_{\eta \in (0,1)} \max\left\{s^{1-\eta}, F(\eta)\right\}$$

converges to 1 if we start from an initial value $s > 1$. This will show that $\inf_S e(S) = 1$.

It remains to prove that the group G has exponential growth. L. Bartholdi shows that in his example G contains a free semigroup on 2 generators. J. Wilson constructs a larger group, G_{large}, which contains a free subgroup, satisfies the condition $G_{large} = A \wr G_{large}$ and for which the above arguments can be applied again. See also the paper [**Wil04a**], where J. Wilson constructs an uncountable family of groups of non-uniform exponential growth.

1.8.5. Tree-wreath products.
Tree-wreath products were defined by Said Sidki and Andrew Brunner in [**BS02b**] and were used by S. Sidki and J. Wilson in [**SW03**] to construct the first example of a branch group with free subgroups.

Let us describe a more general construction of S. Sidki from [**Sid04b**]. Let (G, X) be a faithful self-similar action over $\mathsf{X} = \{0, 1\}$. Denote by $\left(\dot{G}, \mathsf{X}^2\right)$ the self-similar action such that if $g \cdot x = y \cdot h$ in (G, X), then

$$g \cdot (x0) = (y0) \cdot h, \quad \text{and} \quad g \cdot (x1) = (y1) \cdot 1$$

in $\left(\dot{G}, \mathsf{X}^2\right)$.

If $g \in G$, then denote by $\widetilde{g}$ the automorphism of X^* defined recurrently by

$$\widetilde{g}(1w) = 1w, \quad \widetilde{g}(00w) = 00\widetilde{g}(w), \quad \widetilde{g}(01w) = 01g(w).$$

Then $\widetilde{G} = \{\widetilde{g} \,:\, g \in G\}$ is an isomorphic copy of G.

S. Sidki and J. Wilson consider in [**SW03**] the group $G = \left\langle \dot{\mathbb{Z}}, \widetilde{H}, H \right\rangle$, where (H, X) is a self-similar group and $\mathbb{Z}$ acts on X^* by the adding machine action. Then $\left(G, \mathsf{X}^2\right)$ is also a self-similar action. They show that if the abelianization H/H' of H has finite exponent, then G is a branch group.

S. Sidki and A. Brunner use in [**BS02b, Sid04b**] the *tree wreath product*, i.e., the groups of the form $\left\langle \dot{G}, \widetilde{H} \right\rangle$, to construct finite-state faithful actions of different solvable groups (for example, the free metabelian groups of any rank).

1.9. Other examples

1.9.1. Affine groups.
The following self-similar action of the affine group $\mathbb{Z}^n \rtimes \mathrm{GL}(n, \mathbb{Z})$ was constructed by A. Brunner and S. Sidki in [**BS98**].

The group $\mathrm{Affine}\,(\mathbb{Z}^n) = \mathbb{Z}^n \rtimes \mathrm{GL}(n, \mathbb{Z})$ is the group of affine transformations $v \mapsto A(v) + b$ of $\mathbb{Z}^n$, where $A \in \mathrm{GL}(n, \mathbb{Z})$ and $b \in \mathbb{Z}^n$. This action on $\mathbb{Z}^n$ extends to a continuous action of $\mathrm{Affine}\,(\mathbb{Z}^n)$ on the set $\mathbb{Z}_2^n$ of n-tuples of dyadic integers.

Let us identify the n-tuple

$$\left(\sum_{k=0}^{\infty} a_{k,1} 2^k, \sum_{k=0}^{\infty} a_{k,2} 2^k, \ldots, \sum_{k=0}^{\infty} a_{k,n} 2^k \right) \in \mathbb{Z}_2^n,$$

where $a_{k,i} \in \{0, 1\}$, with the infinite word

$$(a_{0,1}, a_{0,2}, \ldots, a_{0,n})(a_{1,1}, a_{1,2}, \ldots, a_{1,n})(a_{2,1}, a_{2,2}, \ldots, a_{2,n}) \ldots$$

over the alphabet $\mathsf{X} = \{0, 1\}^n$. Then we get a continuous action of $\mathrm{Affine}\,(\mathbb{Z}^n)$ on X^ω.

One can prove that this action is self-similar and finite-state. This construction actually defines a self-similar action of the affine group $\mathbb{Z}_2^n \rtimes \mathrm{GL}(n, \mathbb{Z}_2)$ over the ring of dyadic integers $\mathbb{Z}_2$. It is proved in [**BS98**] that the subgroup $\mathbb{Z}_{(2)}^n \rtimes \mathrm{GL}(n, \mathbb{Z}_{(2)})$, where $\mathbb{Z}_{(2)}$ is the ring of rational numbers with odd denominators, acts by finite automata.

See a general treatment of analogous actions of affine groups and their subgroups in [**NS04**].

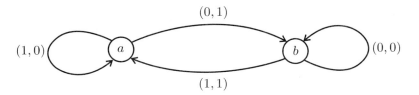

FIGURE 1.8. The lamplighter group

1.9.2. Lamplighter group and generalizations. Consider the group generated by the automaton shown in Figure 1.8 over the alphabet $\mathsf{X} = \{0, 1\}$.

The following proposition is due to R. Grigorchuk and A. Żuk [**GŻ01**]. Here we present a different proof from [**GNS00**].

PROPOSITION 1.9.1. *The group generated by the transformations a and b is isomorphic to the "lamplighter group", i.e., to the semi-direct product $(\mathbb{Z}/2\mathbb{Z})^{\mathbb{Z}} \rtimes \mathbb{Z}$, where $\mathbb{Z}$ acts on $(\mathbb{Z}/2\mathbb{Z})^{\mathbb{Z}}$ by the shift, or equivalently, to the wreath product $(\mathbb{Z}/2\mathbb{Z}) \operatorname{wr} \mathbb{Z}$.*

PROOF. The generators can be written as
$$a = \sigma(b, a), \qquad b = (b, a).$$

Let us identify the alphabet $\mathsf{X} = \{0, 1\}$ with the field $\mathbb{F}_2 = \mathbb{Z}/2\mathbb{Z}$. Then $ab^{-1} = \sigma$ acts according to the rule
$$\sigma(x_1 x_2 \ldots) = (x_1 + 1) x_2 x_3 x_4 \ldots.$$

A direct verification shows that b acts according to the rule
$$b(x_1 x_2 \ldots) = x_1 (x_2 + x_1)(x_3 + x_2)(x_4 + x_5) \ldots.$$

Let us identify every $x_1 x_2 \ldots \in \mathsf{X}^\omega$ with the formal power series $x_1 + x_2 t + x_3 t^2 + \ldots \in \mathbb{F}_2[[t]]$. It follows that this identification conjugates σ with the mapping $\phi_\sigma : F(t) \mapsto F(t) + 1$ and b with $\phi_b : F(t) \mapsto (1+t) F(t)$. Therefore, the group generated by a and b is isomorphic to the group generated by the transformations ϕ_σ and ϕ_b. This group obviously consists of the transformations of the form

$$(1.18) \qquad F(t) \mapsto (1+t)^n F(t) + \sum_{s=-\infty}^{+\infty} k_s (1+t)^s,$$

where $n \in \mathbb{Z}$, and all but a finite number of the coefficients $k_s \in \mathbb{F}_2$ are equal to zero. Indeed, transformations of this type form a group containing ϕ_σ and ϕ_b, and on the other hand,
$$F(t) + (1+t)^s = \left(F(t)(1+t)^{-s} + 1 \right)(1+t)^s = \phi_b^s \cdot \phi_\sigma \cdot \phi_b^{-s}(F(t));$$

therefore, all transformations of the type (1.18) belong to the group generated by ϕ_σ and ϕ_b.

It implies that the group generated by a and b is isomorphic to the group $(\mathbb{Z}/2\mathbb{Z}) \operatorname{wr} \mathbb{Z}$, where the base of the wreath product $(\mathbb{Z}/2\mathbb{Z})^{\mathbb{Z}}$ is identified with the normal subgroup of the transformations
$$F(t) \mapsto F(t) + \sum_{s=-\infty}^{+\infty} k_s (1+t)^s$$

on which ϕ_b acts by conjugation as a shift:

$$(1+t)^{-1}\left((1+t)F(t) + \sum_{s=-\infty}^{+\infty} k_s(1+t)^s\right) = F(t) + \sum_{s=-\infty}^{+\infty} k_{s+1}(1+t)^s.$$

$\square$

In the paper [**GŻ01**] this representation was used to compute the spectrum of the Markov operator on the lamplighter group. It was also used in [**GLSŻ00**] to construct a counterexample to the strong Atiyah conjecture.

The following generalization of the described group was defined by P. V. Silva and B. Steinberg in [**SS**].

If G is a finite group, then its *Cayley machine* $\mathcal{C}(G)$ is the automaton with the set of states and the alphabet both identified with G, whose transition and output functions are defined by the equalities

$$g(h) = gh, \quad g|_h = gh.$$

The following is Theorem 3.1 of [**SS**].

THEOREM 1.9.2. *Let G be a non-trivial finite group.*
- *The states of $\mathcal{C}(G)$ generate a free semigroup of transformations of X^*.*
- *If G is abelian, then the group generated by the automaton $\mathcal{C}(G)$ is isomorphic to $G \operatorname{wr} \mathbb{Z}$.*
- *In general, the group generated by $\mathcal{C}(G)$ is isomorphic to $N \rtimes \mathbb{Z}$ where N is a locally finite group.*

1.9.3. Stabilizers of vertex-transitive actions. Suppose that Γ is a locally finite graph and suppose that a group G acts by automorphisms of the graph Γ and that the action is faithful and transitive on the set of vertices. Then the stabilizer G_0 of a vertex v_0 of Γ has a natural faithful self-similar action on a rooted tree, which is constructed in the following way.

Let $\{v_1, v_2, \ldots, v_d\}$ be the vertices adjacent to v_0. Choose elements $g_i \in G$ such that $g_i(v_0) = v_i$. The set $\{g_1, \ldots, g_d\} = \mathsf{X}$ will be our alphabet. Let $g \in G_0$ and $g_i \in \mathsf{X}$ be arbitrary. Then $gg_i(v_0) = g(v_i)$ is a vertex adjacent to v_0. Let $v_j = gg_i(v_0)$. Then $g_j^{-1}gg_i(v_0) = v_0$; i.e., $h = g_j^{-1}gg_i \in G_0$.

We see that for every $g \in G_0$ and $g_i \in \mathsf{X}$ there exist unique $g_j \in \mathsf{X}$ and $h \in G_0$ such that

(1.19) $$g \cdot g_i = g_j \cdot h$$

in G.

We define a self-similar action of G_0 on X^* by (1.19). Such actions are completely described in terms of virtual endomorphisms in the paper [**Nek00**].

1.10. Bi-reversible automata and free groups

1.10.1. Bi-reversible automata. A finite invertible automaton (A, X) is said to be *bi-reversible* if its dual $(\mathsf{A}, \mathsf{X})'$ and dual of its inverse $(\mathsf{A}^{-1}, \mathsf{X})'$ both are invertible (see [**MNS00**]).

The dual of the automaton (A, X) is invertible if and only if the transformation $q \mapsto q|_x$ is a permutation of A for every $x \in \mathsf{X}$.

For example, the dual $(\mathsf{A}, \mathsf{X})'$ of the automaton shown in Figure 1.8 is not invertible, while the dual $(\mathsf{A}^{-1}, \mathsf{X})'$ of the inverse is.

The *abstract commensurator* Comm G of a group G is the set of equivalence classes of *virtual automorphisms* of G. A virtual automorphism of G is an isomorphism between two subgroups of finite index. Two virtual automorphisms are equivalent if their restrictions onto some subgroup of finite index are equal.

The set of all automorphisms of X^* which are defined by states of bi-reversible automata is a group called the *group of bi-reversible automata*. This group is a subgroup of Comm $F(\mathsf{X})$, where $F(\mathsf{X})$ is the free group generated by X. Namely, it is isomorphic to the group of the virtual automorphisms which are extendable to automorphisms of the directed Cayley graph of the group $F(\mathsf{X})$ (see [**MNS00**], Theorems 4 and 5).

The notion of bi-reversible automata is closely related to the theory of lattices in the automorphism groups of regular (non-rooted) trees.

An example of such a lattice is the free group $F(\mathsf{X})$ acting in the natural way on its Cayley graph T. Following [**GM03**], let us denote by C the *commensurator* of this lattice in Aut T; i.e., the set of elements $g \in \operatorname{Aut} T$ such that $g^{-1} \cdot F(\mathsf{X}) \cdot g \cap F(\mathsf{X})$ has finite index both in $F(\mathsf{X})$ and in $g^{-1} \cdot F(\mathsf{X}) \cdot g$. Let C_O be the stabilizer of the vertex $1 \in T$. Denote by $\operatorname{Aut}^+ T$ the group of *orientation preserving* automorphisms of T. Here the orientation is given on the Cayley graph T by the generators X of $F(\mathsf{X})$. Then the mentioned isomorphism of the group of bi-reversible automata and a subgroup of the abstract commensurator of $F(\mathsf{X})$ can be formulated in the following way (see [**GM03**], Theorem 2.16).

THEOREM 1.10.1. *The group of bi-reversible automata is isomorphic to the group $C_O \cap \operatorname{Aut}^+ T$. Moreover, the action of the group of bi-reversible automata on X^* coincides with the action of $C_O \cap \operatorname{Aut}^+ T$ on $\mathsf{X}^* \subset T$.*

For more on lattices in automorphism groups of regular trees and their commensurators, see the works [**LMZ94, BL01**].

1.10.2. Free groups. The first example of a self-similar free group (i.e., a faithful self-similar action of a free group) was constructed by Y. Glasner and S. Mozes in [**GM03**] using bi-reversible automata.

Their construction of the free group is the following. We take a pair of different primes p, l both congruent to 1 modulo 4. Then there exist exactly $p + 1$ integral quaternions $x = a + bi + cj + dk$ such that a is odd and positive, b, c, d are all even, and the norm $N(x) = a^2 + b^2 + c^2 + d^2$ is equal to p. (See, for example, [**LPS88**] for proofs.) Denote these quaternions by $x_1, x_2, \ldots, x_{p+1}$. Similarly there are $l + 1$ quaternions $q_1, q_2, \ldots, q_{l+1}$ associated to the prime l.

It is known (see [**GM03, LPS88**] and the bibliography therein) that for any two quaternions q_i, x_j there is a unique pair q_k, x_m satisfying

$$q_i \cdot x_j = \pm x_m \cdot q_k.$$

We interpret these equations (discarding $\pm$) as a definition of an automaton over the alphabet $\mathsf{X} = \{x_1, x_2, \ldots, x_{p+1}\}$ with the set of states $\mathsf{A} = \{q_1, q_2, \ldots, q_{l+1}\}$. Then A generates a free group of rank $l + 1$. The smallest example is therefore an automaton (A, X) with $\{|\mathsf{A}|, |\mathsf{X}|\} = \{6, 14\}$.

A possibly simpler example is the group generated by the automaton over the alphabet $\mathsf{X} = \{0, 1\}$ shown on the left-hand side of Figure 1.9.

This automaton is one of the automata described in [**Ale83**]. There was posed a conjecture in [**Sid00**] (Section 4, Problem 2) that the group generated by a, b, c

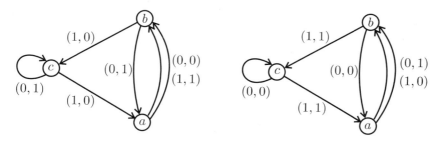

FIGURE 1.9. Bi-reversible automata

is free. It is not known yet if this is true. Note that this automaton is bi-reversible (as are the Glasner-Mozes examples). It is interesting that bi-reversibility was also used in the paper [**Ale83**], though the proof there is not complete and it is not known yet if the statement of the main theorem is true.

1.10.3. Free product $C_2 * C_2 * C_2$. Consider the automaton shown on the right-hand side of Figure 1.9. The following is a result of E. Muntyan and D. Savchuk. We will give a complete proof, since it contains several useful techniques (like, for example, two different ways to prove that some action is level-transitive).

THEOREM 1.10.2. *The group generated by the transformations*
$$a = \sigma(b,b), \quad b = (a,c), \quad c = (c,a)$$
(i.e., the group generated by the automaton shown on the right-hand side of Figure 1.9) is isomorphic to the free product $C_2 * C_2 * C_2$ of three groups of order 2.

PROOF. We see that $a^2 = (b^2, b^2)$, $b^2 = (a^2, c^2)$ and $c^2 = (c^2, a^2)$; hence the transformations a, b, c are of order two. In particular, every element of the group $G = \langle a, b, c \rangle$ can be written as a word in the alphabet $\mathsf{A} = \{a, b, c\}$ without equal consecutive letters.

Let us prove at first that the group G is infinite. It is sufficient to prove that the element ab generates a level-transitive cyclic group (i.e., that it is *level-transitive*). We will use the following lemma.

LEMMA 1.10.3. *Suppose that $|\mathsf{X}| = 2$. Then an automorphism $g \in \operatorname{Aut} \mathsf{X}^*$ is level-transitive if and only if for every $n \geq 0$ the number of words $v \in \mathsf{X}^n$ with active restriction $g|_v$ is odd.* □

Recall that an automorphism is said to be active if it acts non-trivially on the first level of the tree. Proof of Lemma 1.10.3 is an easy induction on the level number. See for example [**GNS00**], Lemma 4.4; or [**Sid00**], Corollary 21.

LEMMA 1.10.4. *The element $ab \in G$ is level-transitive; hence the group G is infinite.*

PROOF. Let $p : \operatorname{Aut} \mathsf{X}^* \longrightarrow \mathbb{F}_2^{\mathbb{N}}$ be defined as $p(g) = (p_0, p_1, \ldots)$, where p_n is the parity of the number of words $v \in \mathsf{X}^n$ such that the restriction $g|_v$ is active. We have to prove that $p(ab) = (1, 1, \ldots)$.

It is straightforward that $p((g_0, g_1)) = (0, p(g_0) + p(g_1))$ and $p(\sigma(g_0, g_1)) = (1, p(g_0) + p(g_1))$. Here, for $\xi = (p_0, p_1, \ldots) \in \mathbb{F}_2^{\mathbb{N}}$, we denote $(1, \xi) = (1, p_0, p_1, \ldots)$ and $(0, \xi) = (0, p_0, p_1, \ldots)$.

It is not hard to see that p is a homomorphism of groups (it is, actually, the abelianization).

We have $ab = \sigma(ba, bc)$ and $bc = (ac, ca)$. Therefore
$$p(bc) = (0, p(ac) + p(ca)) = (0, 0, 0, \ldots)$$
and
$$p(ab) = (1, p(ba) + p(bc)) = (1, p(ab));$$
hence $p(ab) = (1, 1, 1, \ldots)$ and ab is level-transitive. □

Let us denote the letters of the alphabet X by x_0, x_1. Then the recursions defining the generators are written
$$a \cdot x_0 = x_1 \cdot b \qquad a \cdot x_1 = x_0 \cdot b$$
$$b \cdot x_0 = x_0 \cdot a \qquad b \cdot x_1 = x_1 \cdot c$$
$$c \cdot x_0 = x_0 \cdot c \qquad c \cdot x_1 = x_1 \cdot a.$$

Let us interpret now the recursions above as a definition of an automaton (X, A) with the set of states $\mathsf{X} = \{x_0, x_1\}$ over the alphabet $\mathsf{A} = \{a, b, c\}$ (so, for example, $x_0 \cdot a = b \cdot x_0$). We see that this automaton is invertible. (It is actually the inverse of the automaton dual to A.) Let $\widetilde{H}$ be the automorphism group of the tree A^* generated by the automaton (X, A).

Let T be the subtree of A^* consisting of the words which do not have equal consecutive letters. The empty word is the root of T. All vertices of the tree T have degree 3. So, every vertex of T, except for the root, is adjacent to 2 vertices of the next level. The tree T is the Cayley graph of the group $C_2 * C_2 * C_2$.

LEMMA 1.10.5. *The tree T is invariant under the action of $\widetilde{H}$.*

Moreover, for every $g \in \widetilde{H}$, $t \in \mathsf{A}$ and $u, v \in \mathsf{A}^*$ the word $g(uttv)$ has the form $u't't'v'$, where $g(uv) = u'v'$ and $|u| = |u'|$, $|v| = |v'|$.

PROOF. It follows from the equalities
$$x_0 \cdot aa = bb \cdot x_0, \quad x_0 \cdot bb = aa \cdot x_0, \quad x_0 \cdot cc = cc \cdot x_0$$
$$x_1 \cdot aa = cc \cdot x_1, \quad x_1 \cdot bb = aa \cdot x_1, \quad x_1 \cdot cc = bb \cdot x_1,$$
which are checked directly. □

Let H be the automorphism group of the tree T generated by $\mathsf{X} = \{x_0, x_1\}$ (one can prove that H is the same as $\widetilde{H}$, i.e., that the action of $\widetilde{H}$ is faithful on T, but we do not need this fact).

LEMMA 1.10.6. *The group H is infinite.*

PROOF. Suppose that H is finite. Then every orbit of its action on the vertex set of T has not more than $|H|$ elements. Let $g \in G$ be arbitrary. It can be written as a word without equal consecutive letters, i.e., as a vertex of T. Let $v \in \mathsf{X}^*$ be arbitrary. We have $g \cdot v = u \cdot h$ for $u = g(v)$ and $h = g|_v$. Let us consider the elements $u, v \in \mathsf{X}^*$ as elements of H and the elements g, h as words of equal length belonging to T. Then we get $u^{-1} \cdot g = h \cdot v^{-1}$, and thus h is the image of g under the action of u^{-1} on T. Therefore, for a given $g \in G$ there is not more than $|H|$ different restrictions $h = g|_v$, and thus every element of G is defined by an automaton with at most $|H|$ states. But there are only finitely many such automata and the group G is infinite. Contradiction. □

Let $\mathsf{St}(n)$ be the stabilizer of the nth level of the action of H on T.

LEMMA 1.10.7. *The stabilizers $\mathsf{St}(n)$ are pairwise different.*

PROOF. We have to find for every $n \geq 0$ an element $g \in H$ such that $g \in \mathsf{St}(n) \setminus \mathsf{St}(n+1)$.

By Lemma 1.10.6 the group H is infinite and thus the stabilizer $\mathsf{St}(n)$ is non-trivial. Take any non-trivial $h \in \mathsf{St}(n)$ and let $m \geq n$ be the smallest number such that $h \notin \mathsf{St}(m+1)$. Let $\widetilde{h}$ be a preimage of h in $\widetilde{H}$. There exists $v = a_1 a_2 \ldots a_{m+1} \in T$ be such that $h(v) \neq v$. Since m is smallest, we have $h \in \mathsf{St}(m) \setminus \mathsf{St}(m+1)$. Take the element $\widetilde{g} = \widetilde{h}|_{a_1 \ldots a_{m-n}}$ of $\widetilde{H}$ and denote by g the image of $\widetilde{g}$ in H. Let us prove that $g \in \mathsf{St}(n) \setminus \mathsf{St}(n+1)$.

Let $v = t_1 t_2 \ldots t_n$ be an arbitrary vertex of the nth level of T. Then we have

$$\widetilde{h}(a_1 a_2 \ldots a_{m-n} t_1 t_2 \ldots t_n) = \widetilde{h}(a_1 a_2 \ldots a_{m-n}) \widetilde{g}(t_1 t_2 \ldots t_n).$$

It is possible that the word $a_1 a_2 \ldots a_{m-n} t_1 t_2 \ldots t_n$ does not belong to T. Then it can be written in the form $u_1 v^{-1} v u_2$, where $u_1 v^{-1} = a_1 \ldots a_{m-n}$, $v u_2 = t_1 \ldots t_n$, where v^{-1} is the word v written in the opposite order and the word $u_1 u_2$ belongs to T (here v and v^{-1} are the parts which cancel out if we reduce the word $a_1 a_2 \ldots a_{m-n} t_1 t_2 \ldots t_n$ in $C_2 * C_2 * C_2$).

Then Lemma 1.10.5 implies that

$$\widetilde{h}(a_1 a_2 \ldots a_{m-n} t_1 t_2 \ldots t_n) = \widetilde{h}\left(u_1 v^{-1} v u_2\right) = u_1' w^{-1} w u_2',$$

where u_1', w, u_2' are such that $|u_1'| = |u_1|$, $|u_2'| = |u_2|$ and $\widetilde{h}(u_1 u_2) = u_1' u_2'$. But $h \in \mathsf{St}(m)$; hence $u_1 v^{-1} = \widetilde{h}(u_1 v^{-1}) = u_1' w^{-1}$ and $u_1 u_2 = \widetilde{h}(u_1 u_2) = u_1' u_2'$. Therefore, $u_1 = u_1'$, $u_2 = u_2'$ and $w = v$. Then

$$\widetilde{g}(t_1 t_2 \ldots t_n) = w u_2' = v u_2 = t_1 t_2 \ldots t_n$$

and $g \in \mathsf{St}(n)$.

We also have

$$a_1 \ldots a_{m+1} \neq \widetilde{h}(a_1 \ldots a_{m+1})$$
$$= \widetilde{h}(a_1 \ldots a_{m-n}) \widetilde{g}(a_{m-n+1} \ldots a_{m+1})$$
$$= a_1 \ldots a_{m-n} \widetilde{g}(a_{m-n+1} \ldots a_{m+1});$$

hence $\widetilde{g}(a_{m-n+1} \ldots a_{m+1}) \neq a_{m-n+1} \ldots a_{m+1}$ and thus $g \notin \mathsf{St}(n+1)$. $\square$

LEMMA 1.10.8. *The group H is level-transitive on T.*

PROOF. Let us prove by induction that the action of H is transitive on the nth level of the tree T. The statement is true for $n = 1$ (since x_1 acts transitively on A). Suppose that it is true for $n = k$; let us prove it for $n = k+1$. There exists $g \in \mathsf{St}(k) \setminus \mathsf{St}(k+1)$. Take a vertex $v \in \mathsf{A}^k \cap T$ such that g permutes the two vertices $v t_1$ and $v t_2$ of the level number $k+1$, which are adjacent to v.

Let $wx \in T$ be an arbitrary vertex, where $w \in \mathsf{A}^k$ and $x \in \mathsf{A}$. By the inductive assumption, there exists $h \in H$ such that $h(w) = v$. Then $h(wx) = vt_1$ or $h(wx) = vt_2$. In the second case we have $gh(wx) = vt_1$. We see that every vertex of the level number $k+1$ can be mapped by an element of H to vt_1; i.e., H acts transitively on the level number $k+1$. $\square$

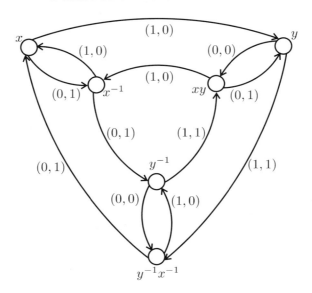

FIGURE 1.10. Automaton generating F_2

The proofs of Lemmas 1.10.7 and 1.10.8 are based on an idea communicated to the author by E. Muntyan and D. Savchuk.

The statement of the theorem follows now from the last lemma. Suppose that on the contrary, there exist non-empty words $v \in T$ representing trivial automorphisms of the tree X^*. Let K be the set of such words (i.e., the kernel of the homomorphism $C_2 * C_2 * C_2 \longrightarrow \operatorname{Aut} X^*$).

If v belongs to K, then $v \cdot x_0 = x_0 \cdot v_0$ and $v \cdot x_1 = x_1 \cdot v_1$ for some $v_0, v_1 \in K$. We get then that $x_0^{-1} \cdot v = v_0 \cdot x_0^{-1}$ and $x_1^{-1} \cdot v = v_1 \cdot x_1^{-1}$. This shows that $x_0^{-1}(K) \subseteq K$ and $x_1^{-1}(K) \subseteq K$, where K is seen as a subset of the tree T. But the group H preserves the levels of the tree T, which are finite sets. Therefore, the set K is H-invariant. This implies, by Lemma 1.10.8, that K is a union of levels of the tree T. But every level, except for the root, has obviously non-trivial elements (like $ab^{n-1} = \sigma(ba^{n-1}, bc^{n-1})$, for example). Hence K contains only the empty word (the root of T). $\square$

1.10.4. Free group of rank 2. Let a, b, c be the generators of $C_2 * C_2 * C_2$ from the previous example. Then $x = ab$ and $y = bc$ generate a free subgroup of index 2 in $\langle a, b, c \rangle \simeq C_2 * C_2 * C_2$. This subgroup is self-similar, since

$$x = \sigma(ba, bc) = \sigma(x^{-1}, y)$$
$$y = (ac, ca) = (xy, y^{-1}x^{-1}).$$

Thus, $\langle x, y \rangle$ is an example of a free self-similar group of rank 2. This group is generated by a finite automaton with six states $\{x, y, x^{-1}, y^{-1}, xy, y^{-1}x^{-1}\}$. This automaton is shown in Figure 1.10. Note that $\langle x, y \rangle$ is also a subgroup of the group generated by the automaton shown on the left-hand side of Figure 1.9. We have $x = a^{-1}b$ and $y = b^{-1}c$, where a, b and c are the states of this automaton.

On the other hand, there are many examples of non-self-similar faithful actions of free groups on the rooted tree X^*. M. Bchattacharjee has shown in [**Bha95**] that almost any k-tuple of automorphisms of X^* are free generators of a free group. It was harder though to give an example of a finite-state free automorphism group of X^*. The first attempt was made in [**Ale83**], but as we already mentioned, a proof that the proposed automata generate a free group is still not known. The first example of a finite-state free group (for $|\mathsf{X}| = 4$) is contained in [**BS98**], where affine groups were represented by finite automata (see Subsection 1.9.1 in our book). The first example of a free finite-state subgroup of $\operatorname{Aut} \mathsf{X}^*$ for $|\mathsf{X}| = 2$ was constructed by A. Oliĭnyk [**Oli99, Oli98**].

CHAPTER 2

Algebraic Theory

We will study in this chapter algebraic aspects of self-similarity of group actions. We will interpret the notation $g \cdot x = y \cdot h$ as a bimodule structure on the direct product $\mathsf{X} \times G$. The notion of a *permutational bimodule* together with the natural notion of the *tensor product* will give us a convenient algebraic formalism for working with self-similar groups.

We will also start to study an important class of *contracting* self-similar actions. The next chapters will be devoted to geometric and dynamical aspects of contracting actions, while here we collect their basic algebraic properties.

2.1. Permutational bimodules

2.1.1. Definitions.

DEFINITION 2.1.1. Let G be a group. A *permutational G-bimodule* is a set $\mathfrak{M}$ together with commuting left and right actions of G on $\mathfrak{M}$. In other words, we have two maps $G \times \mathfrak{M} \longrightarrow \mathfrak{M} : (g, m) \mapsto g \cdot m$ and $\mathfrak{M} \times G \longrightarrow \mathfrak{M} : (m, g) \mapsto m \cdot g$ such that

(1) $1 \cdot m = m \cdot 1 = m$ for all $m \in \mathfrak{M}$;
(2) $(g_1 g_2) \cdot m = g_1 \cdot (g_2 \cdot m)$ and $m \cdot (g_1 g_2) = (m \cdot g_1) \cdot g_2$ for all $g_1, g_2 \in G$ and $m \in \mathfrak{M}$;
(3) $(g_1 \cdot m) \cdot g_2 = g_1 \cdot (m \cdot g_2)$ for all $g_1, g_2 \in G$ and $m \in \mathfrak{M}$.

Two G-bimodules $\mathfrak{M}_1, \mathfrak{M}_2$ are *isomorphic* if there exists a bijection $f : \mathfrak{M}_1 \longrightarrow \mathfrak{M}_2$ which agrees with the left and the right actions, i.e., such that $g \cdot f(m) \cdot h = f(g \cdot m \cdot h)$ for all $g, h \in G$ and $m \in \mathfrak{M}_1$.

For a given permutational G-bimodule we denote by $\mathfrak{M}_G$ and $_G\mathfrak{M}$ the respective right and left G-modules (i.e., the set $\mathfrak{M}$ with the right and the left actions respectively).

DEFINITION 2.1.2. A permutational bimodule $\mathfrak{M}$ is called a *covering bimodule* if the right module $\mathfrak{M}_G$ is *free*, i.e., if $x \cdot g = x$ implies $g = 1$ for any $x \in \mathfrak{M}$.

We say that $\mathfrak{M}$ is *d-fold* if the set of right G-orbits on $\mathfrak{M}$ has cardinality d.

Permutational G-bimodules can be seen as *correspondences*. A correspondence between groups G_1 and G_2 is a $(G_1$–$G_2)$-bimodule, i.e., a set $\mathfrak{M}$ with a left action of G_1 and a right action of G_2 which commute. One can "compose" correspondences taking their *tensor products* (see 2.3.1).

We will see below that self-similarity of a group can be interpreted as an auto-correspondence on it. The study of iterations of these auto-correspondences is an essential part of the study of self-similar groups.

2.1.2. Bimodules associated to self-similar actions.
Suppose that we have a self-similar action (G, X).

The *associated bimodule* of the action (or the *self-similarity bimodule*) is the direct product $\mathfrak{M} = \mathsf{X} \times G$ with the right action given by

$$(x, g) \cdot h = (x, gh),$$

and the left action

$$h \cdot (x, g) = (h(x), h|_x g).$$

We will identify a letter $x \in \mathsf{X}$ with the pair $(x, 1) \in \mathsf{X} \times G$. Then the pair (x, g) is written $x \cdot g$, by definition of the right action.

In other words, we define the bimodule $\mathfrak{M}$ in such a way that the equality $g \cdot x = y \cdot h$ holds in $\mathfrak{M}$ if and only if $g(xw) = yh(w)$ for all $w \in \mathsf{X}^*$, i.e., if it holds in the complete automaton of the action.

It follows directly from the definition of the right action that $\mathfrak{M} = \mathsf{X} \cdot G$ is a d-fold covering bimodule for $d = |\mathsf{X}|$.

If we identify an element $x \cdot g \in \mathfrak{M}$ with the map

$$w \mapsto xg(w)$$

on X^*, then both the left and the right actions of G on $\mathfrak{M}$ coincide with composition of maps:

$$x \cdot g(h(w)) = x \cdot gh(w), \qquad h(x \cdot g(w)) = h(x) \cdot h|_x g(w).$$

The axioms of a bimodule easily follow from this interpretation (or from (1.3)).

2.2. Bases of a covering bimodule and wreath recursions

Suppose that $\mathfrak{M}$ is a d-fold covering bimodule over G. Then a *basis* of $\mathfrak{M}$ is, by definition, an orbit transversal of the right action of G on $\mathfrak{M}$. It follows from the definition that if X is a basis, then $|\mathsf{X}| = d$ and every element $m \in \mathfrak{M}$ is written in a unique way in the form $m = x \cdot g$, where $x \in \mathsf{X}$ and $g \in G$.

If $\mathsf{X} \cdot G = \mathfrak{M}$ is the bimodule associated to a self-similar action (G, X), then the set $\mathsf{X} = \{x = x \cdot 1\}$ is a natural basis of $\mathfrak{M}$.

It also follows that if $\mathsf{X} = \{x_1, \ldots, x_d\}$ is a basis of $\mathfrak{M}$, then a collection $\mathsf{Y} = \{y_1, \ldots, y_d\}$ is a basis of $\mathfrak{M}$ if and only if there exists a permutation $\pi \in \mathfrak{S}(d)$ and elements $g_i \in G$ such that

$$y_i = x_{\pi(i)} \cdot g_i.$$

A bijection $\alpha : \mathfrak{M} \longrightarrow \mathfrak{M}$ is an *automorphism of the right module* $\mathfrak{M}_G$ if

$$\alpha(m \cdot g) = \alpha(m) \cdot g$$

for all $m \in \mathfrak{M}$ and $g \in G$. We denote the automorphism group of the right module $\mathfrak{M}_G$ by $\operatorname{Aut} \mathfrak{M}_G$.

PROPOSITION 2.2.1. *Let $\mathfrak{M}$ be a d-fold covering G-bimodule with a basis $\mathsf{X} = \{x_1, \ldots, x_d\}$. Then the map putting in correspondence to an automorphism $\alpha \in \operatorname{Aut} \mathfrak{M}_G$ the element*

$$\pi(g_1, \ldots, g_d) \in \mathfrak{S}(\mathsf{X}) \wr G,$$

where $g_i \in G$ and $\pi \in \mathfrak{S}(\mathsf{X})$ are such that

$$\alpha(x_i) = \pi(x_i) \cdot g_i,$$

is an isomorphism between $\operatorname{Aut} \mathfrak{M}_G$ and $\mathfrak{S}(\mathsf{X}) \wr G$.

PROOF. For any (ordered) basis $\mathsf{Y} = \{y_1, \ldots, y_d\}$ of $\mathfrak{M}$ the map
$$\alpha : x_i \mapsto y_i$$
extends to a unique automorphism $\alpha \in \operatorname{Aut} \mathfrak{M}_G$. It is the automorphism given by
$$\alpha(x_i \cdot g) = y_i \cdot g.$$

On the other hand, if α is an automorphism of $\mathfrak{M}_G$, then $\{\alpha(x_1), \ldots, \alpha(x_d)\}$ is a basis of $\mathfrak{M}_G$.

Consequently, the set $\operatorname{Aut} \mathfrak{M}_G$ is in a bijective correspondence with the set of ordered bases $\mathsf{Y} = \{y_1, \ldots, y_d\}$. The latter, as we know, is in a bijective correspondence with the group $\mathfrak{S}(\mathsf{X}) \wr G$, so that $\mathsf{Y} = \{y_1 = \pi(x_1) \cdot g_1, \ldots, y_d = \pi(x_d) \cdot g_d\}$ corresponds to $\pi(g_1, \ldots, g_d) \in \mathfrak{S}(\mathsf{X}) \wr G$.

We have to check that the obtained bijection $\alpha \mapsto \pi \cdot (g_1, \ldots, g_d)$ is a homomorphism of groups.

Let α and $\beta \in \operatorname{Aut} \mathfrak{M}_G$ correspond to $\pi(g_1, \ldots, g_d)$ and $\sigma(h_1, \ldots, h_d)$. Then
$$\alpha(\beta(x_i)) = \alpha(\sigma(x_i) \cdot h_i) = \alpha(\sigma(x_i)) \cdot h_i = \pi(\sigma(x_i)) g_{\sigma(i)} h_i.$$
Hence, $\alpha\beta$ corresponds to $\pi\sigma(g_{\sigma(1)}h_1, \ldots, g_{\sigma(d)}h_d)$, which agrees with the multiplication in $\mathfrak{S}(\mathsf{X}) \wr G$. $\square$

The left action of G on $\mathfrak{M}$ commutes with the right action, so that we get for every $g \in G$ an automorphism $\psi(g) \in \operatorname{Aut} \mathfrak{M}_G$:
$$\psi(g)(m) = g \cdot m.$$
The bimodule $\mathfrak{M}$ is then uniquely determined by this *structural homomorphism*
$$\psi : G \longrightarrow \operatorname{Aut} \mathfrak{M}_G \cong \mathfrak{S}(\mathsf{X}) \wr G,$$
which is called the *wreath recursion*.

On the other hand, suppose that we have a homomorphism
$$\psi : G \longrightarrow \mathfrak{S}(\mathsf{X}) \wr G.$$
Let $\mathfrak{M}$ be the set $\mathsf{X} \times G$. For $g_1 \in G$, let $\psi(g_1) = \pi \cdot \overline{g}_1$, where $\overline{g}_1 \in G^\mathsf{X}$ is a function $\mathsf{X} \longrightarrow G$ and $\pi \in \mathfrak{S}(\mathsf{X})$ is a permutation. We define then the actions of G on $\mathfrak{M}$ by
$$g_1 \cdot (x, g) \cdot g_2 = (\pi(x), \overline{g}_1(x) g g_2),$$
where $\overline{g}_1(x) \in G$ is the value of the function $\overline{g}_1 \in G^\mathsf{X}$ at the point $x \in \mathsf{X}$.

It is easy to check then that $\mathfrak{M}$ is a G-bimodule with the structural homomorphism ψ.

2.3. Tensor products and self-similar actions

2.3.1. Tensor products of bimodules.
Let $\mathfrak{M}_1$ and $\mathfrak{M}_2$ be permutational G-bimodules. Then their *tensor product* $\mathfrak{M}_1 \otimes \mathfrak{M}_2$ is the quotient of the set $\mathfrak{M}_1 \times \mathfrak{M}_2$ by the equivalence relation
$$(x_1 \cdot g) \otimes x_2 = x_1 \otimes (g \cdot x_2),$$
where $g \in G$, $x_1 \in \mathfrak{M}_1$, $x_2 \in \mathfrak{M}_2$ and $x \otimes y = (x, y) \in \mathfrak{M}_1 \times \mathfrak{M}_2$.

The proof of the following proposition is straightforward.

PROPOSITION 2.3.1. *The quotient $\mathfrak{M}_1 \otimes \mathfrak{M}_2$ is well defined, and the actions*
$$g \cdot (x_1 \otimes x_2) = (g \cdot x_1) \otimes x_2, \quad (x_1 \otimes x_2) \cdot g = x_1 \otimes (x_2 \cdot g)$$
give a well defined bimodule structure on $\mathfrak{M}_1 \otimes \mathfrak{M}_2$.

If $\mathfrak{M}_1, \mathfrak{M}_2, \mathfrak{M}_3$ are permutational G-bimodules, then the mapping
$$(x_1 \otimes x_2) \otimes x_3 \mapsto x_1 \otimes (x_2 \otimes x_3)$$
induces an isomorphism of the bimodules $(\mathfrak{M}_1 \otimes \mathfrak{M}_2) \otimes \mathfrak{M}_3$ and $\mathfrak{M}_1 \otimes (\mathfrak{M}_2 \otimes \mathfrak{M}_3)$.

In particular, the nth tensor power
$$\mathfrak{M}^{\otimes n} = \underbrace{\mathfrak{M} \otimes \mathfrak{M} \otimes \cdots \otimes \mathfrak{M}}_{n \text{ times}}$$
of a G-bimodule $\mathfrak{M}$ is defined.

We set $\mathfrak{M}^{\otimes 0}$ equal to the group G with the natural G-bimodule structure. The bimodules $G \otimes \mathfrak{M}$ and $\mathfrak{M} \otimes G$ are obviously isomorphic to $\mathfrak{M}$.

If $\mathfrak{M}$ is a G-bimodule and $\mathfrak{M}'$ is a right (or left) G-module, then the right module $\mathfrak{M}' \otimes \mathfrak{M}$ (resp. the left module $\mathfrak{M} \otimes \mathfrak{M}'$) is defined. Also if $\mathfrak{M}_1$ is a right module and $\mathfrak{M}_2$ is a left module, then the tensor product $\mathfrak{M}_1 \otimes \mathfrak{M}_2$ is also defined but is just a set.

PROPOSITION 2.3.2. *Let $\mathfrak{M}_1$ and $\mathfrak{M}_2$ be covering bimodules and let $\mathsf{X}_1, \mathsf{X}_2$ be their bases. Then $\mathfrak{M}_1 \otimes \mathfrak{M}_2$ is a covering bimodule and the set $\mathsf{X}_1 \otimes \mathsf{X}_2 = \{x_1 \otimes x_2 : x_1 \in \mathsf{X}_1, x_2 \in \mathsf{X}_2\}$ is its basis.*

PROOF. Suppose that $(m_1 \otimes m_2) \cdot g = m_1 \otimes m_2$. This means that there exists $h \in G$ such that $m_1 = m_1 \cdot h$ and $m_2 \cdot g = h \cdot m_2$. The right action in $\mathfrak{M}_1$ is free therefore $h = 1$. But then $m_2 \cdot g = m_2$, which implies that $g = 1$. Consequently, the bimodule $\mathfrak{M}_1 \otimes \mathfrak{M}_2$ has a free right action, i.e., is a covering bimodule.

Let $m_1 \otimes m_2$ be an arbitrary element of $\mathfrak{M}_1 \otimes \mathfrak{M}_2$. Then $m_1 = x_1 \cdot g_1$ and $g_1 \cdot m_2 = x_2 \cdot g_2$ for some $g_1, g_2 \in G$ and $x_1 \in \mathsf{X}_1, x_2 \in \mathsf{X}_2$. Then $m_1 \otimes m_2 = (x_1 \cdot g_1) \otimes m_2 = x_1 \otimes (g_1 \cdot m_2) = x_1 \otimes (x_2 \cdot g_2)$. Hence, the set $\mathsf{X}_1 \otimes \mathsf{X}_2$ intersects every right orbit of the bimodule $\mathfrak{M}_1 \otimes \mathfrak{M}_2$.

Suppose that $x_1 \otimes x_2 \cdot g = y_1 \otimes y_2$ for some $x_1, y_1 \in \mathsf{X}_1$, $x_2, y_2 \in \mathsf{X}_2$ and $g \in G$. Then there exists $h \in G$ such that $y_1 \cdot h = x_1$ and $y_2 = h \cdot x_2 \cdot g$. But X_1 is a basis of the bimodule $\mathfrak{M}_1$; thus the first equality implies that $h = 1$ and $x_1 = y_1$. But then $y_2 = x_2 \cdot g$; thus $g = 1$ and $x_2 = y_2$. So, every right orbit of $\mathfrak{M}_1 \otimes \mathfrak{M}_2$ contains not more than one element of $\mathsf{X}_1 \otimes \mathsf{X}_2$. □

2.3.2. Associated self-similar action on X^*. As a corollary of Proposition 2.3.2 we get that if X is a basis of a covering bimodule $\mathfrak{M}$, then
$$\mathsf{X}^n = \{x_1 \otimes x_2 \otimes \cdots \otimes x_n \ : \ x_i \in \mathsf{X}\}$$
is a basis of the bimodule $\mathfrak{M}^{\otimes n}$. We will use the short-hand notation
$$x_1 x_2 \ldots x_n = x_1 \otimes x_2 \otimes \cdots \otimes x_n.$$

Every element of $\mathfrak{M}^{\otimes n}$ is uniquely written in the form $v \cdot g$, where $v \in \mathsf{X}^n$ and $g \in G$. In particular, for every pair $g \in G$, $v \in \mathsf{X}^n$ there exists a pair $h \in G$, $u \in \mathsf{X}^n$ such that $g \cdot v = u \cdot h$ in $\mathfrak{M}^{\otimes n}$. The pair u, v is uniquely defined due to Proposition 2.3.2, and we denote $u = g(v)$ and $h = g|_v$. The following proposition follows directly from the uniqueness and the definitions of permutational bimodules and their tensor products.

2.3. TENSOR PRODUCTS AND SELF-SIMILAR ACTIONS

PROPOSITION 2.3.3. *The equalities*

$$g \cdot v = g(v) \cdot g|_v$$

in the bimodules $\mathfrak{M}^{\otimes n}$ define a self-similar action $g : v \mapsto g(v)$ of G on X^*.

It is the original action of G on X^* if $\mathfrak{M}$ is the bimodule $\mathsf{X} \cdot G$ associated to a self-similar action (G, X).

The restriction map $g \mapsto g|_v$ satisfies

(2.1) $$g|_{v_1 v_2} = (g|_{v_1})|_{v_2}, \qquad (g_1 g_2)|_v = g_1|_{g_2(v)} g_2|_v.$$

The action of G on X^* is defined by the automaton (G, X) whose output and transition functions are given by the condition that

$$g \cdot x = g(x) \cdot g|_x$$

in $\mathfrak{M}$.

PROOF. The proof is straightforward. The only thing to check is (2.1), which follows directly from the axioms of a (covering) bimodule. $\square$

The action described in Proposition 2.3.3 is called the *self-similar action defined by the bimodule $\mathfrak{M}$ and its basis X*. It is denoted $(G, \mathfrak{M}, \mathsf{X})$ or just (G, X) if it is clear which bimodule is being considered.

2.3.3. Tensor power of self-similar actions. Recall that every self-similar action (G, X) can be identified with its *complete automaton* (also denoted (G, X)) with the set of states G over the alphabet X (see Definition 1.5.2).

Using the complete automaton (G, X) we can construct the automaton (G, X^n) (see 1.3.4 on page 5). It will also satisfy the conditions of Definition 1.5.2 and thus will correspond to a self-similar action of G over the alphabet X^n.

This action is just the restriction of the action of G on X^* onto the subset $(\mathsf{X}^n)^*$ of words of length divisible by n if we identify the word

$$(x_1 x_2 \ldots x_n)(x_{n+1} x_{n+2} \ldots x_{2n}) \ldots (x_{kn+1} x_{kn+2} \ldots x_{(k+1)n}) \in (\mathsf{X}^n)^*$$

with the word $x_1 x_2 \ldots x_{(k+1)n} \in \mathsf{X}^*$.

The corresponding actions on X^ω and $(\mathsf{X}^n)^\omega$ are conjugate, and the conjugating homeomorphism is the continuous extension of the above natural identification of finite words.

It follows directly from Proposition 2.3.2 that the bimodule associated to the action (G, X^n) is isomorphic to $\mathfrak{M}^{\otimes n}$. Moreover, the isomorphism is in a sense the tautological map

$$x_1 x_2 \ldots x_n \cdot g \mapsto x_1 \otimes x_2 \otimes \cdots \otimes x_n \cdot g,$$

where on the left-hand side we have an element of $\mathsf{X}^n \cdot G$ and on the right-hand side is an element of $\mathfrak{M}^{\otimes n}$.

Since passing to (G, X^n) corresponds to passing to the nth tensor power of the associated bimodule, the self-similar action (G, X^n) is called the *nth tensor power* of the action (G, X).

2.3.4. Conjugacy of associated actions.

PROPOSITION 2.3.4. *Let $\mathfrak{M}$ be a d-fold covering bimodule over G. Let X, Y be bases of $\mathfrak{M}$. Then the self-similar actions (G, X) and (G, Y) are conjugate and the conjugating isomorphism is the map $\alpha : \mathsf{X}^* \longrightarrow \mathsf{Y}^*$ such that for every $v \in \mathsf{X}^*$ there exists $\alpha_v \in G$ for which*
$$v = \alpha(v) \cdot \alpha_v$$
in $\mathfrak{M}$. The map α is defined by the recurrent formula

(2.2) $$\alpha(xw) = y h_x \alpha(w),$$

where $h_x \in G$ and $y \in \mathsf{Y}$ are such that $x = y \cdot h_x$ and $w \in \mathsf{X}^*$ is arbitrary.

PROOF. Let $v = x_{i_1} x_{i_2} \ldots x_{i_n}$ be an arbitrary element of X^*. Proposition 2.3.2 implies that there exists a unique $\alpha(v) \in \mathsf{Y}^*$ such that $v = \alpha(v) \cdot \alpha_v$ for some (also uniquely defined) $\alpha_v \in G$.

We have $g \cdot v = g(v) \cdot g|_v$ (where $g(v)$ is computed using the associated action of G on X^*); therefore $g \cdot v = \alpha(g(v)) \cdot \alpha_{g(v)} g|_v$. On the other hand,
$$g \cdot v = g \cdot \alpha(v) \cdot \alpha_v(v) = g(\alpha(v)) \cdot g|_{\alpha(v)} h(v),$$
where $g(\alpha(v))$ is computed using the self-similar action (G, Y). Hence, $\alpha(g(v)) = g(\alpha(v))$.

Recurrent formula (2.2) follows directly from the definition of α. $\square$

Let us index the bases $\mathsf{X} = \{x_1, \ldots, x_d\}$ and $\mathsf{Y} = \{y_1, \ldots, y_d\}$ and consider the bijections $x_i \leftrightarrow i$ and $y_i \leftrightarrow i$ of X and Y with $\mathsf{D} = \{1, 2, \ldots, d\}$. Let $\tilde{\alpha}$ be the conjugator α, interpreted as an automorphism of the tree D^*, i.e., such that $\tilde{\alpha}(i_1 \ldots i_n) = j_1 \ldots j_n$ if and only if $\alpha(x_{i_1} \ldots x_{i_n}) = y_{j_1} \ldots y_{j_n}$. Then the recurrent definition (2.2) of the conjugator α can be written in terms of the wreath recursion as

(2.3) $$\alpha = \pi \left(h_1 \alpha, h_2 \alpha, \ldots, h_d \alpha \right),$$

where $\pi \in \mathfrak{S}(d)$ and h_i are such that $x_i = y_{\pi(i)} \cdot h_i$.

2.4. The left G-space $\mathfrak{M}^{\otimes \omega}$

Let $\mathfrak{M}$ be a d-fold covering bimodule over G and let X be its basis. Then we get the associated self-similar action (G, X). It is an action by automorphisms of the tree X^*, and thus it induces an action of G by homeomorphisms of the boundary X^ω.

The space X^ω and the action of G on it can be naturally interpreted as the infinite tensor power $\mathfrak{M}^\omega = \mathfrak{M}^{\otimes \omega}$ of bimodules.

Namely, we say that two infinite sequences $a_1, a_2, \ldots$ and $b_1, b_2, \ldots$ in $\mathfrak{M}$ define equal infinite products $a_1 \otimes a_2 \otimes \cdots$ and $b_1 \otimes b_2 \otimes \cdots$ if and only if there exists a sequence $g_n \in G$ such that
$$a_1 \otimes a_2 \otimes \cdots \otimes a_n \cdot g_n = b_1 \otimes b_2 \otimes \cdots \otimes b_n$$
for every $n \geq 1$.

The defined set of expressions $a_1 \otimes a_2 \otimes \cdots$ (i.e., the quotient of the set of the infinite sequences by the described equivalence relation) is denoted $\mathfrak{M}^{\otimes \omega}$, or just $\mathfrak{M}^\omega$.

The equivalence relation agrees with the left action of G, so that
$$g \cdot (a_1 \otimes a_2 \otimes \cdots) = (g \cdot a_1) \otimes a_2 \otimes \cdots$$
is a well defined left action of G on $\mathfrak{M}^\omega$.

PROPOSITION 2.4.1. *Let X be a basis of $\mathfrak{M}_G$. Then every element of $\mathfrak{M}^\omega$ can be written in a unique way in the form $x_1 \otimes x_2 \otimes \ldots$ for $x_i \in \mathsf{X}$.*

PROOF. A direct corollary of Proposition 2.3.2 and the definition of $\mathfrak{M}^\omega$. □

Proposition 2.4.1 shows that we have a natural bijection between X^ω and $\mathfrak{M}^\omega$ given by
$$x_1 x_2 \ldots \mapsto x_1 \otimes x_2 \otimes \cdots.$$
We will write the infinite tensor product $a_1 \otimes a_2 \otimes \cdots$ just as a sequence $a_1 a_2 \ldots$ (so that the above bijection becomes tautological).

The left action of G on $\mathfrak{M}^\omega = \mathsf{X}^\omega$ will coincide with the associated self-similar action of G on X^ω.

It follows from Proposition 2.3.4 that if X and Y are two bases of $\mathfrak{M}$, then $x_1 x_2 \ldots \in \mathsf{X}^\omega$ and $y_1 y_2 \ldots \mathsf{Y}^\omega$ represent the same point of $\mathfrak{M}^\omega$ if and only if $\alpha(x_1 x_2 \ldots) = y_1 y_2 \ldots$, for α as in Proposition 2.3.4 (more precisely, α is the action of the conjugator on the boundaries of the trees X^* and Y^*).

But we know that $\alpha : \mathsf{X}^\omega \longrightarrow \mathsf{Y}^\omega$ is a homeomorphism; therefore we get a well defined topology on $\mathfrak{M}^\omega$ if we pull back the topology from X^ω to $\mathfrak{M}^\omega$ by the natural bijection.

2.5. Virtual endomorphisms

2.5.1. Definitions.

DEFINITION 2.5.1. A *virtual homomorphism* $\phi : G_1 \dashrightarrow G_2$ is a homomorphism of groups $\phi : \mathrm{Dom}\,\phi \longrightarrow G_2$, where $\mathrm{Dom}\,\phi \leq G_1$ is a subgroup of finite index called the *domain* of the virtual homomorphism.

A *virtual endomorphism* of a group G is a virtual homomorphism $\phi : G \dashrightarrow G$.

The index $[G_1 : \mathrm{Dom}\,\phi]$ is called the *index of the virtual homomorphism* ϕ and is denoted $\mathrm{ind}\,\phi$.

By $\mathrm{Ran}\,\phi$ we denote the image of $\mathrm{Dom}\,\phi$ under ϕ.

We say that a virtual homomorphism ϕ *is defined* on an element $g \in G_1$ if $g \in \mathrm{Dom}\,\phi$.

A composition of two virtual homomorphisms $\phi_1 : G_1 \dashrightarrow G_2$, $\phi_2 : G_2 \dashrightarrow G_3$ is defined on an element $g \in G_1$ if and only if ϕ_1 is defined on g and ϕ_2 is defined on $\phi_1(g)$. Thus, the domain of the composition $\phi_2 \circ \phi_1$ is the subgroup
$$\mathrm{Dom}\,(\phi_2 \circ \phi_1) = \{g \in \mathrm{Dom}\,\phi_1 : \phi_1(g) \in \mathrm{Dom}\,\phi_2\} \leq G_1.$$

PROPOSITION 2.5.2. *Let $\phi_1 : G_1 \dashrightarrow G_2$ and $\phi_2 : G_2 \dashrightarrow G_3$ be two virtual homomorphisms. Then*
$$[\mathrm{Dom}\,\phi_1 : \mathrm{Dom}\,(\phi_2 \circ \phi_1)] \leq [G_2 : \mathrm{Dom}\,\phi_2] = \mathrm{ind}\,\phi_2.$$
If ϕ_1 is onto, then
$$[\mathrm{Dom}\,\phi_1 : \mathrm{Dom}\,(\phi_2 \circ \phi_1)] = [G_2 : \mathrm{Dom}\,\phi_2].$$

PROOF. We have $[\operatorname{Ran}\phi_1 : \operatorname{Dom}\phi_2 \cap \operatorname{Ran}\phi_1] \leq \operatorname{ind}\phi_2$ and we have here equality in the case when ϕ_1 is onto. Let $T = \{\phi_1(h_1), \phi_1(h_2), \ldots \phi_1(h_d)\}$ be a left coset transversal for $\operatorname{Dom}\phi_2 \cap \operatorname{Ran}\phi_1$ in $\operatorname{Ran}\phi_1$. Then for every $g \in \operatorname{Dom}\phi_1$ there exists a unique $\phi_1(h_i) \in T$ such that $\phi_1(h_j)^{-1}\phi_1(g) = \phi_1(h_i^{-1}g) \in \operatorname{Dom}\phi_2$. This is equivalent to $h_i^{-1}g \in \operatorname{Dom}(\phi_2 \circ \phi_1)$, and hence the set $\{h_1, h_2, \ldots, h_d\}$ is a left coset transversal of $\operatorname{Dom}(\phi_2 \circ \phi_1)$ in G_1. Thus,

$$[G_2 : \operatorname{Dom}\phi_2] = [\operatorname{Ran}\phi_1 : \operatorname{Dom}\phi_2 \cap \operatorname{Ran}\phi_1] \leq \operatorname{ind}\phi_2.$$

□

COROLLARY 2.5.3. *A composition of two virtual homomorphisms is again a virtual homomorphism.*

□

2.5.2. Virtual endomorphisms associated to covering bimodules.
Let $\mathfrak{M}$ be a d-fold covering G-bimodule and take $x \in \mathfrak{M}$. Then the *associated* virtual endomorphism ϕ_x is defined by the condition

$$g \cdot x = x \cdot \phi_x(x).$$

The domain of ϕ_x is the subgroup G_x of those elements $g \in G$ for which x and $g \cdot x$ belong to the same right orbit.

Let us show that $[G : G_x] \leq d$. If $g_1, \ldots, g_n$ are elements of G and $n > d$, then some elements $g_i \cdot x$ belong to one right orbit, since we have only d of them. But if $g_i \cdot x = g_j \cdot x \cdot h$, then $g_j^{-1} g_i \cdot x = x \cdot h$, which implies that $g_j^{-1} g_i \in \operatorname{Dom}\phi_x$, i.e., that g_i and g_j belong to one left $\operatorname{Dom}\phi_x$-coset.

It follows directly from the definitions that $\phi_x : \operatorname{Dom}\phi_x \longrightarrow G$ is a homomorphism.

In particular, if $\mathfrak{M} = \mathsf{X} \cdot G$ is the associated bimodule of a self-similar action (G, X), then the map $\phi_x : G_x \to G$ defined by the formula

$$\phi_x(g) = g|_x$$

is a virtual endomorphism of G. Here G_x is the stabilizer of the one-letter word x in G. This virtual endomorphism $\phi_x : G \dashrightarrow G$ is called the *endomorphism associated to the self-similar action*.

For example, the associated virtual endomorphism of the adding machine action is the map $\mathbb{Z} \dashrightarrow \mathbb{Z} : n \mapsto n/2$ with the domain equal to the set of even numbers. This follows from the equality $a^2 \cdot 0 = 0 \cdot a$, where a is the adding machine $a = \sigma(1, a)$ defined over the alphabet $\mathsf{X} = \{0, 1\}$.

2.5.3. Conjugate virtual endomorphisms.

DEFINITION 2.5.4. We say that virtual homomorphisms $\phi_1, \phi_2 : G_1 \dashrightarrow G_2$ are *conjugate* if there exist $g_1 \in G_1, g_2 \in G_2$ such that $\operatorname{Dom}\phi_1 = g_1^{-1} \cdot \operatorname{Dom}\phi_2 \cdot g_1$ and

$$\phi_2(x) = g_2^{-1}\phi_1(g_1^{-1}xg_1)g_2$$

for all $x \in \operatorname{Dom}\phi_2$.

DEFINITION 2.5.5. We say that a permutational bimodule $\mathfrak{M}$ is *irreducible* if for any $x_1, x_2 \in \mathfrak{M}$ there exist $g_1, g_2 \in G$ such that $g_1 \cdot x_1 \cdot g_2 = x_2$.

The bimodule associated to a self-similar action is irreducible if and only if the action is transitive on the first level X^1 of the tree X^*.

PROPOSITION 2.5.6. *Let $\mathfrak{M}$ be an irreducible d-fold covering G-bimodule. Then every two associated virtual endomorphisms ϕ_x and ϕ_y are conjugate. If ϕ is conjugate to an associated virtual endomorphism ϕ_x, then it is also associated to $\mathfrak{M}$; i.e., there exists $y \in \mathfrak{M}$ such that $\phi = \phi_y$.*

PROOF. There exist $g, h \in G$ such that $y = g \cdot x \cdot h$. Then for every $f \in \text{Dom}\,\phi_y$ we have $f \cdot y = y \cdot \phi_y(f)$, which is equivalent to the condition $fg \cdot x \cdot h = g \cdot x \cdot h\phi_y(f)$, i.e., $g^{-1}fg \cdot x = x \cdot h\phi_y(f)h^{-1}$. It follows that $\phi_y(f) = h^{-1}\phi_x(g^{-1}fg)h$.

Similar arguments show that if $\phi(f) = h^{-1}\phi_x(g^{-1}fg)h$, then ϕ is the virtual endomorphism associated to $\mathfrak{M}$ and $g \cdot x \cdot h \in \mathfrak{M}$. □

2.5.4. The bimodule $\phi(G)G$. Let us show that every irreducible covering bimodule is determined uniquely by the associated virtual endomorphism.

Let ϕ be a virtual endomorphism of a group G. Let us define the set $\phi(G)G$ of expressions of the form $\phi(g_1)g_0$, where $g_1, g_0 \in G$. Two expressions $\phi(g_1)g_0$ and $\phi(h_1)h_0$ are considered to be equal if and only if $g_1^{-1}h_1 \in \text{Dom}\,\phi$, and
$$\phi(g_1^{-1}h_1) = g_0 h_0^{-1}.$$

DEFINITION 2.5.7. Let $v = \phi(g_1)g_0 \in \phi(G)G$ and $g \in G$. The *right action* of G on $\phi(G)G$ is defined by $v \cdot g = \phi(g_1)g_0 g$, and the *left action* is defined by $g \cdot v = \phi(gg_1)g_0$.

The left and the right actions are well defined, since $\phi(g_1)g_0 = \phi(h_1)h_0$ implies
$$\phi\left(g_1^{-1}h_1\right) = \phi\left((gg_1)^{-1}(gh_1)\right) = g_0 h_0^{-1} = (g_0 g)(h_0 g)^{-1};$$
hence $\phi(gg_1)g_0 = \phi(gh_1)h_0$ and $\phi(g_1)g_0 g = \phi(h_1)h_0 g$.

It follows directly from the definitions that the right and the left actions commute, and thus we get the *bimodule* $\phi(G)G$.

It is easy to see that the bimodule $\phi(G)G$ is irreducible and has free right action with the number of orbits equal to the index $\text{ind}\,\phi = [G : \text{Dom}\,\phi]$.

PROPOSITION 2.5.8. *Let $\mathfrak{M}$ be an irreducible d-fold covering G-bimodule and let ϕ be an associated virtual endomorphism. Then the bimodules $\mathfrak{M}$ and $\phi(G)G$ are isomorphic.*

PROOF. Let $\phi = \phi_{x_0}$ for $x_0 \in \mathfrak{M}$. Let us define a map $F : \phi(G)G \longrightarrow \mathfrak{M}$ by the formula
$$(2.4) \qquad F(\phi(g_1)g_0) = g_1 \cdot x_0 \cdot g_0.$$

The equality $\phi(g_1)g_0 = \phi(h_1)h_0$ holds if and only if $g_1^{-1}h_1 \cdot x_0 = x_0 \cdot g_0 h_0^{-1}$, which is equivalent to $h_1 \cdot x_0 \cdot h_0 = g_1 \cdot x_0 \cdot g_0$. Hence the map F is well defined and injective.

Since the bimodule $\mathfrak{M}$ is irreducible, one can find for every $x \in G$ elements $g_1, g_0 \in G$ such that $x = g_1 \cdot x_0 \cdot g_0$; hence the map F is onto.

We have $F(\phi(g \cdot g_1)g_0 \cdot h) = gg_1 \cdot x_0 \cdot g_0 h = g \cdot F(\phi(g_1)g_0) \cdot h$; hence F is an isomorphism of the bimodules. □

We will usually identify an element $\phi_{x_0}(g)h \in \phi_{x_0}(G)G$ with its image $g \cdot x_0 \cdot h \in \mathsf{X} \cdot G$ under F.

Propositions 2.5.6 and 2.5.8 imply the next corollary.

COROLLARY 2.5.9. *The G-bimodules $\phi_1(G)G$ and $\phi_2(G)G$ are isomorphic if and only if the virtual endomorphisms ϕ_1 and ϕ_2 are conjugate.* □

2.5.5. Self-similar action in terms of ϕ. Let us describe the bases of the bimodule $\phi(G)G$ and the associated self-similar action in terms of the virtual endomorphism ϕ.

It is easy to see that a set $\{\phi(g_i)h_i\}_{i=1,\ldots,d}$ is a basis of the bimodule $\phi(G)G$ if and only if the set $T = \{g_i\}$ is a left coset transversal of $\operatorname{Dom}\phi$, i.e., if G is the disjoint union of the cosets $g_i \operatorname{Dom}\phi$. The sequence $C = \{h_i\}$ may be arbitrary.

PROPOSITION 2.5.10. *If $\mathsf{X} = \{x_i = \phi(g_i)h_i\}_{i=1,\ldots,d}$ is a basis of the bimodule $\phi(G)G$, then the associated self-similar action $(G, \phi(G)G, \mathsf{X})$ is defined by the formula:*

$$g \cdot x_i = x_j \cdot h_j^{-1} \phi(g_j^{-1} g g_i) h_i, \tag{2.5}$$

where j is such that $g_j^{-1} g g_i \in \operatorname{Dom}\phi$ (i.e., $g g_i \in g_j \operatorname{Dom}\phi$).

PROOF. A direct computation in $\phi(G)G$. □

Equation (2.5) can be interpreted as a "ϕ"-adic adding machine, so that we get in some sense generalized numeration systems (compare with 1.7).

We get the next corollary of Proposition 2.3.4 and Corollary 2.5.9.

COROLLARY 2.5.11. *The associated virtual endomorphism determines the associated self-similar action uniquely up to a conjugacy.* □

If we start from a given self-similar action (G, X), then it may be convenient to know how one gets the elements g_i, h_i such that $\{x_i = \phi(g_i)h_i\} = \mathsf{X}$. For example, one can then use (2.5) to compute the action of the group elements on the words. The answer is actually given in the proof of Proposition 2.5.8. Namely, if ϕ is associated to $x_0 \in \mathsf{X}$ (i.e., defined by the condition $g \cdot x_0 = x_0 \cdot \phi(g)$), then g_i and h_i are such elements of the group that

$$g_i \cdot x_0 = x_i \cdot h_i^{-1}, \tag{2.6}$$

since $\phi(g_i)h_i$ corresponds to $g_i \cdot x_0 \cdot h_i$ in the proof of Proposition 2.5.8.

2.6. The linear recursion

Let A, B be algebras over a field $\Bbbk$. An $(A - B)$-*bimodule* is a right B-module Φ together with a homomorphism ψ from A to the endomorphism algebra of the right B-module Φ. The $(A - A)$-bimodules are called A-bimodules.

We write $\psi(a) \cdot \xi = a \cdot \xi$ for $a \in A, \xi \in \Phi$. Hence Φ is also a left A-module and the left multiplication by A commutes with the right multiplication by B.

It is required in many definitions of a bimodule that the homomorphism ψ is injective. We need to consider a more general definition.

On the rôle of (Hilbert) bimodules (or *correspondences*) in C^*-algebras, see the papers [**Con94, CJ85, JS97**].

If $\mathfrak{M}$ is a permutational G-bimodule and $\Bbbk$ is a field, then the left and the right actions of G on $\mathfrak{M}$ are extended to a structure of a $\Bbbk[G]$-bimodule on the linear space $\langle \mathfrak{M} \rangle_\Bbbk$. Here $\langle \mathfrak{M} \rangle_\Bbbk$ denotes the linear $\Bbbk$-space with the basis $\mathfrak{M}$ and $\Bbbk[G]$ is the group algebra. The obtained $\Bbbk[G]$-bimodule $\langle \mathfrak{M} \rangle_\Bbbk$ is called the *linear span* of the permutational bimodule $\mathfrak{M}$.

Now let $\mathfrak{M}$ be a d-fold covering G-bimodule. Let X be its basis. Then the right module of the span $\Phi = \langle \mathfrak{M} \rangle_\Bbbk$ is isomorphic to the free d-dimensional right $\Bbbk[G]$-module $(\Bbbk[G])^d$ and X is a basis of this right module.

Therefore the left module structure is defined by a homomorphism of the $\Bbbk$-algebras

$$\psi : \Bbbk[G] \longrightarrow M_{d\times d}(\Bbbk[G]),$$

where $M_{d\times d}(\Bbbk[G])$ is the algebra of the $d \times d$ matrices over $\Bbbk[G]$, i.e., the algebra $\Bbbk[G] \otimes_\Bbbk M_{d\times d}(\Bbbk)$.

Note that the homomorphism ψ need not be injective even if the self-similar action is faithful.

The homomorphism ψ is the *linear recursion* associated to the bimodule $\mathfrak{M}$ (or to the self-similar action if $\mathfrak{M}$ is the self-similarity bimodule).

The linear recursion is given by

$$(2.7) \quad \psi(g) = (a_{xy})_{x,y \in \mathsf{X}}, \quad \text{where} \quad a_{xy} = \begin{cases} h, & \text{if } g \cdot y = x \cdot h, \\ 0, & \text{if such } h \text{ does not exist.} \end{cases}$$

For instance, for the adding machine action $a = \sigma(1, a)$ we have

$$\psi(a) = \begin{pmatrix} 0 & a \\ 1 & 0 \end{pmatrix}.$$

In principle, the linear recursion when restricted to the group G is nothing more than just another way to write the wreath recursion $G \longrightarrow \mathfrak{S}(\mathsf{X}) \wr G$. It becomes, however, very important and convenient in the cases when the group algebra, measures on groups, or linear representations of G are considered.

For example, R. Grigorchuk, L. Bartholdi and A. Żuk used linear recursions in [**BG00b, GŻ01**] to compute the spectra of Markov operators on Schreier graphs and Hecke type operators of representations for some self-similar groups.

For more on linear recursions see the papers [**Sid97, Nek02, Nek04**].

2.7. Invariant subgroups and the kernel of a self-similar action

2.7.1. (Semi-)Invariant subgroups. Let $\phi : G \dashrightarrow G$ be a virtual endomorphism and let $\mathfrak{M} = \phi(G)G$ be the respective bimodule.

DEFINITION 2.7.1. A subgroup $H \leq G$ is said to be ϕ-*semi-invariant* if

$$\phi(H \cap \mathrm{Dom}\,\phi) \leq H.$$

A subgroup $H \leq G$ is said to be ϕ-*invariant* if $H \leq \mathrm{Dom}\,\phi$ and $\phi(H) \leq H$.

If H is ϕ-semi-invariant, then the map $\phi_H : \mathrm{Dom}\,\phi \cap H \longrightarrow H$ is a virtual endomorphism of H. Its index $[H : \mathrm{Dom}\,\phi_H]$ is not greater than the index $d = [G : \mathrm{Dom}\,\phi]$ of ϕ. We get a natural embedding $\phi_H(H)H \longrightarrow \phi(G)G$ given by

$$\phi_H(h_1)h_2 \mapsto \phi(h_1)h_2,$$

which is well defined and injective by semi-invariance of H. Hence, the bimodule $\mathfrak{M}(H) = \phi_H(H)H$ is a sub-bimodule of the H-bimodule $\mathfrak{M}$.

We have the following description of semi-invariant groups in terms of self-similar actions.

LEMMA 2.7.2. *A subgroup $H \leq G$ is ϕ-semi-invariant if and only if there exists a self-similar action (G, X) defined by a basis X of $\phi(G)G$ and a subset $\mathsf{Y} \subseteq \mathsf{X}$ such that $\phi(1)1 \in \mathsf{Y}$, the sub-tree Y^* of X^* is H-invariant and $H|_y \subseteq H$ for every $y \in \mathsf{Y}$.*

PROOF. Suppose that H is ϕ-semi-invariant. Let $\mathfrak{M}(H) = \phi_H(H)H$ be the corresponding sub-bimodule of $\mathfrak{M} = \phi(G)G$.

The elements of $\mathfrak{M}(H)$ are of the form $\phi(g)h$ for $g, h \in H$. Two elements $m_1, m_2 \in \mathfrak{M}(H) \subset \mathfrak{M}$ belong to one right H-orbit if and only if they belong to one right G-orbit. In particular, any basis Y of the H-bimodule $\mathfrak{M}(H)$ can be extended to a basis $\mathsf{X} \supseteq \mathsf{Y}$ of the G-bimodule $\mathfrak{M}$.

Then it follows directly from Proposition 2.5.10 that the set $\mathsf{Y}^* \subseteq \mathsf{X}^*$ is H-invariant and that $H|_y \subseteq H$ for every $y \in \mathsf{X}$.

Suppose now that X and $\mathsf{Y} \subseteq \mathsf{X}$ are such that $H|_y \subset H$ for all $y \in \mathsf{Y}$. Let ϕ_y be the virtual endomorphism associated to $y \in \mathsf{Y}$. Take an arbitrary $g \in \text{Dom}\,\phi_y$. Then by the definition of the associated virtual endomorphism

$$g \cdot y = y \cdot \phi_y(g);$$

hence $\phi_y(g) \in H$ for all $g \in \text{Dom}\,\phi_y$. □

Thus a subgroup $H \leq G$ is semi-invariant if and only if the restriction of its action on a subtree Y^* of X^* is *self-similar*. Note that a subgroup may be self-similar with respect to the action defined by one basis and not be self-similar with respect to another action.

Note also that $|\mathsf{Y}| = [H : \text{Dom}\,\phi|_H]$ and $|\mathsf{X}| = [G : \text{Dom}\,\phi]$; therefore $\mathsf{Y} = \mathsf{X}$ if and only the indices of ϕ and ϕ_H coincide. If it is so, then $\mathsf{Y}^* = \mathsf{X}^*$ and we say that the subgroup H *is transitive on the first level*, for obvious reasons.

Another remark is that the action (H, Y) may be non-faithful even if the action (G, X) is faithful.

2.7.2. Quotients of virtual endomorphisms. Let a subgroup $H \trianglelefteq G$ be normal and ϕ-semi-invariant. Take some virtual endomorphism

$$\phi_1(x) = g_1^{-1} \cdot \phi(g_2^{-1} x g_2) \cdot g_1$$

conjugate to ϕ. Then $g \in H \cap \text{Dom}\,\phi_1$ implies that $g_2^{-1} g g_2 \in \text{Dom}\,\phi$, but then $g_2^{-1} g g_2 \in H \cap \text{Dom}\,\phi$. Therefore $\phi\left(g_2^{-1} g g_2\right) \in H$ by ϕ-semi-invariance; hence $\phi_1(g) = g_1^{-1} \cdot \phi(g_2^{-1} g g_2) \cdot g_1 \in H$. Thus, any normal ϕ-semi-invariant subgroup is also ϕ_1-semi-invariant.

It is also straightforward to check that the H-bimodule $\mathfrak{M}(H) = \phi_H(H)H$ does not depend on the choice of the associated virtual endomorphism ϕ, i.e., depends only on H and the conjugacy class of ϕ.

PROPOSITION 2.7.3. *If $H \trianglelefteq G$ is a normal ϕ-semi-invariant subgroup, then the formula*

$$\psi(gH) = \phi(g)H$$

for $g \in \text{Dom}\,\phi$ gives a well defined virtual endomorphism ψ of the quotient G/H.

PROOF. The domain of the map ψ is the image of the subgroup of finite index $\text{Dom}\,\phi$ under the canonical homomorphism $G \to G/H$ and thus has finite index in G/H. Suppose that $g_1 H = g_2 H$ for some $g_1, g_2 \in \text{Dom}\,\phi$. Then $g_1^{-1} g_2 \in H \cap \text{Dom}\,\phi$, so $\phi(g_1^{-1} g_2) \in H$; thus $\phi(g_1)H = \phi(g_2)H$. □

The virtual endomorphism ψ is called the *quotient of ϕ by the subgroup H* and is denoted ϕ/H.

2.7.3. The kernel of a self-similar action.

PROPOSITION 2.7.4. *Let $\mathfrak{M}$ be an irreducible d-fold covering bimodule over G. Let ϕ be an associated virtual endomorphism and choose some basis X of $\mathfrak{M}$. If N is a normal subgroup of G, then the following conditions are equivalent:*

(1) *for all $g \in N$ and $m \in \mathfrak{M}$ we have $g \cdot m = m \cdot h$ for some $h \in N$,*
(2) *for all $g \in N$ and $x \in \mathsf{X}$ we have $g(x) = x$ and $g|_x \in N$,*
(3) *N is ϕ-invariant;*

and they imply that

(4) *N is contained in the kernel of the associated self-similar action.*

PROOF. Condition (1) obviously implies (2). The converse implication follows from the definition of a basis, since if (2) holds, then every $m \in \mathfrak{M}$ can be written in the form $m = x \cdot f$ for some $f \in G$ and then

$$g \cdot m = g \cdot x \cdot f = x \cdot g|_x f = x \cdot f \cdot f^{-1} g|_x f = m \cdot f^{-1} g|_x f.$$

Condition (3) implies (2) by Proposition 2.5.10, since $g_i^{-1} g g_i \in N \le \operatorname{Dom} \phi$ and $h_i^{-1} \phi(g_i^{-1} g g_i) h_i \in N$ for every $g \in N$ if N is normal and ϕ-invariant.

On the other hand, (1) implies (3), since if ϕ is associated to $\mathfrak{M}$ and $m \in \mathfrak{M}$, then for every $g \in G$ we have $g \cdot m = m \cdot \phi(g)$ for $\phi(g) \in N$.

Implication (2)$\Rightarrow$(4) follows directly from the definition of the associated self-similar action by induction on the length of words in X^*. $\square$

As a corollary we get that if $H \triangleleft G$ is a normal subgroup invariant with respect to a virtual endomorphism ϕ associated to $\mathfrak{M}$, then it is invariant with respect to every associated virtual endomorphism.

Therefore, we will say that a normal subgroup $H \triangleleft G$ is $\mathfrak{M}$-*invariant* if it is invariant with respect to any virtual endomorphism associated to $\mathfrak{M}$.

Another corollary of Proposition 2.7.4 is the following description of the kernel of a self-similar action.

PROPOSITION 2.7.5. *The kernel of a self-similar action of a group G with an associated virtual endomorphism ϕ is equal to the subgroup*

(2.8) $$\mathcal{K}(\phi) = \bigcap_{n \ge 1} \bigcap_{g \in G} g^{-1} \cdot \operatorname{Dom} \phi^n \cdot g$$

and is the maximal one among the normal ϕ-invariant subgroups.

PROOF. It follows from the definition of the virtual endomorphism $\phi = \phi_{x_0}$ that the subgroup $\operatorname{Dom} \phi^n$ is the stabilizer of the word $x_0^n \in \mathsf{X}^*$; thus the group $\bigcap_{g \in G} g^{-1} \cdot \operatorname{Dom} \phi^n \cdot g$ is the stabilizer of all the vertices of the nth level of the tree X^*. Therefore, the subgroup $\mathcal{K}(\phi)$ is the kernel of the action.

The subgroup $\mathcal{K}(\phi)$ is obviously ϕ-invariant and normal. If N is a normal ϕ-invariant subgroup of G, then it is contained in the kernel of the action by Proposition 2.7.4. $\square$

2.7.4. Homomorphisms of self-similar groups.

PROPOSITION 2.7.6. *Let ϕ_i be a virtual endomorphism of a group G_i for $i = 1, 2$ and suppose that we have a homomorphism $f : G_1 \longrightarrow G_2$ such that*

$$f^{-1}(\operatorname{Dom} \phi_2) \le \operatorname{Dom} \phi_1, \qquad f(\operatorname{Dom} \phi_1) \le \operatorname{Dom} \phi_2$$

and
$$f(\phi_1(g)) = \phi_2(f(g))$$
for every $g \in f^{-1}(\mathrm{Dom}\,\phi_2)$.

Then the subgroup $f(G_1)$ of G_2 is ϕ_2-semi-invariant and there exists an injective homomorphism $f^* : G_1/\mathcal{K}(\phi_1) \longrightarrow G_2/\mathcal{K}(\phi_2)$ (where $\mathcal{K}(\phi_i)$ are the kernels of the associated self-similar actions) such that the diagram

$$\begin{array}{ccc} G_1 & \xrightarrow{f} & G_2 \\ \downarrow & & \downarrow \\ G_1/\mathcal{K}(\phi_1) & \xrightarrow{f^*} & G_2/\mathcal{K}(\phi_2) \end{array}$$

is commutative, where the vertical arrows are the canonical epimorphisms. In particular, if f is surjective, then the groups $G_1/\mathcal{K}(\phi_1)$ and $G_2/\mathcal{K}(\phi_i)$ are isomorphic and the respective self-similar actions are conjugate.

PROOF. Choose some basis $\mathsf{X}_1 = \{x_i = \phi_1(r_i) \cdot 1\}_{i=1,\ldots,d}$ of the bimodule $\phi_1(G_1)G_1$. Here $\{r_i\}_{i=1,\ldots,d}$ is a left coset transversal of the subgroup $\mathrm{Dom}\,\phi_1$.

Consider the set $\mathsf{X}'_2 = \{x'_i = \phi_2(f(r_i)) \cdot 1\}_{i=1,\ldots,d} \subset \phi_2(G_2)G_2$. If
$$f(r_i)\,\mathrm{Dom}\,\phi_2 = f(r_j)\,\mathrm{Dom}\,\phi_2,$$
then $f(r_i^{-1}r_j) \in \mathrm{Dom}\,\phi_2$; therefore $r_i^{-1}r_j \in \mathrm{Dom}\,\phi_1$ and thus $i = j$. Consequently, the elements of $\mathsf{X}'_2 \subset \phi_2(G_2)G_2$ belong to different orbits of the right action and the set X'_2 can be extended to a basis X_2 of $\phi_2(G_2)G_2$. Let $F : \mathsf{X}_1^* \longrightarrow \mathsf{X}_2^*$ be the natural extension of the map
$$F : \phi_1(r_i) \cdot 1 = x_i \mapsto x'_i = \phi_2(f(r_i)) \cdot 1;$$
i.e., we put $F(a_1 a_2 \ldots a_n) = F(a_1)F(a_2)\ldots F(a_n)$ for $a_1 a_2 \ldots a_n \in \mathsf{X}_1^*$.

Suppose that $g \cdot x_i = x_j \cdot h$ for $g, h \in G_1$. Then $r_j^{-1} g r_i \in \mathrm{Dom}\,\phi_1$ and $h = \phi_1(r_j^{-1} g r_i)$; hence $f(r_j^{-1} g r_i) = f(r_j)^{-1} f(g) f(r_i) \in \mathrm{Dom}\,\phi_2$ and $f(h) = \phi_2(f(r_j)^{-1} f(g) f(r_i))$. Thus $f(g) \cdot F(x_i) = F(x_j) \cdot f(h)$, and we get by induction on the length of the words that the equality $g \cdot v = u \cdot h$ for $v, u \in \mathsf{X}_1^*$ and $g, h \in G_1$ implies $f(g) \cdot F(v) = F(u) \cdot f(h)$.

Thus, the map $F : \mathsf{X}_1^* \longrightarrow \mathsf{X}_2^*$ semi-conjugates the action of G_1 on X_1^* to the action of $f(G_1)$ on X_2^* and $f(G_1)$ is a self-similar subgroup of G_2. This implies the statement of the proposition. $\square$

Note that in conditions of Proposition 2.7.6 we have
$$\ker f \leq \mathrm{Dom}\,\phi_1, \quad \phi_1(\ker f) \leq \ker f.$$

The first inclusion follows from the condition $f^{-1}(\mathrm{Dom}\,\phi_2) \leq \mathrm{Dom}\,\phi_1$, and the second one from $f(\phi_1(g)) = \phi_2(f(g))$. Thus $\ker f$ is ϕ_1-invariant.

On the other hand, if $N \leq G_1$ is a normal ϕ-invariant subgroup, then the canonical homomorphism $f : G_1 \longrightarrow G_2 = G_1/N$ and the virtual endomorphisms $\phi_1 = \phi$ and $\phi_2 = \phi_1/N$ satisfy the conditions of Proposition 2.7.6.

2.8. Recurrent actions

2.8.1. If ϕ is onto, then every basis of $\phi(G)G$ is of the form $\{\phi(g_i) \cdot 1\}_{i=1,\ldots,d}$, where $\{g_i\}$ is a left coset transversal, since $\phi(g_i)h_i = \phi(g_i s_i) \cdot 1$, where $s_i \in \mathrm{Dom}\,\phi$ are such that $\phi(s_i) = h_i$.

2.8. RECURRENT ACTIONS

DEFINITION 2.8.1. A self-similar action is said to be *recurrent* (or *fractal*) if it is transitive on the first level X^1 of the tree X^* and the associated virtual endomorphism ϕ_x is onto, i.e., if $\operatorname{Ran}\phi_x = G$.

We have the following description of recurrent actions in terms of permutational bimodules.

PROPOSITION 2.8.2. *A self-similar action is recurrent if and only if the left action of the associated bimodule is transitive.*

PROOF. Let ϕ be the associated virtual endomorphism. Then the associated bimodule and $\phi(G)G$ are isomorphic. If ϕ is onto, then for every $\phi(g_1)h_1$ and $\phi(g_2)h_2$ we can find $r \in G$ such that $\phi(rg_1)h_1 = \phi(g_2)h_2$; i.e., $\phi(g_2^{-1}rg_1) = h_2h_1^{-1}$. One can take any $r \in g_2\phi^{-1}(h_2h_1^{-1})g_1^{-1}$.

On the other hand, if $\phi(G)G$ has a transitive right action, then for every $h \in G$ there exists $g \in G$ such that $\phi(g) = h$, i.e., $g \cdot (\phi(1)1) = \phi(1)h$ in $\phi(G)G$. □

As a corollary of Proposition 2.8.2 we get that the definition of a recurrent action does not depend on the choice of the associated virtual endomorphism.

If the action is recurrent, then every element of the bimodule $\mathfrak{M} = \mathsf{X} \cdot G$ can be written in the form $\phi(g) \cdot 1 \in \phi(G)G = \mathfrak{M}$, where $g \in G$ and $\phi = \phi_{x_0}$ is the associated virtual endomorphism. In particular, the basis X is a set of elements of the form $x_i = \phi(r_{x_i}) \cdot 1$, where r_{x_i} are such that $r_{x_i} \cdot x_0 = x_i \cdot 1$.

DEFINITION 2.8.3. Let (G, X) be a recurrent action. A set $\{r_{x_0}, r_{x_1}, \ldots, r_{x_{d-1}}\}$ is a *digit system* (associated to the *initial letter* x_0) if $r_{x_i} \cdot x_0 = x_i \cdot 1$ for any $x_i \in \mathsf{X}$, i.e., if $x_i = \phi_{x_0}(r_{x_i}) \cdot 1$.

2.8.2. Tensor powers of a recurrent action.

PROPOSITION 2.8.4. *Let $\mathfrak{M}_1$ and $\mathfrak{M}_2$ be irreducible covering G-bimodules. Let ϕ_i be a virtual endomorphism associated to $\mathfrak{M}_i$.*

If the bimodule $\mathfrak{M}_1 \otimes \mathfrak{M}_2$ is irreducible, then $\phi_2 \circ \phi_1$ is its associated virtual endomorphism.

If the bimodules $\mathfrak{M}_1$ and $\mathfrak{M}_2$ have transitive left actions, then the bimodule $\mathfrak{M}_1 \otimes \mathfrak{M}_2$ also has a transitive left action.

PROOF. Let ϕ_i be associated to $\mathfrak{M}_i$ and $x_i \in \mathfrak{M}_i$, $i = 1, 2$. Let ϕ be the virtual endomorphism associated to $\mathfrak{M}_1 \otimes \mathfrak{M}_2$ and $x_1 \otimes x_2$. Then an element $g \in G$ belongs to $\operatorname{Dom}\phi$ if and only if $g \cdot x_1 \otimes x_2 = x_1 \otimes x_2 \cdot h$ for some $h \in G$ and then $h = \phi(g)$. But then $g \in \operatorname{Dom}\phi_1$, $g \cdot x_1 = x_1 \cdot \phi_1(g)$, and $\phi_1(g) \in \operatorname{Dom}\phi_2$ and $\phi_1(g) \cdot x_2 = x_2 \cdot \phi_2 \circ \phi_1(g)$; therefore $\phi(g) = \phi_2 \circ \phi_1(g)$. The converse implications show that $\operatorname{Dom}\phi = \operatorname{Dom}\phi_2 \circ \phi_1$ and thus $\phi = \phi_2 \circ \phi_1$.

Suppose that the left actions are transitive on $\mathfrak{M}_i$. Then every element of $\mathfrak{M}_i$ can be written in the form $g \cdot x_i$, where $g \in G$. Therefore, every element of $\mathfrak{M}_1 \otimes \mathfrak{M}_2$ can be written in the form $g_1 \cdot x_1 \otimes g_2 \cdot x_2$ for $g_1, g_2 \in G$. We can write $x_1 \cdot g_2 = h \cdot x_1$ for some $h \in G$. Then $g_1 \cdot x_1 \otimes g_2 \cdot x_2 = g_1 h \cdot x_1 \otimes x_2$, which proves that the left action of G on $\mathfrak{M}_1 \otimes \mathfrak{M}_2$ is transitive. □

COROLLARY 2.8.5. *If a self-similar action (G, X) is recurrent and ϕ is its associated virtual endomorphism, then the action is level-transitive and its nth tensor power is also recurrent with the associated virtual endomorphism ϕ^n.*

PROOF. Let $\mathfrak{M}$ be the associated G-bimodule. Proposition 2.8.4 implies that the bimodule $\mathfrak{M}^{\otimes n}$ has transitive left action. It follows that the action of G on X^n is transitive and that the nth tensor power (G, X^n) is recurrent.

If ϕ is associated to $x \in \mathsf{X}$, then the virtual endomorphism associated to $\mathfrak{M}^{\otimes n}$ and $x^{\otimes n} \in \mathfrak{M}^{\otimes n}$ is obviously ϕ^n. $\square$

An interesting observation of Y. Muntyan and D. Savchuk is that if (G, X) is a self-similar action of an *infinite* group G over a *two-letter* alphabet X, then it is necessary level-transitive. See Lemmas 1.10.7 and 1.10.8 for the idea of the proof.

2.9. Example: free abelian groups

2.9.1. Let us illustrate the developed technique and study self-similar actions of free abelian groups. The results of this section where obtained (for the case $|\mathsf{X}| = 2$) jointly with S. Sidki in [**NS04**]. We use additive notation here.

We know (by Propositions 2.5.10 and 2.3.4) that every self-similar and transitive on X^1 action is determined (up to a conjugacy) by the associated virtual endomorphism. Consider such an action of $G = \mathbb{Z}^n$ and let $\phi : \mathbb{Z}^n \dashrightarrow \mathbb{Z}^n$ be the associated virtual endomorphism. The map $\phi : \mathrm{Dom}\,\phi \longrightarrow \mathbb{Z}^n$ can be extended in a unique way to a linear map

$$A = \mathbb{Q} \otimes \phi : \mathbb{Q} \otimes \mathbb{Z}^n = \mathbb{Q}^n \to \mathbb{Q}^n,$$

since $\mathbb{Q} \otimes \mathrm{Dom}\,\phi = \mathbb{Q}^n$.

Let $\mathbf{A}$ be the matrix of the linear map $A = \mathbb{Q} \otimes \phi$ in the standard basis of the group $\mathbb{Z}^n \subset \mathbb{Q}^n$ (seen as a basis of the vector space $\mathbb{Q}^n$). The matrix $\mathbf{A}$ has rational entries. Moreover, if $k \in \mathbb{N}$ is such that $k\mathbb{Z}^n \leq \mathrm{Dom}\,\phi$, then $k \cdot \mathbf{A}$ has integral entries.

If the self-similar action is recurrent (i.e., if ϕ is onto) and ϕ is invertible (we will see that the last condition is satisfied for faithful actions of $\mathbb{Z}^n$), then the map ϕ^{-1} is defined on the whole group $\mathbb{Z}^n$ and is injective; therefore $\mathbf{A}^{-1}$ is a matrix with integral entries.

Let $\mathsf{X} = \{x_0 = \phi(r_0) + h_0, x_1 = \phi(r_1) + h_1, \ldots, x_{d-1} = \phi(r_{d-1}) + h_{d-1}\}$ be a basis of the bimodule $\phi(\mathbb{Z}^n) + \mathbb{Z}^n$; i.e., $\{r_i\}$ is a coset transversal of $\mathrm{Dom}\,\phi$. Recall that if we start from a given self-similar action, then r_i and h_i are chosen so that $r_i \cdot x_0 = x_i \cdot (-h_i)$ for every $0 \leq i \leq d-1$ (see (2.6) on page 40).

Then by (2.5) on page 40, the equality $g \cdot x_i = x_j \cdot h$ is equivalent to the conditions $g + r_i - r_j \in \mathrm{Dom}\,\phi$ and

$$(2.9) \qquad h = A(g + r_i - r_j) + h_i - h_j.$$

We will consider only recurrent actions in this section. Then every basis of the bimodule $\phi(\mathbb{Z}^n) + \mathbb{Z}^n$ is of the form $\{x_0 = \phi(r_0), x_1 = \phi(r_1), \ldots, x_{d-1} = \phi(r_{d-1})\}$, where $\{r_0, r_1, \ldots, r_{d-1}\}$ is a coset transversal of $A^{-1}(\mathbb{Z}^n)$. We say that $\{r_0, r_1, \ldots, r_{d-1}\}$ is a *digit system*; see Definition 2.8.3.

Then (2.9) is written

$$(2.10) \qquad h = A(g + r_i - r_j),$$

where j is such that $g + r_i - r_j \in A^{-1}(\mathbb{Z}^n)$.

Let us rewrite Proposition 2.7.5 for the case of a commutative group.

2.9. EXAMPLE: FREE ABELIAN GROUPS

PROPOSITION 2.9.1. *Let ϕ be a surjective virtual endomorphism of an abelian group G. Then the kernel of the self-similar action defined by ϕ is the subgroup*

$$\mathcal{K}(\phi) = \bigcap_{n=1}^{\infty} \phi^{-n}(G).$$

We have the following criterion (see [**NS04, BJ99**]).

PROPOSITION 2.9.2. *Let A be a linear operator on $\mathbb{Q}^n$. Consider the virtual endomorphism $\phi : v \mapsto A(v)$ of the group $\mathbb{Z}^n$.*

Then the subgroup $\mathcal{K}(\phi)$ is trivial if and only if the characteristic polynomial of A is not divisible by a monic polynomial with integral coefficients (or, in other words, if and only if no eigenvalue of A is an algebraic integer).

PROOF. Let $\mathbf{A}$ be the matrix of A. Suppose that $U = \mathcal{K}(\phi)$ is non-trivial. Then $A(U) \le U$. Let C be the restriction of the linear operator A onto $U \otimes \mathbb{Q}$. Then the characteristic polynomial of C is monic, has integral coefficients and is a factor of the characteristic polynomial of $\mathbf{A}$.

On the other hand, suppose that $f(x) = x^k + a_1 x^{k-1} + \cdots + a_k \in \mathbb{Z}[x]$ is an irreducible factor of the characteristic polynomial of $\mathbf{A}$. Let $\widehat{U} \le \mathbb{Q}^n$ be the kernel of $f(A)$. Then for an arbitrary non-zero element $v \in \widehat{U}$ the vectors $v, A(v), A^2(v), \ldots A^{k-1}(v)$ form a basis of the space $\widehat{U}$ such that the matrix of the operator $A|_{\widehat{U}}$ with respect to it has integral entries. Take some $q \in \mathbb{Z}$ not equal to zero such that the vectors $qv, qA(v), qA^2(v), \ldots qA^{k-1}(v)$ belong to $\mathbb{Z}^n$. Then they also form a basis of $\widehat{U}$ giving the same matrix of $A|_{\widehat{U}}$. Then the subgroup of $\mathbb{Z}^n$ generated by $qv, qA(v), qA^2(v), \ldots qA^{k-1}(v)$ is a non-trivial ϕ-invariant subgroup and $\mathcal{K}(\phi)$ is not trivial. $\square$

2.9.2. "Sausage" automaton.
As an example, consider the $n \times n$ matrix

$$\begin{pmatrix} 0 & 1 & 0 & \ldots & 0 \\ 0 & 0 & 1 & \ldots & 0 \\ \vdots & \vdots & \vdots & \ddots & \vdots \\ 0 & 0 & \ldots & \ldots & 1 \\ 1/2 & 0 & \ldots & \ldots & 0 \end{pmatrix}.$$

It defines a virtual endomorphism $\phi : \mathbb{Z}^n \dashrightarrow \mathbb{Z}^n$ whose domain is $\langle 2e_1 = (2, 0, \ldots, 0), e_2 = (0, 1, \ldots, 0), \ldots e_n = (0, 0, \ldots, 1) \rangle = 2\mathbb{Z} \oplus \mathbb{Z}^{n-1}$ and whose action is given on the generators of the domain by

$$\phi(2e_1) = e_m, \quad \phi(e_i) = e_{i-1}, \text{ for } i = 2, \ldots, n.$$

The characteristic polynomial of this matrix is $f(x) = x^n - 1/2$, and therefore the virtual endomorphism ϕ defines a faithful self-similar action of $\mathbb{Z}^n$ on the binary tree. Let us choose the coset transversal $R = \{r_0 = 0, r_1 = e_1\}$ and let $\mathsf{X} = \{0 = \phi(0) + 0, 1 = \phi(e_1) + 0\}$ be the respective basis of the bimodule $\phi(\mathbb{Z}^n) + \mathbb{Z}^n$. Let us compute the action $(\mathbb{Z}^n, \phi(\mathbb{Z}^n) + \mathbb{Z}^n, \mathsf{X})$.

The only generator which does not belong to $\operatorname{Dom}\phi$ is e_1. Then

$$e_1 = \sigma(id, e_n),$$

since $e_1 \cdot 0 = 1 \cdot \phi(e_1 + r_0 - r_1) = 1 \cdot 0$ and $e_1 \cdot 1 = 0 \cdot \phi(e_1 + r_1 - r_0) = 0 \cdot e_m$ by (2.10).

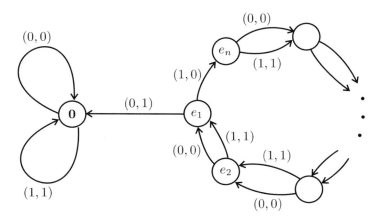

FIGURE 2.1. Automaton generating $\mathbb{Z}^n$

The action of e_i on X^* for $i \geq 2$ is given by the recursion
$$e_i = (e_{i-1}, e_{i-1}),$$
since $e_i \cdot 0 = 0 \cdot \phi(e_i + r_0 - r_0) = 0 \cdot e_{i-1}$ and $e_i \cdot 1 = 1 \cdot \phi(e_i + r_1 - r_1) = 1 \cdot e_{i-1}$.

Thus the defined action of $\mathbb{Z}^n$ on X^* is generated by the automaton, shown in Figure 2.1. It coincides with the adding machine action for the case $n = 1$.

Actually, this construction is a general method to construct a self-similar action of G^n having a self-similar action of G.

Suppose that we have a faithful self-similar action (G, X). If $g \cdot x = y \cdot h$ for $g, h \in G$ and $x, y \in \mathsf{X}$, then we set

(2.11) $\qquad (g, g_2, \ldots, g_n) \cdot x = y \cdot (g_2, \ldots, g_n, h)$

for $(g, g_2, \ldots, g_n) \in G^n$.

PROPOSITION 2.9.3. *The recurrent relation* (2.11) *defines a faithful self-similar action of the direct power* G^n.

PROOF. If we prove that (2.11) is a well defined bimodule structure on $\mathsf{X} \times G^n$, then we will show that (2.11) defines an action of G^n on X^*. It is sufficient to check that $(g'g'') \cdot x = g' \cdot (g'' \cdot x)$, which is easily done.

Note that if ϕ_x is the virtual endomorphism associated to the action, then ϕ_x^n acts by the rule $\phi_x^n(g_1, \ldots, g_n) = (\psi_x(g_1), \ldots, \psi_x(g_n))$, where ψ_x is the virtual endomorphism associated to the action (G, X). Since the action (G, X) is faithful, every normal ψ_x-invariant subgroup of G is trivial. This implies that every normal ϕ^n-invariant subgroup of G^n is trivial; hence the action of G^n on X^* is also faithful (see Proposition 2.7.5). □

2.9.3. "A-adic" groups.

LEMMA 2.9.4. *Let* $G \leq \operatorname{Aut} \mathsf{X}^*$ *be self-similar and let* $\widehat{G}$ *be its closure in* $\operatorname{Aut} \mathsf{X}^*$. *Then* $\widehat{G}$ *is also self-similar.*

PROOF. An automorphism $g \in \operatorname{Aut} \mathsf{X}^*$ belongs to $\widehat{G}$ if and only if for every $n \in \mathbb{N}$ the action of g on the first n levels of X^* coincides with the action of some element $g' \in G$ (where g' depends on n). Therefore, if $g \in \widehat{G}$, then for every $x \in \mathsf{X}$

the restriction $g|_x$ also belongs to $\widehat{G}$, since the actions of $g|_x$ and $g'|_x \in G$ coincide on the first $n-1$ levels of the tree. □

Consider some faithful recurrent action of $\mathbb{Z}^n$ on X^*. Denote $G_k = \phi^{-k}(\mathbb{Z}^n) = \mathrm{Dom}\,\phi^k$, where ϕ is the associated virtual endomorphism.

LEMMA 2.9.5. *The group G_k coincides with the kth level stabilizer $\mathsf{St}_G(k)$.*

PROOF. Stabilizer of the kth level of a self-similar group G is equal to
$$\bigcap_{g\in G} g^{-1}\cdot \mathrm{Dom}\,\phi^k \cdot g,$$
where ϕ is the associated virtual endomorphism. We have in our case
$$g^{-1}\cdot \mathrm{Dom}\,\phi^k \cdot g = \mathrm{Dom}\,\phi^n = \phi^{-k}(\mathbb{Z}^n).$$
□

Let A be the linear operator $\mathbb{Q}\otimes\phi$. Denote by $\widehat{\mathbb{Z}}_A^n$ the completion of the group $\mathbb{Z}^n$ with respect to the sequence of finite-index subgroups
$$\mathbb{Z}^n > A^{-1}(\mathbb{Z}^n) > A^{-2}(\mathbb{Z}^n) > \cdots > A^{-k}(\mathbb{Z}^n)\cdots.$$
In other words, it is the closure of $\mathbb{Z}^n$ with respect to the metric
$$\|g_1 - g_2\| = d^{-k},$$
where k is maximal among such that $g_1 - g_2 \in A^{-k}(\mathbb{Z}^n)$. Then $\widehat{\mathbb{Z}}_A^n$ is a profinite abelian group. We call $\widehat{\mathbb{Z}}_A^n$ the *group of integral A-adic vectors*.

Now let $R = \{r_0, r_1, \ldots, r_{d-1}\}$ be a digit system of the self-similar action, i.e., a coset transversal of $A^{-1}(\mathbb{Z}^n)$ in $\mathbb{Z}^n$. Then every letter x_i of the alphabet is identified with the element $\phi(r_i)$ of the bimodule $\phi(\mathbb{Z}^n) + \mathbb{Z}^n$. Every element of $\widehat{\mathbb{Z}}_A^n$ is written uniquely in the form
$$(2.12) \qquad a = r_{i_0} + A^{-1}(r_{i_1}) + A^{-2}(r_{i_2}) + \cdots.$$

If we have an A-adic number (2.12), then its partial sums are given by
$$a_m = r_{i_0} + A^{-1}(r_{i_1}) + A^{-2}(r_{i_2}) + \cdots + A^{-m}(r_{i_m}).$$

The sequence a_m converges in $\widehat{\mathbb{Z}}_A^n$, since it is a Cauchy sequence with respect to the defined metric.

PROPOSITION 2.9.6. *If $a = \sum_{k=0}^{\infty} A^{-k}(r_{i_k})$ is an A-adic vector, then the sequence of its partial sums a_m seen as automorphisms of the tree X^* converge in $\mathrm{Aut}\,X^*$. The set of the limits coincides with the closure $\widehat{G}$ of $G = \mathbb{Z}^n$ in $\mathrm{Aut}\,X^*$. The map $\Psi : X^\omega \longrightarrow \widehat{G}$ given by*
$$\Psi(x_{i_0}x_{i_1}x_{i_2}\ldots) = r_{i_0} + A^{-1}(r_{i_1}) + A^{-2}(r_{i_2}) + \cdots$$
is a homeomorphism.

PROOF. It is sufficient to prove that the sequence a_m is a Cauchy sequence in $\mathrm{Aut}\,X^*$ in order to prove the convergence. But this follows from the condition $a_{m_1} - a_{m_2} \in A^{-\min(m_1,m_2)}(G) \subset \mathsf{St}(\min(m_1, m_2))$.

Thus Ψ is well defined. If we prove that for every m and every $g \in G$ there exists a unique element of the form $g_m = r_{i_0} + A^{-1}(r_{i_1}) + \cdots + A^{-m}(r_{i_m})$ such that $g - g_m \in G_{m+1}$, then this will imply that the map Ψ is bijective. Let us prove it by induction on m. For every $g \in G$ there exists a unique index i_1 such that

$g - r_{i_1} \in A^{-1}(\mathbb{Z}^n)$. Hence, the statement is true for $m = 1$. Suppose that it is true for $m = k - 1$, where $k \geq 2$. Since $g - g_{k-1} \in G_k = A^{-k}(G)$ and the map A is injective, there exists a unique $h \in G$ such that $g - g_{k-1} = A^{-k}(h)$. Then there exists a unique i_k such that $h - r_{i_k} \in A^{-1}(\mathbb{Z}^n)$, i.e., $g - (g_{k-1} + A^{-k}(r_{i_k})) \in A^{-k-1}(r_{i_k})$. Then $g_k = g_{k-1} + A^{-k}(\dot{r}_{i_k})$.

Continuity of Ψ and its inverse follows now easily from the definitions of the topologies on X^ω and $\operatorname{Aut} \mathsf{X}^*$. $\square$

THEOREM 2.9.7. *The closure $\widehat{G}$ of $\mathbb{Z}^n$ in $\operatorname{Aut} \mathsf{X}^*$ coincides with the group $\widehat{\mathbb{Z}}_A^n$ of A-adic vectors. They are both homeomorphic to X^ω, and the natural homeomorphism $\Psi : \mathsf{X}^\omega \longrightarrow \widehat{G}$ (defined in Proposition 2.9.6) has the property that*

$$g(w) = \Psi^{-1}\left(\Psi(w) + g\right)$$

for all $w \in \mathsf{X}^\omega$ and $g \in \mathbb{Z}^n$.

In particular, the natural action of $\mathbb{Z}^n$ on its completion $\widehat{\mathbb{Z}}_A^n = \widehat{G}$ is conjugate to the action of $\mathbb{Z}^n$ on X^ω.

PROOF. A straightforward corollary of (2.10) on page 46 and uniqueness of the map Ψ. $\square$

2.10. Rigidity

2.10.1. Level-transitive rooted trees. The aim of this section is to investigate when the group structure uniquely determines the action on a rooted tree. It is based on the joint work [**LN02**] with Y. Lavreniuk.

It is not natural to formulate the result of this section only for the case of regular rooted trees X^* and self-similar actions. We need therefore some more general definitions.

Let $\mathsf{X} = (X_1, X_2, \ldots)$ be a sequence of finite sets (we assume that $|X_i| \geq 2$ for all i). We denote then by X^* the set of all finite words $x_1 x_2 \ldots x_n$, where $x_i \in X_i$. We include the empty word $\varnothing$ in X^*. The set X^* has a structure of a rooted tree defined in the same way as in the regular case: a vertex $x_1 \ldots x_{n-1}$ is adjacent to a vertex $x_1 \ldots x_{n-1} x_n$. The set $\mathsf{X}^n = \{x_1 x_2 \ldots x_n \;:\; x_i \in X_i\}$ is the nth level of the tree X^*. We put $\mathsf{X}^0 = \{\varnothing\}$.

The tree X^* is called the *level-transitive* (or the *spherically homogeneous*) rooted tree of the *spherical index* $(|X_1|, |X_2|, \ldots)$. It is easy to see that a level-transitive rooted tree is uniquely determined up to an isomorphism by the spherical index.

A *regular rooted tree* is the tree of a constant spherical index $(d, d, d, \ldots)$, i.e., the tree X^* defined by one alphabet X (or $\mathsf{X} = (X, X, \ldots)$ in the terminology of this section). Most rooted trees appearing in this book are regular.

The *boundary* of the tree X^* is the set X^ω of infinite sequences $x_1 x_2 \ldots$, where $x_i \in X_i$. The boundary X^ω comes with the natural topology of the direct product $\prod_{i=1}^\infty X_i$ of discrete sets and with the *Bernoulli* measure equal to the direct product of the uniform probability measures on X_i.

If g is an automorphism of the level-transitive rooted tree X^* defined by the sequence $\mathsf{X} = (X_1, X_2, \ldots)$ and $v = x_1 x_2 \ldots x_n$ is a vertex of the tree, then the restriction $g|_v$ is the automorphism of the tree X_n^*, where $\mathsf{X}_n = (X_{n+1}, X_{n+2}, \ldots)$ and is defined by the condition

$$g(x_1 \ldots x_n y_{n+1} \ldots y_m) = g(x_1 \ldots x_n) g|_v (y_{n+1} \ldots y_m) \quad \text{for all } y_{n+1} \ldots y_m \in \mathsf{X}_n^*.$$

So in general, $g|_v$ acts on a different tree than g does. This is the reason why self-similar actions are defined only on regular trees (when $\mathsf{X}_n = \mathsf{X}$).

The notions of a level-stabilizer $\mathsf{St}_G(n)$, rigid stabilizers $G[v]$ and $\mathsf{RiSt}_G(n)$ are defined in the same way as they were defined for the regular case in Definition 1.2.2.

2.10.2. Topological rigidity. We are going to prove the following theorem and its corollaries. For the notion of a (weakly) branch group, see Definition 1.2.4.

THEOREM 2.10.1. *Let G_1 and G_2 be weakly branch automorphism groups of level-transitive rooted trees X^* and Y^* respectively. If $\varphi : G_1 \longrightarrow G_2$ is an isomorphism of abstract groups, then there exists a measure-preserving homeomorphism $F : \mathsf{X}_1^\omega \longrightarrow \mathsf{X}_2^\omega$ such that*

$$\varphi(g)\left(F(w)\right) = F\left(g(w)\right)$$

for all $w \in \mathsf{X}_1^\omega$ and $g \in G_1$, i.e., such that φ is induced by F.

This theorem follows from a more general result of M. Rubin (see [**Rub89**]). However, we present here a more accessible proof from [**LN02**].

LEMMA 2.10.2. *Let G be an automorphism group of X^* and let $w \in \mathsf{X}^*$. If $h \in G[w]$ is non-trivial and $g(w) \neq w$, then*

$$[h, g] \neq 1.$$

Here $[h, g] = h^{-1} g^{-1} h g$.

PROOF. We have

$$(gh)|_w = g|_w h|_w,$$

since h fixes w, and

$$(hg)|_w = g|_w,$$

since g moves w and h acts trivially outside $w\mathsf{X}_{|w|}^*$. Therefore $hg \neq gh$. □

The next is the main technical lemma used in the proof of Theorem 2.10.1. It essentially coincides with Proposition 6.2 from [**LN02**]; however we reproduce here a much shorter proof found by C. Röver (see Lemma 5.7 in [**Röv02**]).

LEMMA 2.10.3. *Suppose that $G_1 \leq \mathrm{Aut}\,\mathsf{X}^*$ and $G_2 \leq \mathrm{Aut}\,\mathsf{Y}^*$ are weakly branch groups. Let $\varphi : G_1 \longrightarrow G_2$ be an isomorphism. If $\varphi^{-1}(G_2[v])$ moves a vertex $u \in \mathsf{X}^*$, where $v \in \mathsf{Y}^*$, then*

$$\varphi\left(G_1[u]\right) \cap \mathsf{St}_{G_2}\left(|v|\right) \leq G_2[v].$$

PROOF. Suppose that the lemma is false. Choose

$$g \in \left(G_1[u] \cap \varphi^{-1}\left(\mathsf{St}_{G_2}(|v|)\right)\right) \setminus \varphi^{-1}\left(G_2[v]\right).$$

Then $\varphi(g) \notin G_2[v]$; hence there exists a vertex $w \notin v\mathsf{Y}_{|v|}^*$ moved by $\varphi(g)$. We have $|w| > |v|$ and $v \notin w\mathsf{Y}_{|w|}^*$, since $\varphi(g) \in \mathsf{St}_{G_2}(|v|)$.

The subgroup $G_2[w] \cap \varphi\left(\mathsf{St}_{G_1}(|u|)\right)$ is non-trivial, since $\mathsf{St}_{G_1}(|u|)$ has finite index in G_1 and $G_2[w]$ is infinite. Take a non-trivial element $h \in G_2[w] \cap \varphi\left(\mathsf{St}_{G_1}(|u|)\right)$.

We have then, by Lemma 2.10.2, $[\varphi(g), h] \neq 1$, i.e., $[g, \varphi^{-1}(h)] \neq 1$, since $\varphi(g)$ moves w and $h \in G_2[w]$. This implies that there exists a vertex $z \in \mathsf{X}^*$ moved both by g and $\varphi^{-1}(h)$. The vertex z must belong to $u\mathsf{X}_{|u|}^*$, since $g \in G_1[u]$. The image $\varphi^{-1}(h)(z)$ also belongs to $u\mathsf{X}_{|u|}^*$, since $\varphi^{-1}(h) \in \mathsf{St}_{G_1}(|u|)$.

Let $\hat{g}$ be a non-trivial element of $G_1[z] \cap \varphi^{-1}(\mathsf{St}_{G_2}(|v|)) \subset G_1[u]$ and let $\hat{h} \in G_2[v]$ be such that $\varphi^{-1}(\hat{h})$ moves u ($\hat{h}$ exists by the condition of the lemma). Then we again have $1 \neq [\hat{g}, \varphi^{-1}(h)]$, by Lemma 2.10.2. We know that $\hat{g} \in G_1[z] \leq G_1[u]$ and $\varphi^{-1}(h) \in \mathsf{St}_{G_1}(|u|)$; therefore $[\hat{g}, \varphi^{-1}(h)] = \hat{g}^{-1} \cdot (\hat{g})^{\varphi^{-1}(h)} \in G_1[u]$. It follows (again by Lemma 2.10.2) that $\left[[\hat{g}, \varphi^{-1}(h)], \varphi^{-1}(\hat{h})\right] \neq 1$.

On the other hand, $[\varphi(\hat{g}), h]|_v = 1$ and $[\varphi(\hat{g}), h] \in \mathsf{St}_{G_2}(|v|)$, since $h|_v = 1$ and $h, \varphi(\hat{g}) \in \mathsf{St}_{G_2}(|v|) \supset G_2[w]$. Hence $\left[[\varphi(\hat{g}), h], \hat{h}\right] = 1$; i.e.,

$$\left[[\hat{g}, \varphi^{-1}(h)], \varphi^{-1}(\hat{h})\right] = 1.$$

We get a contradiction. $\square$

PROPOSITION 2.10.4. *Let $G_1 \leq \mathsf{Aut}\,\mathsf{X}^*$ and $G_2 \leq \mathsf{Aut}\,\mathsf{Y}^*$ be weakly branch groups. Then for every pair of positive integers n_1, n_2 such that $|\mathsf{X}^{n_1}| \geq |\mathsf{Y}^{n_2}|$ and for every isomorphism $\varphi : G_1 \longrightarrow G_2$, we have*

$$\varphi(\mathsf{RiSt}_{G_1}(n_1)) \leq \mathsf{St}_{G_2}(n_2).$$

Recall that if $\mathsf{X} = (X_1, X_2, \ldots)$, then X^n denotes the set $X_1 \times \cdots \times X_n$.

PROOF. For every $v \in \mathsf{X}^{n_1}$ put W_v to be the set of vertices $u \in \mathsf{Y}^{n_2}$ moved by $\varphi(G_1[v])$.

If $u \in W_{v_1} \cap W_{v_2}$ for different $v_1, v_2 \in \mathsf{X}^{n_1}$, then Lemma 2.10.3 implies that $\varphi^{-1}(G_2[u]) \cap \mathsf{St}_{G_1}(n_1) \leq G_1[v_1] \cap G_1[v_2]$. But then we get the contradiction $\{1\} \neq \varphi^{-1}(G_2[u]) \cap \mathsf{St}_{G_1}(n_1) \leq G_1[v_1] \cap G_1[v_2] = \{1\}$. Therefore the sets W_v are disjoint for $v \in \mathsf{X}^{n_1}$. If a set W_v is not empty, then it contains more than one element. The union of the sets W_v has cardinality not greater than $|\mathsf{Y}^{n_2}| \leq |\mathsf{X}^{n_1}|$, while we have $|\mathsf{X}^{n_1}|$ of them. Consequently, W_{v_0} is empty for some $v_0 \in \mathsf{X}^{n_1}$. This means that $\varphi(G_1[v_0]) \leq \mathsf{St}(n_2)$.

Since G_1 is level-transitive, for every $v \in \mathsf{X}^{n_1}$ there exists $g \in G_1$ such that $v_0 = g(v)$. Then

$$\varphi(G_1[v]) = \varphi(g^{-1} \cdot G_1[v_0] \cdot g) = \varphi(g)^{-1} \cdot \varphi(G_1[v_0]) \cdot \varphi(g)$$
$$\leq \varphi(g)^{-1} \cdot \mathsf{St}(n_2) \cdot \varphi(g) = \mathsf{St}(n_2).$$

Thus $\varphi(G_1[v]) \leq \mathsf{St}_{G_2}(n_2)$ for all $v \in \mathsf{X}^{n_1}$; consequently $\varphi(\mathsf{RiSt}_{G_1}(n_1)) \leq \mathsf{St}_{G_2}(n_2)$. $\square$

Lemma 2.10.3 and Proposition 2.10.4 imply

COROLLARY 2.10.5. *Let $G_1 \leq \mathsf{Aut}\,\mathsf{X}^*$ and $G_2 \leq \mathsf{Aut}\,\mathsf{Y}^*$ be weakly branch groups and let $\varphi : G_1 \longrightarrow G_2$ be an isomorphism. If $|\mathsf{X}^{n_1}| \geq |\mathsf{Y}^{n_2}|$, then for every $v \in \mathsf{Y}^{n_2}$ and every vertex $u \in \mathsf{X}^{n_1}$ moved by $\varphi^{-1}(G_2[v])$ we have*

$$\varphi(G_1[u]) \leq G_2[v].$$

We are ready to prove Theorem 2.10.1.

PROOF OF THEOREM 2.10.1. Consider an infinite word $w = x_1 x_2 \ldots \in \mathsf{X}^\omega$ and a number $n \in \mathbb{N}$. The group $\varphi^{-1}(G_2[v])$ is non-trivial for every $v \in \mathsf{Y}^n$; thus it moves some vertex $u \in \mathsf{X}^*$. The group G_1 is level-transitive; therefore there exists

2.10. RIGIDITY

$g \in G_1$ such that $g(u)$ is of the form $x_1 x_2 \ldots x_{n_1}$. We may assume that n_1 is such that $|\mathsf{X}^{n_1}| \geq |\mathsf{Y}^n|$. We have then, by Corollary 2.10.5, the inclusion

$$\varphi(G_1[u]) \leq G_2[v];$$

hence

$$\varphi(G_1[x_1 \ldots x_{n_1}]) = \varphi(G[g(u)]) = \varphi(G[u])^{\varphi(g)^{-1}} \leq G_2[v]^{\varphi(g)^{-1}} = G_2[\varphi(g)(v)].$$

Let us denote $\varphi(g)(v)$ by v_n. Note that if $m \geq n_1$, then we also have

$$\varphi(G_1[x_1 \ldots x_m]) \leq G_2[v_n].$$

Thus we have proved that for every $n \in \mathbb{N}$ there exists a word $v_n \in \mathsf{Y}^n$ and a number n_1 such that $\varphi(G_1[x_1 \ldots x_m]) \leq G_2[v_n]$ for all $m \geq n_1$. If v'_n and $v''_n \in \mathsf{Y}^*$ are two such words, then $G_2[v'_n] \cap G_2[v''_n]$ is non-trivial; therefore $v'_n = v''_n$. This implies that the sequence $\{v_n\}_{n \in \mathbb{N}}$ is unique. If $n_1 \leq n_2$, then also $\varphi(G_1[x_1 \ldots x_m]) \leq G_2[v_{n_1}] \cap G_2[v_{n_2}]$ for m big enough. Therefore $G_2[v_{n_1}] \cap G_2[v_{n_2}]$ is non-trivial; hence v_{n_1} is a beginning of v_{n_2} and there exists an infinite word $y_1 y_2 \ldots \in \mathsf{Y}^\omega$ such that $v_n = y_1 \ldots y_n$ for every n.

We thus get a map $F_\varphi : \mathsf{X}^\omega \longrightarrow \mathsf{Y}^\omega : x_1 x_2 \ldots \mapsto y_1 y_2 \ldots$ such that for every $n \in \mathbb{N}$ there exists $m \in \mathbb{N}$ such that

(2.13) $$\varphi(G_1[x_1 \ldots x_m]) \leq G_2[y_1 \ldots y_n].$$

It follows directly from the definition that F_φ is continuous for every isomorphism $\varphi : G_1 \longrightarrow G_2$. We can define in the same way a continuous map $F_{\varphi^{-1}} : \mathsf{Y}^\omega \longrightarrow \mathsf{X}^\omega$ using the isomorphism φ^{-1}. Then $\varphi^{-1}(y_1 y_2 \ldots) = a_1 a_2 \ldots$ is equivalent to the condition that for every $k \in \mathbb{N}$ there exists $n \in \mathbb{N}$ such that

$$\varphi^{-1}(G_2[y_1 \ldots y_n]) \leq G_1[a_1 \ldots a_k];$$

i.e.,

$$G_2[y_1 \ldots y_n] \leq \varphi(G_1[a_1 \ldots a_k]).$$

Let us find $m \in \mathbb{N}$ such that (2.13) holds. Then

$$G_1[x_1 \ldots x_m] \leq G_1[a_1 \ldots a_k];$$

hence $a_1 \ldots a_k$ is a beginning of $x_1 \ldots x_m$. Thus $a_1 a_2 \ldots = x_1 x_2 \ldots$; i.e., $F_\varphi^{-1} = F_{\varphi^{-1}}$ and F_φ is a homeomorphism.

Let $g \in G_1$, $w = x_1 x_2 \in \mathsf{X}^\omega$ and $y_1 y_2 \ldots = F(x_1 x_2 \ldots)$. Then for every $n \in \mathbb{N}$ there exists $n_1 \in \mathbb{N}$ such that for all $m \geq n_1$

$$\varphi(G_1[g(x_1 \ldots x_m)]) = \varphi(G_1[x_1 \ldots x_m])^{\varphi(g)^{-1}}$$

$$\leq G_2[y_1 \ldots y_n]^{\varphi(g)^{-1}} = G_2[\varphi(g)(y_1 \ldots y_n)];$$

hence $F(g(w)) = \varphi(g)(y_1 y_2 \ldots) = \varphi(g)(F(w))$.

It remains to prove that the homeomorphism F is measure-preserving. Let μ be the measure on Y^ω equal to the image under F of the Bernoulli measure on X^ω. Since the action of G_1 on X^ω is ergodic with respect to the Bernoulli measure, the action of G_2 on Y^ω is ergodic with respect to μ. But the only probabilistic measure on Y^ω with respect to which the action of G_2 is ergodic is the Bernoulli measure (see [**GNS00**]). Thus μ coincides with the Bernoulli measure on Y^ω. □

An action of a group G on a measure space $\mathcal{X}$ is said to be *ergodic* if every measurable G-invariant subset of $\mathcal{X}$ has either zero or full measure.

2.10.3. Combinatorial rigidity. It would be good to know when the homeomorphism F is induced by an isomorphism of the rooted trees, i.e., when the group actions are rigid on the trees (and not only on their boundaries).

We also consider here the general case of a level-transitive rooted tree.

DEFINITION 2.10.6. An isomorphism $\varphi : G_1 \longrightarrow G_2$ of level-transitive automorphism groups of a rooted tree X^* is called *saturated* if there exists a sequence of subgroups $H_n \leq G_1$ such that

(1) $H_n \leq \mathsf{St}_{G_1}(n)$ and $\varphi(H_n) \leq \mathsf{St}_{G_2}(n)$,
(2) the actions of H_n and $\varphi(H_n)$ on $v\mathsf{X}^k$ are transitive for every $k \geq 1$ and $v \in \mathsf{X}^n$.

A level-transitive group $G \leq \mathrm{Aut}\,\mathsf{X}^*$ is *saturated* if there exists a sequence $H_n \leq G$ of characteristic subgroups for which (1) and (2) hold.

If a group is saturated, then every one of its automorphisms is saturated.

PROPOSITION 2.10.7. *Let $G_i \leq \mathrm{Aut}\,\mathsf{X}^*$ be weakly branch groups and let $\varphi : G_1 \longrightarrow G_2$ be a saturated isomorphism. Then φ is induced by an automorphism F_* of the rooted tree X^*.*

PROOF. Let $v \in \mathsf{X}^n$ be an arbitrary word of length n. The action of H_n on $v\mathsf{X}^\omega$ is minimal (i.e., every orbit is dense), since the action of H_n is transitive on $v\mathsf{X}^k$ for every $k \geq 1$. Let $F : \mathsf{X}^\omega \longrightarrow \mathsf{X}^\omega$ be the homeomorphism inducing the isomorphism φ. Then for every $v \in \mathsf{X}^n$ the set $F(v\mathsf{X}^\omega)$ is closure of a $\varphi(H_n)$-orbit on X^ω and therefore is of the form $u\mathsf{X}^\omega$ for $u \in \mathsf{X}^n$. Put $F_*(v) = u$. It is easy to see that so defined map $F_* : \mathsf{X}^* \longrightarrow \mathsf{X}^*$ is a bijection. If $v \in \mathsf{X}^n$ and $vx \in \mathsf{X}^{n+1}$ are adjacent vertices, then $v\mathsf{X}^\omega \supset vx\mathsf{X}^\omega$; hence $F(v\mathsf{X}^\omega) \supset F(vx\mathsf{X}^\omega)$ and therefore $F_*(v)$ and $F_*(vx)$ are also adjacent. □

2.10.4. Wreath branch groups.

DEFINITION 2.10.8. We say that a group $G \leq \mathrm{Aut}\,\mathsf{X}^*$ is *wreath branch* if it is level transitive and $\mathsf{RiSt}_G(n) = \mathsf{St}_G(n)$ for every $n \in \mathbb{N}$.

Examples of wreath branch groups include the full automorphism group $\mathrm{Aut}\,\mathsf{X}^*$ of the tree X^* and every self-similar group G for which the decomposition $G = H \wr G$ is true for some $H \leq \mathfrak{S}(\mathsf{X})$. So, the P. Neumann- and J. Wilson-type examples are wreath branch (see Section 1.8).

THEOREM 2.10.9. *If G_1, G_2 are wreath branch automorphism groups of X^* and $\varphi : G_1 \longrightarrow G_2$ is an isomorphism, then φ is induced by an automorphism of the tree X^*.*

PROOF. A direct corollary of Propositions 2.10.7 and 2.10.4. □

We get the following result of Y. Lavreniuk [**Lav99**].

COROLLARY 2.10.10. *Every automorphism of the group $\mathrm{Aut}\,\mathsf{X}^*$ is inner. Hence $\mathrm{Aut}\,\mathrm{Aut}\,\mathsf{X}^* = \mathrm{Aut}\,\mathsf{X}^*$.* □

Rigidity may be used to distinguish different groups acting on rooted trees. As an example consider the following criterion. For the definition of the groups $\mathcal{W}(A_i)$ see 1.8.3.

PROPOSITION 2.10.11. *Let $A_1, A_2 \le \mathfrak{S}(\mathsf{X})$ be perfect 2-transitive permutation groups. Then the groups $\mathcal{W}(A_1)$ and $\mathcal{W}(A_2)$ are isomorphic as abstract groups if and only if A_1 and A_2 are conjugate in $\mathfrak{S}(\mathsf{X})$.*

PROOF. We have $\mathcal{W}(A_i) = A_i \wr \mathcal{W}(A_i)$, by Proposition 1.8.2. Consequently, $\mathcal{W}(A_i)$ are wreath branch (see Definition 2.10.8). If $\varphi : \mathcal{W}(A_1) \longrightarrow \mathcal{W}(A_2)$ is an isomorphism, then it is induced by a conjugation in $\operatorname{Aut} \mathsf{X}^*$, due to Theorem 2.10.9. But A_i as a permutation group of X coincides with the set of permutations defined by $\mathcal{W}(A_i)$ on the first level X of the tree X^*. □

2.10.5. Grigorchuk groups. Let us show also how rigidity of weakly branch groups can be used to distinguish the *Grigorchuk groups* G_w (and thus answer Question 6 from [**BGŠ03**]).

DEFINITION 2.10.12. Let $w = w_1 w_2 \ldots \in \{0,1,2\}^\omega$. The group G_w is the automorphism group of the binary tree $\mathsf{X}^* = \{0,1\}^*$ generated by the transformations a, b_w, c_w, d_w, where
$$a(0v) = 1v, \quad a(1v) = 0v$$
for all $v \in \mathsf{X}^*$,

$$b_w(1v) = 1 b_{\sigma(w)}(v), \qquad b_w(0v) = \begin{cases} 0v & \text{if } w_1 = 2, \\ 0a(v) & \text{otherwise,} \end{cases}$$

$$c_w(1v) = 1 c_{\sigma(w)}(v), \qquad c_w(0v) = \begin{cases} 0v & \text{if } w_1 = 1, \\ 0a(v) & \text{otherwise,} \end{cases}$$

$$d_w(1v) = 1 d_{\sigma(w)}(v), \qquad d_w(0v) = \begin{cases} 0v & \text{if } w_1 = 0, \\ 0a(v) & \text{otherwise,} \end{cases}$$

where $\sigma(w) = w_2 w_3 \ldots$ is the shift of the sequence w.

For example, the first Grigorchuk group, considered in Section 1.6 coincides with the group G_w for $w = 012012\ldots$.

THEOREM 2.10.13. *Suppose that the sequences $w_1 = x_1 x_2 \ldots, w_2 = y_1 y_2 \ldots \in \{0,1,2\}^\omega$ are not eventually constant (i.e., are not of the form $v(i)^\omega$ for $v \in \{0,1,2\}^*$ and $i \in \{0,1,2\}$). Then the groups G_{w_1} and G_{w_2} are isomorphic if and only if there exists a permutation $\pi \in \mathfrak{S}(0,1,2)$ such that $\pi(x_n) = y_n$ for all n.*

PROOF. It is easy to see that if such a permutation exists, then the groups G_{w_1} and G_{w_2} even *coincide* as subgroups of $\operatorname{Aut} \mathsf{X}^*$.

It is known (see [**Gri85**]) that if the sequence w is not eventually constant, then the group G_w is branch (otherwise it is virtually abelian). Let us define inductively the subgroups $G_n \le G_w$ by $G_0 = G_w$ and $G_n = G_{n-1}^2$. It is not hard to prove that G_n belongs to the nth level stabilizer and that G_n is level-transitive on the subtrees $v\mathsf{X}^*$ for all $v \in \mathsf{X}^n$. One can prove by induction that the group $G_n|_v$ for $v \in \mathsf{X}^n$ contains either $\langle ac_{\sigma^n(w)}, ad_{\sigma^n(w)} \rangle$ or $\langle ab_{\sigma^n(w)}, ac_{\sigma^n(w)} \rangle$ or $\langle ab_{\sigma^n(w)}, ad_{\sigma^n(w)} \rangle$.

Hence, if G_{w_1} and G_{w_2} are isomorphic, then by Proposition 2.10.7, they are conjugate in $\operatorname{Aut} \mathsf{X}^*$. Let us consider the abelianization map $p : \operatorname{Aut} \mathsf{X}^* \longrightarrow \mathbb{F}_2^\mathbb{N}$ (see the proof of Lemma 1.10.4 on page 25). Recall that $p(g) = (p_0, p_1, \ldots)$, where p_n is the parity of the number of words $v \in \mathsf{X}^*$ such that the restriction $g|_v$ is active, i.e., acts non-trivially on the first level of the tree. If the groups G_{w_1} and G_{w_2}

are isomorphic (thus conjugate), then the images $p(G_{w_1})$ and $p(G_{w_2})$ are equal subgroups of $\mathbb{F}_2^\mathbb{N}$. We have

$$\begin{aligned} p(a) &= (1,0,0,0,\ldots), \\ p(b_{w_1}) &= (0, \beta(x_1), \beta(x_2), \ldots), \\ p(c_{w_1}) &= (0, \gamma(x_1), \gamma(x_2), \ldots), \\ p(d_{w_1}) &= (0, \delta(x_1), \delta(x_2), \ldots), \end{aligned}$$

where

$$\begin{aligned} \beta(0) &= 1, & \beta(1) &= 1, & \beta(2) &= 0 \\ \gamma(0) &= 1, & \gamma(1) &= 0, & \gamma(2) &= 1 \\ \delta(0) &= 0, & \delta(1) &= 1, & \delta(2) &= 1. \end{aligned}$$

Consequently, $p(G_{w_1})$ is equal to

$$\{0, p(a), p(b_{w_1}), p(c_{w_1}), p(d_{w_1}), p(a)+p(b_{w_1}), p(a)+p(c_{w_1}), p(a)+p(d_{w_1})\}.$$

It is easy to check now that we can reconstruct the sequence w_1 from the group $p(G_{w_1}) \leq \mathbb{F}_2^\mathbb{N}$ up to a simultaneous permutation of the letters $0, 1, 2$. This finishes the proof. $\square$

We get thus an uncountable family of 3-generated groups (the generator d_w is superfluous, since $d_w = b_w c_w$) with a complete description of the isomorphism classes.

2.10.6. Theorem of R. Grigorchuk and J. Wilson. Let $\mathsf{X} = (X_1, X_2, \ldots)$ be a sequence of alphabets defining a level-transitive rooted tree X^*, as above. Recall that a group $G \leq \operatorname{Aut} \mathsf{X}^*$ is *branch* if the rigid stabilizer $\operatorname{RiSt}_G(n)$ is a subgroup of finite index in G for every n (see Definition 1.2.4).

The following rigidity theorem is proved in [**GW03**].

THEOREM 2.10.14. *Let $G \leq \operatorname{Aut} \mathsf{X}^*$ be a branch group and suppose that*

(*) *for each vertex $u \in \mathsf{X}^*$ the stabilizer G_u acts as a transitive cyclic group of prime order on the edges descending from u (i.e., on the set $uX_{|u|+1}$);*

(**) *whenever u, u' are incomparable (i.e., neither word is a beginning of the other) and $v = uw \in \mathsf{X}^*$, there is an element $g \in G$ such that $g(u') = u'$ and $g(v) \neq v$.*

Let $\mathsf{Y} = (Y_1, Y_2, \ldots)$ be another sequence defining a level-transitive rooted tree Y^* and suppose that G acts on Y^* faithfully as a branch group. Then there is an isomorphism of the rooted tree Y^* with a tree obtained from X^* by deletion of levels, which conjugates the respective actions of G.

Here deletion of levels is passing from the tree defined by the sequence $\mathsf{X} = (X_1, X_2, \ldots)$ to the tree defined by a sequence of the form

$$(X_1 \times X_2 \times \cdots \times X_{n_1}, \quad X_{n_1+1} \times X_{n_1+2} \times \cdots \times X_{n_2}, \ldots)$$

for some sequence $n_1 < n_2 < \ldots$. The set of vertices of the new tree is the subset of X^* consisting of the words of the lengths $0, n_1, n_2$, etc.

2.11. Contracting actions

2.11.1. Definition and the nucleus.

DEFINITION 2.11.1. A self-similar action (G, X) is called *contracting* if there exists a finite set $\mathcal{N} \subset G$ such that for every $g \in G$ there exists $k \in \mathbb{N}$ such that $g|_v \in \mathcal{N}$ for all words $v \in \mathsf{X}^*$ of length $\geq k$. The minimal set $\mathcal{N}$ with this property is called the *nucleus* of the self-similar action.

REMARK. We allow the action here to be non-faithful. In this case a self-similar action is the action defined by a covering G-bimodule $\mathfrak{M}$ and its basis X. The restrictions $g|_v$ are understood then in terms of permutational bimodules (see Proposition 2.3.3).

It follows from the definition that every contracting action is finite-state. The nucleus of a contracting action is unique and is equal to the set

$$\mathcal{N} = \bigcup_{g \in G} \bigcap_{n \geq 0} \{g|_v \,:\, v \in \mathsf{X}^*, |v| \geq n\}.$$

If an element $g \in G$ belongs to a cycle of the Moore diagram of the complete automaton (G, X), i.e., if $g|_v = g$ for some $v \in \mathsf{X}^* \setminus \{\varnothing\}$, then g belongs to the nucleus by the above equality. Moreover, it is easy to see that the nucleus is precisely the set of all restrictions of elements belonging to the cycles.

If A, B are subsets of a group G and $V \subset \mathsf{X}^*$, then $A \cdot B$ denotes the set $\{ab : a \in A, b \in B\} \subset G$ and $A|_V$ denotes the set of the restrictions $\{a|_v : a \in A, v \in V\}$. We also write A^k as a short-hand notation for $\underbrace{A \cdot A \cdots A}_{k \text{ times}}$.

LEMMA 2.11.2. *A self-similar action of a group G with a generating set $S = S^{-1}$, $1 \in S$ is contracting if and only if there exists a finite set $\mathcal{N}$ and a number $k \in \mathbb{N}$ such that*

$$((S \cup \mathcal{N})^2)|_{\mathsf{X}^k} \subseteq \mathcal{N}.$$

PROOF. Induction on the length of the group's element using (1.2) and (1.3). □

An example of a contracting action is the adding machine action of the group $\mathbb{Z}$. If we take $S = \{-1, 0, 1\}$, then $2S = \{-2, -1, 0, 1, 2\}$. The restrictions of the elements of $2S$ in the words of length > 1 are $\{-1, 0, 1\}$, so the nucleus is the set $\{-1, 0, 1\}$.

Actually, all the examples mentioned in Section 1.8 are contracting. An action of $\mathbb{Z}^n$ described in Section 1.7 is contracting if and only if the matrix $B = A^{-1}$ is expanding, i.e., has all eigenvalues greater than one in absolute value (see Section 2.12 below). The examples from Section 1.9 are all non-contracting.

The Grigorchuk group (see Section 1.6) is contracting. The nucleus of the Grigorchuk group coincides with the automaton defining the generators and is shown in Figure 1.6 on page 13. Most of its properties are proved using contraction. See for example the proof of Theorem 1.6.1.

It follows from Definition 2.11.1 that restrictions of the elements of the nucleus $\mathcal{N}$ also belong to the nucleus. Thus $\mathcal{N}$ is a subautomaton of the complete automaton (G, X) of the action. So we will consider the nucleus of a contracting action as an automaton rather than just a subset of the group. For instance, the nucleus of the adding machine has the diagram shown in Figure 2.2.

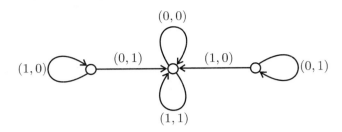

FIGURE 2.2. Nucleus of the adding machine action

Every state of the nucleus has an incoming arrow; i.e., for every $g \in \mathcal{N}$ there exists $x \in \mathsf{X}$ and $h \in \mathcal{N}$ such that $g = h|_x$, since otherwise we can remove g without affecting the conditions of Definition 2.11.1.

PROPOSITION 2.11.3. *Let a self-similar action of a finitely generated group G be recurrent and contracting with the nucleus $\mathcal{N}$. Then $G = \langle \mathcal{N} \rangle$.*

PROOF. Let S be a finite generating set of the group G and let ϕ be an associated virtual endomorphism. There exist n such that for every $g \in S$ and $v \in \mathsf{X}^n$ the restriction $g|_v$ belongs to $\mathcal{N}$. Then the restriction of any element of G in any word of length n is a product of elements of $\mathcal{N}$. Consequently, the range of ϕ^n belongs to the subgroup generated by $\mathcal{N}$. But the action is recurrent, so the range of ϕ^n is equal to G and G is generated by $\mathcal{N}$. □

COROLLARY 2.11.4. *There exists an algorithm which, given two automata over an alphabet X generating recurrent contracting groups G_1, G_2, decides whether G_1 and G_2 are equal subgroups of $\operatorname{Aut} \mathsf{X}^*$.*

PROOF. The equality of two transformations defined by finite automata is algorithmically decidable (see [**Eil74**], for example). Therefore, if an automaton generates a contracting group, then the nucleus of the group can be effectively computed. One has just to find a set $\mathcal{N}$, satisfying the conditions of Lemma 2.11.2; then the nucleus will be a subset of $\mathcal{N}$, which is easy to find. But two recurrent finitely generated contracting groups coincide if and only if their nuclei coincide, by Proposition 2.11.3. □

It is not clear, however, if there exists an algorithm deciding whether an automaton generates a recurrent (or a contracting) group.

2.11.2. Hyperbolic bimodules. The next proposition will be our main technical tool for the study of contracting groups.

PROPOSITION 2.11.5. *Suppose that a self-similar action (G, X) is contracting. Let $\mathsf{Y} \subset \mathfrak{M}$ be a finite set. Then the set of all possible elements $h \in G$ such that*

(2.14) $$y_1 \otimes y_2 \otimes \cdots \otimes y_m = v \cdot h,$$

in $\mathfrak{M}^{\otimes m}$ for some $y_i \in \mathsf{Y}$ and $v \in \mathsf{X}^m$, is finite.

PROOF. It is sufficient to prove the proposition for some set $\mathsf{Y}' \supseteq \mathsf{Y}$, so we may assume that the set Y is of the form $\{x \cdot g \,:\, x \in \mathsf{X}, g \in A\}$, where the set $A \subset G$ contains the nucleus $\mathcal{N}$ of the action and is *state-closed*; i.e., for every $g \in A$ and

$v \in \mathsf{X}^*$ the restriction $g|_v$ also belongs to A. We can do this, since the action is finite-state.

There exists a number k such that $A^2|_v \subseteq \mathcal{N} \subseteq A$ for every word $v \in \mathsf{X}^*$ of length greater or equal to k. It follows that $A^{2n}|_v \subseteq A^n$ for every $v \in \mathsf{X}^k$ and every $n \in \mathbb{N}$.

It is sufficient to find a finite set B such that it contains all h, which appear in (2.14) for the numbers m divisible by k.

We can write
$$y_1 \otimes y_2 \otimes \cdots \otimes y_m = v_1 \cdot h_1 \otimes v_2 \cdot h_2 \otimes \cdots \otimes v_{m/k} \cdot h_{m/k},$$
where $h_i \in G$ and $v_i \in \mathsf{X}^k$ for all i. The elements h_i belong to A^k, since A is state-closed. But then $h_1 \cdot v_2 = h_1(v_2) \cdot h_1|_{v_2}$ and $h_1|_{v_2}$ also belongs to A^k, so $(h_1|_{v_2} h_2)|_{v_3} \in A^{2k}|_{v_3} \subseteq A^k$, and we get an inductive proof of the fact that $v_1 \cdot h_1 \otimes v_2 \cdot h_2 \otimes \cdots \otimes v_{m/k} \cdot h_{m/k} = u \cdot h$ for some $h \in A^{2k}$. $\square$

PROPOSITION 2.11.6. *Suppose that the action of the group G defined by a bimodule $\mathfrak{M}$ and a basis X is contracting. Let $\mathsf{Y} \subset \mathfrak{M}$ be any finite subset. Then there is a finite set $\mathcal{N}(\mathsf{Y}) \subset G$ such that for every $g \in G$ there exists $n_0 \in \mathbb{N}$ such that if $g \cdot y_1 \otimes \cdots \otimes y_n = z_1 \otimes \cdots \otimes z_n \cdot h$ for $y_i, z_i \in \mathsf{Y}$ and $n \geq n_0$, then $h \in \mathcal{N}(\mathsf{Y})$.*

PROOF. Let $\mathcal{N}$ be the nucleus of the action (G, X). Let A be the set of the elements $h \in G$ such that $y_1 \otimes \cdots \otimes y_n = v \cdot h$ for some $n \in \mathbb{N}$, $y_1 \otimes \cdots \otimes y_n \in \mathsf{Y}^n$ and $v \in \mathsf{X}^n$. The set A is finite by Proposition 2.11.5.

Take any $g \in G$. There exists n_0 such that $g|_v \in \mathcal{N}$ for all words $v \in \mathsf{X}^*$ of length $\geq n_0$. Suppose that
$$g \cdot y_1 \otimes \cdots \otimes y_n = z_1 \otimes \cdots \otimes z_n \cdot h$$
for some $y_i, z_i \in \mathsf{Y}$ and $h \in G$. Let $y_1 \otimes \cdots \otimes y_n = v_1 \cdot h_1$ and $z_1 \otimes \cdots \otimes z_n = v_2 \cdot h_2$ for $v_1, v_2 \in \mathsf{X}^n$ and $h_1, h_2 \in G$. The elements h_1, h_2 belong to the finite set A. But then
$$g \cdot y_1 \otimes \cdots \otimes y_n = g \cdot v_1 \cdot h_1 = z_1 \otimes \cdots \otimes z_n \cdot h = v_2 \cdot h_2 h;$$
hence $h_2 h = g|_{v_1} h_1$, so $h = h_2^{-1} g|_{v_1} h_1$, and we can take $\mathcal{N}(\mathsf{Y}) = A^{-1} \cdot \mathcal{N} \cdot A$. $\square$

COROLLARY 2.11.7. *Suppose that the self-similar action (G, X) associated to a bimodule $\mathfrak{M}$ and a basis X is contracting. Let Y be another basis of $\mathfrak{M}$. Then the action (G, Y) is also contracting and the conjugating transformation α defined in Proposition 2.3.4 is finite-state.*

PROOF. A direct corollary of Proposition 2.11.6 and the definition of the conjugator α given in Proposition 2.3.4. $\square$

DEFINITION 2.11.8. We say that a permutational G-bimodule $\mathfrak{M}$ is *hyperbolic* if for some (and thus for all) of its bases X the associated self-similar action (G, X) is contracting.

2.11.3. Contraction coefficient. If the group G is finitely generated, then contraction of the action is equivalent to contraction of the length of the group elements under the restrictions.

If G is a group generated by a finite set $S = S^{-1}$, then by $l(g) = l_S(g)$ we denote the word length of the group element $g \in G$, i.e., the minimal length of a representation of g as a product of elements of S.

DEFINITION 2.11.9. Let G be a finitely generated group with a self-similar action (G, X). The number

$$(2.15) \qquad \rho = \limsup_{n \to \infty} \sqrt[n]{\limsup_{l(g) \to \infty} \max_{v \in \mathsf{X}^n} \frac{l(g|_v)}{l(g)}}$$

is called the *contraction coefficient* of the action.

Let ϕ be a virtual endomorphism of the group G. The number

$$(2.16) \qquad \rho_\phi = \limsup_{n \to \infty} \sqrt[n]{\limsup_{g \in \mathrm{Dom}\, \phi^n, l(g) \to \infty} \frac{l(\phi^n(g))}{l(g)}}$$

is called the *contraction coefficient* (or the *spectral radius*) of the virtual endomorphism ϕ.

LEMMA 2.11.10. *The limits* (2.15) *and* (2.16) *are finite, and they do not depend on the choice of the generating set* S.

PROOF. It is clear that $\rho_\phi \le \rho \le \max_{g \in S, x \in \mathsf{X}} l(g|_x)$, where S is the generating set. This proves that the limits are finite.

If l_1 and l_2 are the length functions on G computed with respect to finite generating sets S_1 and S_2 and if C is any number greater than $l_1(s_2)$ and $l_2(s_1)$ for all $s_1 \in S_1, s_2 \in S_2$, then

$$C^{-1} l_2(g) \le l_1(g) \le C l_2(g)$$

for every $g \in G$. Therefore,

$$\rho_1 = \limsup_{n \to \infty} \sqrt[n]{\limsup_{l(g) \to \infty} \max_{v \in \mathsf{X}^n} \frac{l_1(g|_v)}{l_1(g)}} \le \limsup_{n \to \infty} \sqrt[n]{C^2 \limsup_{l(g) \to \infty} \max_{v \in \mathsf{X}^n} \frac{l_2(g|_v)}{l_2(g)}} = \rho_2.$$

In the same way we prove that $\rho_2 \le \rho_1$; thus $\rho_1 = \rho_2$. Consequently, the value of the contraction coefficient ρ does not depend on the choice of the generating set of the group. The same is obviously true for the contraction coefficient ρ_ϕ of the virtual endomorphism. □

PROPOSITION 2.11.11. *The action is contracting if and only if its contraction coefficient ρ is less than 1.*

Let the action be level-transitive. If it is contracting, then $\rho = \rho_\phi < 1$. If $\rho_\phi < 1$, then the action is contracting.

For example, for the adding machine action and for the Grigorchuk group we have $\rho = \rho_\phi = 1/2$. If a group is finite-state and not contracting, then $\rho = \rho_\phi = 1$.

LEMMA 2.11.12. *Let G be a finitely generated group with a contracting self-similar action. Then there exist a number $M > 0$ and a positive integer n such that for every $g \in G$ and every word $v \in \mathsf{X}^n$ the inequality*

$$(2.17) \qquad l(g|_v) \le \frac{l(g)}{2} + M$$

holds.

Conversely, if there exist M, n and l_0 such that inequality (2.17) *holds for all $v \in \mathsf{X}^n$ and g such that $l(g) > l_0$, then the action is contracting.*

PROOF. Let M be the maximal length of the elements of the nucleus $\mathcal{N}$. There exists a number $n \in \mathbb{N}$ such that for every element $g \in G$ of the length $\leq 2M$ and every word $v \in \mathsf{X}^n$ we have $g|_v \in \mathcal{N}$.

Let $g \in G$ be an arbitrary element. We can write it in the form $g = g_1 \cdots g_k g_{k+1}$, where $k = \lfloor \frac{l(g)}{2M} \rfloor$, $l(g_i) = 2M$ for all $1 \leq i \leq k$ and $l(g_{k+1}) < 2M$. Then for every $v \in \mathsf{X}^n$ the restriction $g|_v$ can be written in the form $h_1 h_2 \cdots h_{k+1}$, where $h_i \in \mathcal{N}$. Consequently

$$l(g|_v) \leq (k+1)M = \left(\left\lfloor \frac{l(g)}{2M} \right\rfloor + 1\right)M \leq \left(\frac{l(g)}{2M} + 1\right)M = \frac{l(g)}{2} + M.$$

Suppose now that $l(g|_v) < \frac{l(g)}{2} + M$ for all g such that $l(g) \geq l_0$ and $v \in \mathsf{X}^n$. Let $L = \max_{l(g) < l_0, v \in \mathsf{X}^n} l(g|_v)$. Then $l(g|_v) < \frac{l(g)}{2} + M + L$ for all $g \in G$. Denote $M_1 = M + L$.

Let $v = v_0 v_1 \ldots v_k$, where $v_i \in \mathsf{X}^*$ are such that $v_i \in \mathsf{X}^n$ for all $1 \leq i \leq k$ and $|v_0| < n$. Then

$$l(g|_{v_0 v_1 \ldots v_k}) < \frac{l(g|_{v_0})}{2^k} + \frac{M_1}{2^{k-1}} + \frac{M_1}{2^{k-2}} + \cdots + M_1 < \frac{l(g|_{v_0})}{2^k} + 2M_1.$$

Therefore, the length of the restrictions $g|_v$ for all the words $v \in \mathsf{X}^*$ of length greater than $n \cdot \max_{v \in \mathsf{X}^*, |v| < n} \frac{\log l(g|_v)}{\log 2}$ is less than $1 + 2M_1$, so the action is contracting with the nucleus contained in the set of the elements of length $< 1 + 2M_1$. $\square$

PROOF OF PROPOSITION 2.11.11. We have $\rho_\phi \leq \rho$, so it is sufficient to prove that the action is contracting if and only if $\rho < 1$ and that in the level-transitive case $\rho_\phi < 1$ implies $\rho_\phi \geq \rho$.

Suppose that $\rho < 1$. Let ρ_1 be an arbitrary number such that $1 > \rho_1 > \rho$. Then there exist n_0 and l_0 such that

$$l(g|_v) < \rho_1^n l(g)$$

for all $g \in G$, and $v \in \mathsf{X}^n$ such that $l(g) > l_0$ and $n > n_0$. Then Lemma 2.11.12 implies that the action is contracting.

Suppose now that the action is contracting. We may assume that the generating set S contains all restrictions of every one of its elements, since there exists only a finite number of them. Then $l(g|_v) \leq l(g)$ for all $g \in G$ and all $v \in \mathsf{X}^*$.

There exist by Lemma 2.11.12 numbers $n_0 \in \mathbb{N}$ and $M > 0$ such that for every word $v \in \mathsf{X}^{n_0}$ and $g \in G$ the inequality

$$l(g|_v) < M + \frac{l(g)}{2}$$

holds.

Suppose that $v \in \mathsf{X}^n$, where $n > n_0$. We can write v as a product $v_0 v_1 \ldots v_k$, where $k = \lfloor \frac{n}{n_0} \rfloor \geq \frac{n}{n_0} - 1$, $|v_i| = n_0$ for $1 \leq i \leq k$ and $|v_0| < n_0$. Then for every $g \in G$

$$l(g|_v) = l(g|_{v_0}|_{v_1} \cdots |_{v_k}) < M + \frac{1}{2}\left(M + \frac{1}{2}\left(\cdots + \left(M + \frac{1}{2}l(g|_{v_0})\right)\right)\right)$$

$$< 2M + \frac{l(g|_{v_0})}{2^k} \leq 2M + \frac{l(g)}{2^k}.$$

Hence,
$$\sqrt[n]{\limsup_{l(g)\to\infty}\max_{v\in \mathsf{X}^n}\frac{l(g|_v)}{l(g)}} \le \sqrt[n]{\limsup_{l(g)\to\infty}\left(2^{-k}+\frac{2M}{l(g)}\right)} = \sqrt[n]{2^{-k}} \le 2^{-\frac{1}{n_0}+\frac{1}{n}}.$$

Consequently
$$\rho \le \lim_{n\to\infty} 2^{-\frac{1}{n_0}+\frac{1}{n}} = 2^{-\frac{1}{n_0}} < 1.$$

Suppose now that the action is level-transitive. Let ρ_ϕ be the contraction coefficient of the associated virtual endomorphism ϕ. Let us prove that $\rho_\phi \ge \rho$ if $\rho_\phi < 1$.

Let ρ_1 be any number such that $\rho_\phi < \rho_1 < 1$. Then there exist $n_0, l_0 \in \mathbb{N}$ such that for all $n > n_0$ and $g \in \mathrm{Dom}\,\phi^n$ such that $l(g) > l_0$, the inequality $l(\phi^n(g)) < \rho_1^n l(g)$ holds.

Let us pass to the nth tensor power of the self-similar action. The associated virtual endomorphism of the nth power action is ϕ^n. The basis X^n of the self-similarity bimodule $\mathfrak{M}^{\otimes n}$ is identified with a basis $\{\phi^n(r_i)h_i\}_{i=1}^{d^n}$ of $\phi^n(G)G$. Let r be an upper bound on the length of the elements r_i and h_i.

Then, by (2.5) on page 40, for every $v \in (\mathsf{X}^n)^k$ and every $g \in G$ long enough, we have
$$l(g|_v) \le l_0 + 2r + \rho_1^n(l_0 + 4r + \rho_1^n(l_0 + 4r + \cdots + \rho_1^n(2r + l(g)))))$$
$$< \frac{4r+l_0}{1-\rho_1^n} + \rho_1^{nk} l(g),$$

and Lemma 2.11.12 implies that the action is contracting. So we may assume that all restrictions of the elements of the generating set also belong to the generating set, so that $l(g|_v) \le l(g)$ for all $g \in G$ and $v \in \mathsf{X}^*$.

Let us denote $C = \frac{4r+l_0}{1-\rho_1^n}$. If $v \in \mathsf{X}^m$ is an arbitrary word, then we can take its beginning v' of the length $n\lfloor \frac{m}{n} \rfloor$ so that
$$l(g|_v) \le l(g|_{v'}) < \rho_1^m l(g) + C.$$

Consequently
$$\rho \le \limsup_{m\to\infty} \sqrt[m]{\limsup_{l(g)\to\infty}\left(\frac{C}{l(g)}+\rho_1^m\right)} = \rho_1.$$

Therefore we have $\rho_1 \ge \rho$ for every ρ_1 such that $1 > \rho_1 > \rho_\phi$. But this is possible only in the case when $\rho_\phi \ge \rho$. $\square$

It is an interesting question to find a method of computation of the contraction coefficient of a self-similar action. For instance, it is not known if the contraction coefficient is always an algebraic number.

2.12. Finite-state actions of $\mathbb{Z}^n$

THEOREM 2.12.1. *Suppose that we have a faithful self-similar recurrent action of $\mathbb{Z}^n$ and let $A = \mathbb{Q}\otimes\phi$ be the associated linear map. Then the following conditions are equivalent:*

(1) *The action is finite-state.*
(2) *The action is contracting.*
(3) *The linear map A is contracting; i.e., its spectral radius is less than one.*

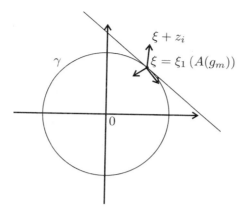

FIGURE 2.3

PROOF. Condition (3) implies (2) by Proposition 2.11.11. Implication (2)⇒(1) is obvious. So we have to prove that (1) implies (3).

Let $\{\varepsilon_1, \varepsilon_2, \ldots, \varepsilon_n\}$ be the basis of $\mathbb{C}^n = \mathbb{C} \otimes \mathbb{Z}^n$ with respect to which A has normal Jordan form. Let $(\xi_1(g), \xi_2(g), \ldots, \xi_n(g))$ denote the coordinates of $g \in \mathbb{Z}^n$ with respect to the basis $\{\varepsilon_i\}$.

Suppose that $\rho(A) \geq 1$. Let λ be the eigenvalue of A such that $|\lambda| \geq 1$. There exists an index i such that $\xi_i(A(g)) = \lambda \cdot \xi_i(g)$ for every $g \in \mathbb{Z}^n$. Let us assume, without loss of generality, that $i = 1$.

There exists $g \in \mathbb{Z}^n$ such that $\xi_1(g) \neq 0$, since $\mathbb{Z}^n$ has rank n. Let us fix such g. We are going to find a sequence $\{g_m = g|_{v_m}\}$ of restrictions of g such that $\{|\xi_1(g_m)|\}$ is non-decreasing. We will define the sequence inductively. Put $g_0 = g$ and suppose that we have defined g_m.

Let $R = \{r_0, r_1, \ldots, r_{d-1}\}$ be the digit system defining the action. Then the restriction of g_m in a one-letter word x_i is equal by (2.10) to $A(g_m + r_i - r'_i)$, where $r'_i \in R$ is defined by the condition $g_m + r_i - r'_i \in A^{-1}(\mathbb{Z}^n)$. Let us denote $d_i = r_i - r'_i$. We have $\sum_{i=0}^{d-1} d_i = 0$, since $\sum_{i=0}^{d-1}(r_i - r'_i) = \sum_{i=0}^{d-1} r_i - \sum_{i=0}^{d-1} r'_i$, and $\{r'_i\}$ is a permutation of $\{r_i\}$. Hence restrictions of g_m in one-letter words are equal to $A(g_m + d_i)$.

Let us denote $z_i = \xi_1(A(d_i))$. We have $\sum_{i=1}^d z_i = \sum_{i=1}^d \xi_1(A(d_i)) = 0$; hence the point $\xi = \xi_1(A(g_m)) = \lambda \cdot \xi_1(g_m)$ is a baricenter of the points $\xi + z_i$. Then either all the numbers z_i are equal to zero or one of the points $\xi + z_i$ is outside the circle $\gamma = \{z \in \mathbb{C} : |z| = |\xi|\}$. Otherwise the baricenter of the points $\xi + z_i$ and γ will be on the same side of the tangent line to γ at ξ (see Figure 2.3).

Hence, either all the numbers $z_i = \xi_1(A(d_i))$ are equal to zero or there exists i_0 such that $|\xi_1(A(g_m + d_{i_0}))| = |\xi_1(A(g_m)) + z_{i_0}| > |\xi_1(A(g_m))|$. We choose in the first case g_{m+1} equal to $A(g_m + d_i)$ for arbitrary i; then

$$|\xi_1(g_{m+1})| = |\xi_1(A(g_m)) + z_i| = |\xi_1(A(g_m))| = |\lambda| \cdot |\xi_1(g_m)|.$$

In the second case we put $g_{m+1} = A(g_m + d_{i_0})$. Then

$$|\xi_1(g_{m+1})| = |\xi_1(A(g_m)) + z_i| > |\xi_1(A(g_m))| = |\lambda| \cdot |\xi_1(g_m)|.$$

Then the sequence $|\xi_1(g_m)|$ is non-decreasing and all its elements are non-zero.

Since the action is finite-state, the set of values of the sequence $\xi_1(g_m)$ is finite. It follows then from the inequalities above that $|\lambda| = 1$ and in all but a finite number of cases all z_i are equal to zero. Then starting from some m we will always have $\xi_1(g_{m+1}) = \lambda \cdot \xi_1(g_m)$. The sequence $\xi_1(g_m)$ has a finite number of different values; hence λ is a root of unity. But this is impossible by Proposition 2.9.2. We get a contradiction proving the implication (1)$\Rightarrow$(3). $\square$

It is a result of [**NS04**] (Theorem 5.2) that for all n and d there exists only a finite number of conjugacy classes of $n \times n$ matrices A which are matrices of virtual endomorphisms associated to a finite state recurrent action of $\mathbb{Z}^n$ over an alphabet X of cardinality d.

For example, the matrix A is conjugate in the case $n = d = 2$ to one of the following matrices:

$$\begin{pmatrix} 0 & 1 \\ 1/2 & 0 \end{pmatrix}, \quad \begin{pmatrix} 1/2 & -1 \\ 1/2 & 0 \end{pmatrix}, \quad \begin{pmatrix} 1/2 & -1/2 \\ 1/2 & 1/2 \end{pmatrix},$$

$$\begin{pmatrix} 0 & 1 \\ -1/2 & 0 \end{pmatrix}, \quad \begin{pmatrix} -1/2 & -1 \\ 1/2 & 0 \end{pmatrix}, \quad \begin{pmatrix} -1/2 & -1/2 \\ 1/2 & -1/2 \end{pmatrix}.$$

If we choose the digit system $\{r_0 = (0,0), r_1 = (1,0)\}$, then the respective actions of the generators $a = (1,0)$, $b = (0,1)$ of $\mathbb{Z}^2$ are given (in the multiplicative notation) by the recurrent relations

$$\begin{cases} a = \sigma(1,b) \\ b = (a,a) \end{cases}, \quad \begin{cases} a = \sigma(1,ab) \\ b = (a^{-1},a^{-1}) \end{cases}, \quad \begin{cases} a = \sigma(1,ab) \\ b = \sigma(a^{-1},b) \end{cases},$$

$$\begin{cases} a = \sigma(1,b^{-1}) \\ b = (a,a) \end{cases}, \quad \begin{cases} a = \sigma(1,a^{-1}b) \\ b = (a^{-1},a^{-1}) \end{cases}, \quad \begin{cases} a = \sigma(1,a^{-1}b) \\ b = \sigma(b^{-1},a^{-1}) \end{cases}.$$

Hence, Proposition 2.3.4 implies that every finite-state recurrent action of $\mathbb{Z}^2$ over a two-letter alphabet is conjugate with one of the six described actions.

For $n = 3$ we will get 14 such conjugacy classes, for $n = 4$ there are 36 of them and for $n = 5$ there are 58 classes (see [**NS04**]).

2.13. Defining relations and word problem

2.13.1. Subgroups $\mathcal{E}_n$. Consider a self-similar action (G, X) and let $\mathfrak{M} = \mathsf{X} \times G$ be the self-similarity bimodule.

The subgroups $\mathcal{E}_n(G) = \mathcal{E}_n$, $n \geq 0$, are defined as

$$\mathcal{E}_n = \{g \in G \,:\, g \cdot v = v \cdot 1, \text{ for all } v \in \mathsf{X}^n\}.$$

In other words, g belongs to $\mathcal{E}_n$ if and only if it belongs to the nth level stabilizer $\mathsf{St}_G(n)$ and all the restrictions $g|_v$ in the words of length n are trivial.

PROPOSITION 2.13.1. *(1) The subgroup $\mathcal{E}_n$ is the kernel of the wreath recursion*

$$\psi_n : G \longrightarrow \mathfrak{S}(\mathsf{X}^n) \wr G$$

associated to the bimodule $\mathfrak{M}^{\otimes n}$ (i.e., to the self-similar action (G, X^n)).

(2) The subgroups $\mathcal{E}_n$ are normal, $\mathfrak{M}$-invariant and $\mathcal{E}_n \geq \mathcal{E}_{n-1}$.

PROOF. The first claim follows directly from the definition of the wreath recursion (see Proposition 2.2.1 and the definition of the wreath recursion after it).

The subgroup $\mathcal{E}_n$ is normal, since it is a kernel. Inclusion $\mathcal{E}_n \geq \mathcal{E}_{n-1}$ is obvious. For $g \in \mathcal{E}_n$ and $x \in \mathsf{X}$ we obviously have $g \cdot x = x \cdot h$ for some $h \in \mathcal{E}_{n-1} \leq \mathcal{E}_n$, which implies that $\mathcal{E}_n$ is invariant with respect to the virtual endomorphism ϕ_x associated to $x \in \mathsf{X}$ and $\mathfrak{M}$. Hence G is $\mathfrak{M}$-invariant (see Proposition 2.7.4). □

We denote by $\mathcal{E}_\infty(G) = \mathcal{E}_\infty$ the union $\bigcup_{n=0}^\infty \mathcal{E}_n$. It is also a normal $\mathfrak{M}$-invariant subgroup of G. Hence, it is a subgroup of the kernel $\mathcal{K}$ of the self-similar action.

PROPOSITION 2.13.2. *If the self-similar action (G, X) is contracting and no element of the nucleus belongs to the kernel $\mathcal{K}$ of the action, then $\mathcal{E}_\infty = \mathcal{K}$.*

PROOF. We have to prove the inclusion $\mathcal{K} \leq \mathcal{E}_\infty$. Let $g \in \mathcal{K}$ be arbitrary. Then there exists $n \in \mathbb{N}$ such that $g|_v$ is an element of the nucleus for every $v \in \mathsf{X}^n$. Note that $g(v) = v$, since g acts trivially on X^*. But then $g|_v \in \mathcal{K} \cap \mathcal{N} = \{1\}$; therefore $g \in \mathcal{E}_n \leq \mathcal{E}_\infty$. □

2.13.2. The covering group. Let us show that hyperbolicity of a bimodule follows from a finite number of group relations.

DEFINITION 2.13.3. Let $\mathfrak{M}$ be a hyperbolic G-bimodule with a basis X, and let $\mathcal{N}$ be the nucleus of the self-similar action (G, X).

The *covering group* $\widetilde{G}$ of the self-similar action (G, X) is the group defined by the presentation whose generators $\widetilde{g}$ are in bijective correspondence with the elements g of $\mathcal{N}$ and the defining relation are all the relations $\widetilde{g}_1 \widetilde{g}_2 \widetilde{g}_3$ of length 3, for $g_i \in \mathcal{N}$, such that $g_1 g_2 g_3 = 1$ in G.

Let $\widetilde{\mathsf{X}}$ be a set together with a bijection $\widetilde{\mathsf{X}} \longrightarrow \mathsf{X} : \widetilde{x} \mapsto x$. If $g \in \mathcal{N}$ and $x \in \mathsf{X}$, then we set

(2.18) $$\widetilde{g} \cdot \widetilde{x} = \widetilde{y} \cdot \widetilde{h} \quad \Leftrightarrow \quad g \cdot x = y \cdot h.$$

PROPOSITION 2.13.4. *Rule (2.18) gives a well defined structure of a hyperbolic $\widetilde{G}$-bimodule on $\widetilde{\mathfrak{M}} = \mathsf{X} \times \widetilde{G}$. The nucleus of the action $\left(\widetilde{G}, \widetilde{\mathsf{X}}\right)$ coincides with the generating set $\widetilde{\mathcal{N}}$ of $\widetilde{G}$.*

PROOF. We have to prove for every word $\widetilde{g}_1 \cdots \widetilde{g}_n$, $g_i \in \mathcal{N}$, which is equal to 1 in $\widetilde{G}$, and for every $x \in \mathsf{X}$ that $\widetilde{h}_1 \cdots \widetilde{h}_n = 1$, where $h_1 = g_1|_{g_2 \cdots g_n(x)}, \ldots, h_{n-1} = g_{n-1}|_{g_n(x)}, h_n = g_n|_x$. This will show that the bimodule structure of $\widetilde{\mathfrak{M}}$ is well defined.

But it is a straightforward corollary of the next two obvious facts:
(1) if $g_1 g_2 g_3 = 1$ in G for $g_i \in \mathcal{N}$, then
$$g_1 g_2 g_3 \cdot x = x \cdot g_1|_{g_2 g_3(x)} g_2|_{g_3(x)} g_3|_x = x \cdot 1;$$
(2) if $g(h(x)) = h(x)$, then $h^{-1} g h \cdot x = x \cdot (h|_x)^{-1} g|_{h(x)} h|_x$.

Let us prove that the self-similar action $\left(\widetilde{G}, \widetilde{\mathsf{X}}\right)$ is contracting. We know that the action of G on X^* is contracting; therefore there exists $n \in \mathbb{N}$ such that $gh|_v = g|_{h(v)} h|_v \in \mathcal{N}$ for all $g, h \in \mathcal{N}$ and $v = x_1 \ldots x_n \in \mathsf{X}^n$. Then, by definition of the bimodule $\widetilde{\mathfrak{M}}$:

$$\widetilde{g}|_{\widetilde{h}(\widetilde{v})} \widetilde{h}|_{\widetilde{v}} \in \widetilde{\mathcal{N}},$$

where $\widetilde{v} = \widetilde{x}_1 \ldots \widetilde{x}_n$, since every relation $gh = f$ between elements of $\mathcal{N}$ implies the relation $\widetilde{gh} = \widetilde{f}$ in $\widetilde{G}$. Hence, $\left(\widetilde{gh}\right)|_{\widetilde{v}} \in \widetilde{\mathcal{N}}$, which implies by Lemma 2.11.2 that the action is contracting. The claim about the nucleus of the action is now straightforward. $\square$

2.13.3. L-presentations. Proposition 2.13.2 describes the kernel of a contracting action in much more convenient terms than the general Proposition 2.7.5 does. However, in many cases a more explicit description of the defining relations of contracting self-similar groups can be found.

In most cases contracting self-similar groups are not finitely presented, but one can usually find a rather simple recursive presentation called a *finite L-presentation*.

The first example of an L-presentation was found by I. Lysionok in [**Lys85**]. He showed that the Grigorchuk group G admits the presentation

$$G = \left\langle a, b, c, d \mid a^2, b^2, c^2, d^2, bcd, \tau^i (ad)^4, \tau^i (adacac)^4, (i \geq 0) \right\rangle,$$

where τ is the endomorphism of the free group $\langle a, b, c, d \rangle$ given by

$$\tau(a) = aca, \qquad \tau(b) = d,$$
$$\tau(c) = b, \qquad \tau(d) = c.$$

This presentation was used later by R. Grigorchuk in [**Gri98**] to construct the first example of a finitely presented amenable, but not an elementary amenable group, thus answering a question of M. Day [**Day57**].

A more general class of recursive presentations generalizing the presentation of I. Lysionok is defined in the following way (see [**Gri99, Bar03a**] and Section 4.2 of [**BGŠ03**]).

DEFINITION 2.13.5. An *endomorphic presentation* is a presentation of the form

$$\left\langle S \;\middle|\; Q \cup \bigcup_{\varphi \in \Phi^*} \varphi(R) \right\rangle,$$

where S is a finite set of generators, Q and R are sets of elements of the free group $F(S)$ generated by S, Φ is a set of endomorphisms of $F(S)$ and Φ^* is the monoid generated by Φ (i.e., the closure of $\Phi \cup \{id\}$ under composition).

The endomorphic presentation is *finite* if the sets Q, R and Φ are finite. It is called *ascending* if the set Q is empty. If Φ consists of one endomorphism, then the endomorphic presentation is called an *L-presentation*.

The following is a result of L. Bartholdi [**Bar03a**].

THEOREM 2.13.6. *Let G be a finitely generated, contracting, regular branch group. Then G has a finite endomorphic presentation. However, G is not finitely presented.*

A self-similar group G is said to be *regular branch* if there exists a finite-index subgroup $K \leq G$ such that $K^{\mathsf{X}} \leq \psi(K)$, where $\psi : G \longrightarrow \mathfrak{S}(\mathsf{X}) \wr G$ is the wreath recursion (see 1.5.2).

See the paper [**Bar03a**] for various examples of groups admitting finite endomorphic presentations.

2.13.4. Growth of orbits.

DEFINITION 2.13.7. Let G be a finitely generated group acting on a set A. *Growth degree* of the G-action is the number
$$\gamma = \sup_{w \in A} \limsup_{r \to \infty} \frac{\log |\{g(w) \, : \, l(g) \leq r\}|}{\log r},$$
where $l(g)$ is the length of the group element g with respect to some fixed finite generating set of G.

One can show, in the same way as in Proposition 2.11.10, that the growth degree γ does not depend on the choice of the generating set of G.

PROPOSITION 2.13.8. *Suppose that a self-similar action (G, X) is contracting. Then the growth degree of the action of G on X^ω is not greater than $\frac{\log |\mathsf{X}|}{-\log \rho}$, where ρ is the contraction coefficient of (G, X).*

PROOF. The statement is more or less classical. See, for instance, similar statements in [**Gro81, BG00b, Fra70**].

Let ρ_1 be such that $\rho < \rho_1 < 1$. Then there exists $C > 0$ and $n \in \mathbb{N}$ such that for all $g \in G$ we have $l(g|_v) < \rho_1^n \cdot l(g) + C$ for all $v \in \mathsf{X}^n$.

Then cardinality of the set $B(w, r) = \{g(w) \, : \, l(g) \leq r\}$, where $w = x_1 x_2 \ldots \in \mathsf{X}^\omega$, is not greater than
$$|\mathsf{X}|^n \cdot |\{B(x_{n+1} x_{n+2} \ldots, \rho_1^n \cdot r + C)|,$$
since the map $\sigma^n : x_1 x_2 \ldots \mapsto x_{n+1} x_{n+2} \ldots$ maps $B(w, r)$ to $B(\sigma^n(w), \rho_1^n \cdot r + C)$ and every point of X^ω has exactly $|\mathsf{X}|^n$ preimages under σ^n. The map σ^n is the nth iteration of the shift map $\sigma(x_1 x_2 \ldots) = x_2 x_3 \ldots$.

Let $k = \left\lfloor \frac{\log r}{-n \log \rho_1} \right\rfloor + 1$. Then $\rho_1^{nk} \cdot r < 1$ and the number of the points in the ball $B(w, r)$ is not greater than
$$|\mathsf{X}|^{nk} \cdot \left| B\left(\sigma^{nk}(w), R\right) \right|,$$
where
$$R = \rho_1^{nk} \cdot r + \rho_1^{n(k-1)} \cdot C + \rho_1^{n(k-2)} \cdot C + \cdots + \rho_1^n \cdot C + C < 1 + \frac{C}{1 - \rho_1^n}.$$

But $|B(u, R)|$ for all $u \in \mathsf{X}^\omega$ is less than $K_1 = |S|^R$, where S is the generating set of G (we assume that $S = S^{-1} \ni 1$). Hence,
$$|B(w, r)| < K_1 \cdot |\mathsf{X}|^{n\left(\frac{\log r}{-n \log \rho_1} + 1\right)}$$
$$= K_1 \cdot \exp\left(\frac{\log |\mathsf{X}| \log r}{-\log \rho_1} + n \log |\mathsf{X}| \right) = K_2 \cdot r^{\frac{\log |\mathsf{X}|}{-\log \rho_1}},$$
where $K_2 = K_1 \cdot |\mathsf{X}|^n$. Thus, the growth degree is not greater than $\frac{\log |\mathsf{X}|}{-\log \rho_1}$ for every $\rho_1 \in (\rho, 1)$, so it is not greater than $\frac{\log |\mathsf{X}|}{-\log \rho}$. $\square$

LEMMA 2.13.9. *Let (G, X) be a contracting faithful self-similar action of an infinite finitely generated group G. Then its contraction coefficient is greater or equal to $1/|\mathsf{X}|$.*

PROOF. Let $\phi = \phi_x$ be the virtual endomorphism of G associated to the action and $x \in \mathsf{X}$. Then the parabolic subgroup $P(\phi) = \bigcap_{n \geq 0} \text{Dom}\, \phi^n$ is the stabilizer of the word $w = xxx\ldots \in \mathsf{X}^\omega$. The subgroup $P(\phi)$ has infinite index in G; otherwise $\bigcap_{g \in G} g^{-1} P g = \mathcal{K}(\phi)$ has finite index, and G does not act faithfully. Consequently, the G-orbit of w is infinite. Then there exists an infinite sequence of generators $s_1, s_2, \ldots$ of the group G such that the elements of the sequence

$$w, s_1(w), s_2 s_1(w), s_3 s_2 s_1(w), \ldots$$

are pairwise different. This implies that the growth degree of the orbit Gw

$$\gamma = \limsup_{r \to \infty} \frac{\log |\{g(w)\ :\ l(g) \leq r\}|}{\log r}$$

is greater than or equal to 1; thus the growth degree of the action of G on X^ω is not less than 1, and by Proposition 2.13.8, $1 \leq \frac{\log |\mathsf{X}|}{-\log \rho}$. □

2.13.5. Word problem algorithm. The proof of the next proposition will give an effective algorithm for solving the word problem in contracting groups. This is essentially the algorithm described already in [**Gri80**] for the Grigorchuk group (see Chapter 3 of [**BGŠ03**] for more information and bibliography). Here we show an interesting relation between the geometry of the self-similar action and its algorithmic properties.

PROPOSITION 2.13.10. *If there exists a faithful contracting action (G, X) of a finitely generated group G, then for any $\epsilon > 0$ there exists an algorithm of polynomial complexity of degree not greater than $\frac{\log |\mathsf{X}|}{-\log \rho} + \epsilon$ solving the word problem in G, where ρ is the contraction coefficient of (G, X).*

PROOF. We assume that the generating set S contains all restrictions of its elements, so that $l(g|_v)$ is always not greater than $l(g)$.

Let us denote by F the free group generated by S. For every $g \in F$ we denote by $\hat{g}$ the image of g in G.

Let $1 > \rho_1 > \rho$. Then $\rho_1 \cdot |\mathsf{X}| > 1$ by Lemma 2.13.9. There exist n_0 and l_0 such that for every word $v \in \mathsf{X}^*$ of length n_0 and every $g \in G$ of length $\geq l_0$ we have

$$l(g|_v) < \rho_1^{n_0} l(g).$$

Let us choose a number $R > l_0$ and assume that we know for every $g \in F$ of length less than R if $\hat{g}$ is trivial or not. Assume also that we have a table of all relations of the form $\hat{g} \cdot v = u \cdot \hat{h}$ where $g, h \in F$, $v, u \in \mathsf{X}^{n_0}$ and $l(g) \leq l_0$. We assume that $l(h) \leq l(g)$ for all g and $l(h) < \rho_1^{n_0} l(g)$, whenever $l(g) = l_0$.

Suppose that $l(g) \geq R$. We can compute in $l(g)$ steps, for any $v \in \mathsf{X}^{n_0}$, an element $h \in F$ and a word $u \in \mathsf{X}^{n_0}$ such that $\hat{g} \cdot v = u \cdot \hat{h}$ and $l(h) < \rho_1^{n_0} l(g) + l_0$. We have just to partition the group word g into pieces of length l_0 and then compute the respective restrictions. If $v \neq u$, then we conclude that $\hat{g}$ is not trivial and stop the algorithm. If for all $v \in \mathsf{X}^{n_0}$ we have $v = u$, then $\hat{g}$ is trivial if and only if all the obtained restrictions $\hat{h} = \hat{g}|_v$ are trivial.

We have

$$l(h) < \rho_1^{n_0} l(g) + l_0 = \left(\rho_1^{n_0} + \frac{l_0}{l(g)}\right) l(g) \leq \left(\rho_1^{n_0} + \frac{l_0}{R}\right) l(g).$$

Denote $\rho_2 = \sqrt[n_0]{\rho_1^{n_0} + \frac{l_0}{R}}$. Changing R, we may assume that $\rho_1 < \rho_2 < 1$ and can make ρ_2 as close to ρ_1 as we wish.

2.13. DEFINING RELATIONS AND WORD PROBLEM

We know whether $\hat{h}$ is trivial if $l(h) < R$. We proceed further, applying the above computations for those h which have length not less than R.

So on each step the length of the elements becomes smaller and the algorithm stops in not more than $\frac{\log l(g) - \log R}{-n_0 \log \rho_2} < \frac{\log l(g)}{-n_0 \log \rho_2}$ steps. On each step the algorithm branches into $|X|^{n_0}$ algorithms. Since $\rho_2 \cdot |X| > 1$, the total time is bounded by

$$l(g)\left(1 + (\rho_2 \cdot |X|)^{n_0} + (\rho_2 \cdot |X|)^{2n_0} + \cdots + (\rho_2 \cdot |X|)^{n_0 \cdot \lfloor -\log l(g)/n_0 \log \rho_2 \rfloor}\right)$$

$$< \frac{l(g)}{(\rho_2 \cdot |X|)^{n_0} - 1}\left((\rho_2 \cdot |X|)^{n_0 - \log l(g)/\log \rho_2} - 1\right)$$

$$= \frac{l(g)(\rho_2 \cdot |X|)^{n_0}}{(\rho_2 \cdot |X|)^{n_0} - 1}\left((\rho_2 \cdot |X|)^{-\log l(g)/\log \rho_2} - (\rho_2 \cdot |X|)^{-n_0}\right)$$

$$= C_1 l(g)\left(\exp\left(\log l(g)\left(\frac{\log |X|}{-\log \rho_2} - 1\right)\right) - C_2\right)$$

$$= C_1 l(g)^{-\log |X|/\log \rho_2} - C_1 C_2 l(g),$$

where $C_1 = \frac{(\rho_2 \cdot |X|)^{n_0}}{(\rho_2 \cdot |X|)^{n_0} - 1}$ and $C_2 = (\rho_2 \cdot |X|)^{-n_0}$. $\square$

CHAPTER 3

Limit Spaces

3.1. Introduction

We show in this chapter how a topological dynamical system called the *limit dynamical system* is associated to every contracting self-similar action. This construction is a bridge between self-similar groups and self-similar topological spaces. The converse construction, called the *iterated monodromy group*, will be studied in the subsequent chapters.

The first observation giving a hint to the construction of the limit space was made by L. Bartholdi and R. Grigorchuk in [**BG00b**]. They noticed that the Schreier graphs (the graphs of the action) of some self-similar groups on the levels of the tree X^* converge (after rescaling) to a fractal space.

The other origin of the limit spaces is the connection between self-similar groups and numeration systems on $\mathbb{Z}^n$ (see Sections 2.9 and 2.12). We have seen that a self-similar action of $\mathbb{Z}^n$ on X^ω can be interpreted as the natural action of $\mathbb{Z}^n$ on the profinite "A-adic" completion of $\mathbb{Z}^n$ (see Proposition 2.9.6), i.e., on the set of the formal power series

$$r_{i_1} + A^{-1}(r_{i_2}) + A^{-2}(r_{i_3}) + \cdots,$$

where r_i are the "digits", i.e., elements of a coset transversal of the subgroup $A^{-1}(\mathbb{Z}^n)$ of $\mathbb{Z}^n$. Here A is the matrix of the virtual endomorphism associated to the action.

If the matrix A is contracting (i.e., if the self-similar action is contracting), then we can use the digits r_i and the matrix A to construct a numeration system on $\mathbb{R}^n$, since then the series

$$A(r_{j_1}) + A^2(r_{j_2}) + A^3(r_{j_3}) \cdots$$

is convergent in $\mathbb{R}^n$ for every sequence $j_1 j_2 \ldots$.

These kinds of numeration systems on $\mathbb{R}^n$ are classical (see the bibliography at the end of Section 6.2), and many of the results and the notions of this chapter were inspired by the results and questions in this field.

Unlike the case of the numeration systems on profinite A-adic groups, an A-adic expansion of a point of $\mathbb{R}^n$ is not unique. For example, the binary real numbers $0.111\ldots$ and $1.000\ldots$ are equal. Identification rules for the points of $\mathbb{R}^n$ may be even more complicated. Such identification is the essence of the definition of a limit space.

DEFINITION 3.1.1. Let (G, X) be a contracting self-similar action. Left-infinite sequences $\ldots x_2 x_1, \ldots y_2 y_1 \in \mathsf{X}^{-\omega}$ are said to be *asymptotically equivalent* (with respect to the action (G, X)) if there exists a bounded sequence $g_k \in G$ such that $g_k(x_k \ldots x_2 x_1) = y_k \ldots y_2 y_1$ for all $k \geq 1$.

The quotient of the topological space $\mathsf{X}^{-\omega}$ by the asymptotic equivalence relation is called the *limit space* of the self-similar action and is denoted $\mathcal{J}_G$.

Here a sequence $\{g_k\}$ is called *bounded* if it assumes only a finite number of different values.

We will prove (Theorem 3.6.3) that two sequences, $\ldots x_2 x_1$ and $\ldots y_2 y_1$, are asymptotically equivalent if and only if there exists a left-infinite directed path $\ldots e_2 e_1$ in the Moore diagram of the nucleus such that the arrow e_i is labeled by (x_i, y_i). Recall that the labels of the right-infinite paths show the action of the elements of the nucleus on the space X^ω. The limit space may be considered, therefore, as a "dual" of the self-similar action.

The adding machine action of $\mathbb{Z}$ (see 1.7.1) corresponds to the usual binary numeration system. If we look carefully at the Moore diagram of its nucleus (Figure 2.2 on page 58), we see that the two sequences $\ldots x_2 x_1$ and $\ldots y_2 y_1$ are equivalent if and only if the binary real numbers $.x_1 x_2 \ldots$ and $.y_1 y_2 \ldots$ are equal (or differ by 1). It follows that the limit space $\mathcal{J}_\mathbb{Z}$ is the circle $\mathbb{R}/\mathbb{Z}$ (obtained from the segment $[0, 1]$ by gluing the endpoints).

The definition of the asymptotic equivalence relation (and its description in the terms of the nucleus) show that it is invariant under the shift $\ldots x_2 x_1 \mapsto \ldots x_3 x_2$. Hence, the shift defines a continuous map $\mathsf{s} : \mathcal{J}_G \longrightarrow \mathcal{J}_G$. The dynamical system $(\mathcal{J}_G, \mathsf{s})$ is called the *limit dynamical system* of the self-similar action (G, X).

The theory becomes nicer if we add some more structure to the limit dynamical system. The limit space $\mathcal{J}_G$ is actually an *orbispace* in a natural way. The shift s with respect to this orbispace structure is a *partial self-covering* of $\mathcal{J}_G$, i.e., a covering by an *open sub-orbispace*. We will see later (Theorem 5.3.1) that the self-similar action is uniquely defined (up to an isomorphism of the self-similarity bimodules) by this partial self-covering.

The orbispace structure of the limit space comes from a presentation of $\mathcal{J}_G$ as a space of orbits of a proper action of G on the *limit G-space* $\mathcal{X}_G$.

The space $\mathcal{X}_G$ is defined as the quotient of the space $\mathsf{X}^{-\omega} \times G$ by the asymptotic equivalence relation. Here $\ldots x_2 x_1 \cdot g$ is equivalent to $\ldots y_2 y_1 \cdot h$ if there exists a bounded sequence g_k such that

$$g_k \cdot x_k \ldots x_2 x_1 \cdot g = y_k \ldots y_2 y_1 \cdot h$$

for all $k \geq 1$, i.e., such that $g_k(x_k \ldots x_1) = y_k \ldots y_1$ and $g_k|_{x_k \ldots x_1} g = h$.

One can also prove (Proposition 3.2.6) that $\ldots x_2 x_1 \cdot g$ is equivalent to $\ldots y_2 y_1 \cdot h$ if and only if there exists a path $\ldots e_2 e_1$ in the Moore diagram of the nucleus which ends in $h g^{-1}$ and is such that e_i is labeled by (x_i, y_i).

It is easy to see that the natural right action of G on $\mathsf{X}^{-\omega} \cdot G$ induces an action of G on $\mathcal{X}_G$.

It follows from the diagram in Figure 2.2 on page 58 that in the case of the adding machine action of $\mathbb{Z} \cong \langle a \rangle$, two sequences $\ldots x_2 x_1 \cdot a^n$ and $\ldots y_2 y_1 \cdot a^m$ are asymptotically equivalent if and only if the real numbers $n.x_1 x_2 \ldots = n + \sum_{k=1}^\infty x_k/2^k$ and $m.y_1 y_2 \ldots = m + \sum_{k=1}^\infty y_k/2^k$ are equal. So, the limit $\mathbb{Z}$-space $\mathcal{X}_\mathbb{Z}$ is $\mathbb{R}$ with the natural action of $\mathbb{Z}$ on it. This also agrees with the fact that the limit space $\mathcal{J}_\mathbb{Z}$ is the orbit space $\mathbb{R}/\mathbb{Z}$ of this action.

We will also prove (see Section 6.2) that in the case of a contracting recurrent action of $\mathbb{Z}^n$, the limit space $\mathcal{X}_{\mathbb{Z}^n}$ is the Euclidean space $\mathbb{R}^n$ with the natural action of $\mathbb{Z}^n$. Thus the G-space $\mathcal{X}_G$ is an even better model of "A-adic real numbers" than

$\mathcal{J}_G$. The points of $\mathcal{X}_G$ are encoded by the sequences $\ldots x_2 x_1 \cdot g$, where $\ldots x_2 x_1$ is the "fractional" part and g is the "integral" part of the point.

The set $\mathcal{T}$ of "fractions", i.e., the set of the points of $\mathcal{X}_G$ with the trivial "integral parts", is a classical object in the case of the numeration systems on $\mathbb{R}^n$. They are called the *digit tiles* and often have interesting fractal shapes (see Figure 6.1 and Figure 6.2 on page 168).

The shifts of the tile $\mathcal{T}$ by the action of G cover the space $\mathcal{X}_G$ and will be the main tool of the study of the topologies on $\mathcal{X}_G$ and $\mathcal{J}_G$. In particular, we prove that the space $\mathcal{X}_G$ can be axiomatically described as the unique proper co-compact right G-space with a contracting self-similarity on it. Here a *self-similarity* of a right G-space $\mathcal{X}$ is a homeomorphism $\Phi : \mathcal{X} \otimes \mathfrak{M} \longrightarrow \mathcal{X}$ conjugating the right actions of G, where $\mathfrak{M}$ is the self-similarity bimodule. The self-similarity is *contracting* if the map $\xi \longrightarrow \Phi(\xi \otimes x)$ contracts the distances on $\mathcal{X}$. More details and an accurate definition are given in Section 3.4.

We start the study of the limit spaces with the limit G-space $\mathcal{X}_G$, since it contains more information than $\mathcal{J}_G$, and most properties of $\mathcal{J}_G$ will be deduced from the properties of $\mathcal{X}_G$.

We also treat the limit spaces from the very beginning in terms of the self-similarity bimodules. This has many technical advantages in further applications.

3.2. The limit G-space $\mathcal{X}_G$

3.2.1. Definition of $\mathcal{X}_G$ in terms of G-bimodules. Let us fix some hyperbolic G-bimodule $\mathfrak{M}$ (recall that it just means that the associated self-similar action is contracting).

We say that a sequence $x_1, x_2, \ldots$ of elements of some set is *bounded* if the set $\{x_i\}$ of values of the sequence is finite. Let $\Omega(\mathfrak{M})$ be the set of all bounded sequences $\ldots \otimes x_2 \otimes x_1$ of elements of $\mathfrak{M}$. We write the left-infinite sequences, since we are going to define the right G-space $\mathcal{X}_G = \mathfrak{M}^{\otimes -\omega} = \ldots \otimes \mathfrak{M} \otimes \mathfrak{M}$.

DEFINITION 3.2.1. Two sequences $\ldots \otimes x_2 \otimes x_1, \ldots \otimes y_2 \otimes y_1 \in \Omega(\mathfrak{M})$ are *asymptotically equivalent* if there exists a bounded sequence $g_n \in G$ such that

$$g_n \cdot x_n \otimes x_{n-1} \otimes \cdots \otimes x_1 = y_n \otimes y_{n-1} \otimes \cdots \otimes y_1$$

in $\mathfrak{M}^{\otimes n}$ for every $n \geq 1$.

The quotient of set $\Omega(\mathfrak{M})$ by the asymptotic equivalence relation is denoted $\mathfrak{M}^{\otimes -\omega}$ or $\mathcal{X}_G$ and is called the *limit G-space*.

Compare the definition of $\mathfrak{M}^{\otimes -\omega}$ with the definition of the G-space $\mathfrak{M}^{\otimes \omega}$ in Section 2.4. We will introduce a topology on $\mathfrak{M}^{\otimes -\omega}$ later.

It is easy to see that the space $\mathfrak{M}^{\otimes -\omega}$ is a right G-space, i.e., that the right action

$$(\ldots \otimes x_2 \otimes x_1) \cdot g = \ldots \otimes x_2 \otimes (x_1 \cdot g)$$

is a well defined action on $\mathfrak{M}^{\otimes -\omega}$.

We will write the sequence $\ldots \otimes x_2 \otimes x_1$ often just as a left-infinite word $\ldots x_2 x_1$.

LEMMA 3.2.2. *Let $\mathsf{Y} \subset \mathfrak{M}$ be a finite subset and let $\mathsf{Y}^{-\omega} \subset \Omega(\mathfrak{M})$ be taken with the topology of a direct product of the discrete sets Y. Then the asymptotic equivalence relation is closed on $\mathsf{Y}^{-\omega}$.*

74   3. LIMIT SPACES

PROOF. We have to prove that if $C \subset \mathsf{Y}^{-\omega}$ is closed, then its saturation $[C]$ is closed in $\mathsf{Y}^{-\omega}$. Saturation of C is the set of all points which are equivalent to some points of C. It is sufficient to prove that the equivalence on $\mathsf{Y}^{-\omega}$ is a closed subset of $\mathsf{Y}^{-\omega} \times \mathsf{Y}^{-\omega}$, since the space $\mathsf{Y}^{-\omega}$ is compact.

Let the set $\mathcal{N}(\mathsf{Y})$ be as in Proposition 2.11.6. Let us construct a labeled directed graph (also denoted $\mathcal{N}(\mathsf{Y})$) whose set of vertices is $\mathcal{N}(\mathsf{Y})$ in which we have an arrow from a vertex g_1 to a vertex g_2 if and only if there exists a pair $(y_1, y_2) \in \mathsf{Y} \times \mathsf{Y}$ such that $g_1 \cdot y_1 = y_2 \cdot g_2$. The respective arrow will be labeled by (y_1, y_2).

We are going to prove the following lemma, which will also be used in the proof of another proposition.

LEMMA 3.2.3. *Two sequences* $\ldots y_2 y_1, \ldots z_2 z_1 \in \mathsf{Y}^{-\omega}$ *are asymptotically equivalent if and only if there exists a directed path* $\ldots e_2 e_1$ *in the graph* $\mathcal{N}(\mathsf{Y})$ *which ends in the vertex* 1 *and is such that the edge* e_i *is labeled by* (y_i, z_i).

PROOF. If such a path exists and if g_i is the beginning of the edge e_i (and the end of the edge e_{i+1}), then, by construction of the graph, we have

$$g_n \cdot y_n y_{n-1} \ldots y_1 = z_n z_{n-1} \ldots z_1,$$

and the sequences $\ldots y_2 y_1$ and $\ldots z_2 z_1$ are asymptotically equivalent.

On the other hand, suppose that the sequences $\ldots y_2 y_1$ and $\ldots z_2 z_1$ are asymptotically equivalent. Let $\{g_n\}$ be a bounded sequence such that $g_n \cdot y_n y_{n-1} \ldots y_1 = z_n z_{n-1} \ldots z_1$. It follows from the definition of the tensor product that for every pair $n > m$ of indices there exists an element $g_{n,m} \in G$ such that

$$g_n \cdot y_n y_{n-1} \ldots y_{m-1} = z_n z_{n-1} \ldots z_{m-1} \cdot g_{n,m},$$
$$g_{n,m} \cdot y_m y_{m-1} \ldots y_1 = z_m z_{m-1} \ldots z_1.$$

It follows from Proposition 2.11.6 that there exists $n_0 \in \mathbb{N}$ such that $g_{n,m} \in \mathcal{N}(\mathsf{Y})$ for every pair n, m such that $n - m \geq n_0$.

Let K_m be the set of the elements $g_{n,m} \in \mathcal{N}(\mathsf{Y})$ for all $n \geq m + n_0$. The set K_m is finite and non-empty. We also have for every $h_m \in K_m$

$$h_m \cdot y_m y_{m-1} \ldots y_1 = z_m z_{m-1} \ldots z_1.$$

If h_m is an element of K_m, then there exists an element h_{m-1} of K_{m-1} such that $h_m \cdot y_m = z_m \cdot h_{m-1}$. Since the inverse limit of a sequence of finite non-empty sets is non-empty, there exists a sequence $h_m \in K_m$ such that $h_m \cdot y_m = z_m \cdot h_{m-1}$ for all $m \geq 1$. This sequence gives the necessary path in $\mathcal{N}(\mathsf{Y})$. $\square$

The set of all left-infinite directed paths in the graph $\mathcal{N}(\mathsf{Y})$ is obviously a compact subset of the space $E^{-\omega}$, where E is the set of edges of $\mathcal{N}(\mathsf{Y})$. The map putting into correspondence the pair $(\ldots y_2 y_1, \ldots z_2 z_1)$ to a path $\ldots e_2 e_1$ with the consecutive labels $\ldots (y_2, z_2)(y_1, z_1)$ is continuous. Hence, the asymptotic equivalence relation on $\mathsf{Y}^{-\omega}$ is a compact, and thus closed, subset of $\mathsf{Y}^{-\omega} \times \mathsf{Y}^{-\omega}$. $\square$

Now we are ready to introduce the topology on $\mathfrak{M}^{\otimes -\omega}$.

DEFINITION 3.2.4. Let $\pi : \Omega(\mathfrak{M}) \longrightarrow \mathfrak{M}^{\otimes -\omega}$ be the quotient map. Then a subset $C \subset \mathfrak{M}^{\otimes -\omega}$ is closed if for every finite set $\mathsf{Y} \subset \mathfrak{M}$ the set $\pi^{-1}(C) \cap \mathsf{Y}^{-\omega}$ is closed in $\mathsf{Y}^{-\omega}$.

In other words, we introduce on $\mathfrak{M}^{\otimes -\omega}$ the coarsest topology for which the restriction of π onto $\mathsf{Y}^{-\omega}$ is continuous for every finite $\mathsf{Y} \subset \mathfrak{M}$.

A more accessible definition of the topology on $\mathfrak{M}^{\otimes -\omega} = \mathcal{X}_G$ will be given in the next subsection.

3.2.2. Definition of $\mathcal{X}_G$ in terms of the action (G, X).

PROPOSITION 3.2.5. *Let X be a basis of $\mathfrak{M}$. Then every element $\ldots a_2 a_1 \in \mathfrak{M}^{\otimes -\omega}$ can be written in the form $\ldots x_2 x_1 \cdot g$ for some $x_i \in \mathsf{X}$ and $g \in G$.*

PROOF. The element $a_n \ldots a_2 a_1 \in \mathfrak{M}^{\otimes n}$ can be written uniquely in the form $v_n \cdot g_n$ for some $v_n \in \mathsf{X}^n$ and $g_n \in G$.

Recall that the space $\mathsf{X}^* \sqcup \mathsf{X}^{-\omega}$ has a natural topology given by the fundamental system of open sets $\mathsf{X}^* v \cup \mathsf{X}^{-\omega} v$, where v runs through X^*.

The set of possible g_n is finite by Proposition 2.11.5. The space $\mathsf{X}^* \sqcup \mathsf{X}^{-\omega}$ is compact; thus there exists a monotone sequence n_k such that v_{n_k} converges to a sequence $\ldots x_2 x_1$ and $g_{n_k} = g$ is constant. Let us prove that $\ldots x_2 x_1 \cdot g = \ldots a_2 a_1$.

Let us fix some n. The sequence v_{n_k} converges to $\ldots x_2 x_1$; hence there exists k_0 such that the end of length n of the word v_{n_k} is equal to $x_n \ldots x_1$ for all $k \geq k_0$. We have for every $k \geq k_0$:

$$a_{n_k} a_{n_k - 1} \ldots a_{n+1} a_n \ldots a_1 = v_{n_k} \cdot g = u_{n_k} x_n \ldots x_2 x_1 \cdot g,$$

where u_{n_k} is the beginning of length $n_k - n$ of the word v_{n_k}.

It follows now from the definition of the tensor product that $a_{n_k} a_{n_k - 1} \ldots a_{n+1} = u_{n_k} \cdot h_{n,k}$ and $h_{n,k} \cdot a_n a_{n-1} \ldots a_1 = x_n \ldots x_1 \cdot g$ for some $h_{n,k}$. The set of possible $h_{n,k}$ is finite by Proposition 2.11.5, which proves that $\ldots a_2 a_1 = \ldots x_2 x_1 \cdot g$ in $\mathcal{X}_G$. $\square$

Let us denote by $\mathsf{X}^{-\omega} \cdot G \subset \Omega(\mathfrak{M})$ the set of sequences $\ldots x_2 x_1 \cdot g$ for $x_i \in \mathsf{X}$ and $g \in G$. We introduce on it the direct product topology, where X and G are discrete.

PROPOSITION 3.2.6. *Two elements $\ldots x_2 x_1 \cdot g$ and $\ldots y_2 y_1 \cdot h$ of $\mathsf{X}^{-\omega} \cdot G$ are asymptotically equivalent if and only if there exists a left-infinite directed path $\ldots e_2 e_1$ in the Moore diagram of the nucleus $\mathcal{N}$ ending in the vertex hg^{-1} such that the edge e_i is labeled by (x_i, y_i).*

The quotient of $\mathsf{X}^{-\omega} \cdot G \subset \Omega(\mathfrak{M})$ by the asymptotic equivalence relation is homeomorphic to $\mathcal{X}_G$.

PROOF. The first part of the proposition follows from Lemma 3.2.3. The set $\mathsf{X}^{-\omega} \cdot G$ intersects every equivalence class, due to Lemma 3.2.5. Consequently, the quotients of $\mathsf{X}^{-\omega} \cdot G$ and of $\Omega(\mathfrak{M})$ by the asymptotic equivalence relation coincide as sets.

Let us prove that the quotient topology coincides with the topology, introduced on $\mathcal{X}_G$ before.

Let $\pi : \Omega(\mathfrak{M}) \longrightarrow \mathcal{X}_G$ be the canonical projection. We have to prove that $C \subset \mathcal{X}_G$ is closed in $\mathcal{X}_G$ if and only if the set $\pi^{-1}(C) \cap \mathsf{X}^{-\omega} \cdot G$ is closed in the product topology on $\mathsf{X}^{-\omega} \cdot G$.

Suppose that $\pi^{-1}(C) \cap \mathsf{X}^{-\omega} \cdot G$ is closed in $\mathsf{X}^{-\omega} \cdot G$. Let $\mathsf{Y} \subset \mathfrak{M}$ be an arbitrary finite subset. Let B be the set of elements $g \in G$ such that there exist asymptotically equivalent sequences $\ldots x_2 x_1 \cdot g \in \mathsf{X}^{-\omega} \cdot G$ and $\ldots y_2 y_1 \in \mathsf{Y}^{-\omega}$. The set B is finite by Proposition 2.11.5. The set $\pi^{-1}(C) \cap \mathsf{X}^{-\omega} \cdot B$ is closed and contains all the elements of $\pi^{-1}(C) \cap \mathsf{X}^{-\omega} \cdot G$, which are asymptotically equivalent to some elements of $\mathsf{Y}^{-\omega}$.

Therefore, applying Lemma 3.2.2 to the finite set $Y \cup (X \cdot B)$, we conclude that the set $\pi^{-1}(C) \cap Y^{-\omega}$ is closed in $Y^{-\omega}$. Hence, the set C is closed in $\mathcal{X}_G$.

Suppose now that the set $C \subset \mathcal{X}_G$ is closed in $\mathcal{X}_G$. It follows that the set $\pi^{-1}(C) \cap X^{-\omega} \cdot B$ is closed for every finite set $B \subset G$. But this implies that the set $\pi^{-1}(C) \cap X^{-\omega} \cdot G$ is closed in the product topology on $X^{-\omega} \cdot G$, since G is discrete. □

The action of G on $\mathcal{X}_G$ is defined in terms of $X^{-\omega} \cdot G$ by

$$(\ldots x_2 x_1 \cdot g) \cdot h = \ldots x_2 x_1 \cdot gh.$$

EXAMPLE. In the case of the adding machine action of $\mathbb{Z} = \langle a \rangle$, one sees on the diagram of the nucleus (Figure 2.2 on page 58) that two sequences are asymptotically equivalent if and only if they are either equal or are of the form

$$\ldots 0001 x_m x_{m-1} \ldots x_1 \cdot a^n \qquad \ldots 1110 x_m x_{m-1} \ldots x_1 \cdot a^n,$$

where $x_m x_{m-1} \ldots x_1 \in X^*$ is an arbitrary finite (possibly empty) word, or are of the form

$$\ldots 000 \cdot a^{n+1} \qquad \ldots 111 \cdot a^n.$$

But this is the usual identification of the dyadic expansions of reals; i.e., two sequences $\ldots x_2 x_1 \cdot a^n, \ldots y_2 y_1 \cdot a^m$ are equivalent if and only if

$$n + \sum_{i=1}^{\infty} x_i \cdot 2^{-i} = m + \sum_{i=1}^{\infty} y_i \cdot 2^{-i}.$$

Consequently, the limit space $\mathcal{X}_G$ is the real line $\mathbb{R}$ with the natural action of $\mathbb{Z}$ on it.

3.2.3. Generation of the asymptotic equivalence relation. Actually one does not need to know the nucleus of the action in order to describe the asymptotic equivalence relation on $X^{-\omega} \cdot G$. It is sufficient to know the automaton generating the action, as the next proposition shows.

PROPOSITION 3.2.7. *Let (G, X) be a contracting action of a finitely generated group. Let (A, X) be a finite automaton generating the action. Denote by $D \subset (X^{-\omega} \cdot G) \times (X^{-\omega} \cdot G)$ the set of the pairs $(\ldots x_2 x_1 \cdot 1, \ldots y_2 y_1 \cdot g)$ such that there exists a path $\ldots e_2 e_1$ in the Moore diagram of A ending in g and such that (x_i, y_i) is the label of e_i. Then the G-invariant equivalence relation on $X^{-\omega} \cdot G$ generated by D (i.e., the smallest G-invariant equivalence relation containing D) coincides with the asymptotic equivalence relation.*

PROOF. If $\ldots e_2 e_1$ is a path in A labeled by $\ldots (x_2, y_2)(x_1, y_1)$ and ending in g, then $\ldots x_2 x_1$ and $\ldots y_2 y_1 \cdot g$ are asymptotically equivalent, since $g_k \cdot x_k \ldots x_1 = y_k \ldots y_1 \cdot g$, where g_k is the beginning of the arrow e_k. Therefore D belongs to the asymptotic equivalence relation.

On the other hand, suppose that $\ldots x_2 x_1 \cdot h, \ldots y_2 y_1 \cdot g$ are asymptotically equivalent. Multiplying by h^{-1} from the right, we may assume that $h = 1$.

There exists a bounded sequence $(g_k)_{k=0,1,\ldots}$, of elements of the group G such that $g_k \cdot x_k = y_k \cdot g_{k-1}$ and $g_0 = g$. If $g_k = h_m \ldots h_2 h_1$ is a representation of g_k as a product of the states of A and their inverses, then

$$g_{k-1} = g_k|_{x_k} = h_m|_{h_{m-1} \ldots h_2 h_1 (x_k)} \cdot h_{m-2}|_{h_{m-3} \ldots h_2 h_1 (x_k)} \cdots h_2|_{h_1(x_k)} \cdot h_1|_{x_k}$$

is also a representation of g_{k-1} as a product of the states of A and their inverses. It follows from the standard argument about inverse limits of finite sets that for some $m \in \mathbb{N}$ there exist decompositions $g_k = h_{m,k} \cdots h_{2,k} h_{1,k}$ for every k, where $h_{i,k} \in \mathsf{A} \cup \mathsf{A}^{-1}$, such that

$$\begin{aligned} g_k \cdot x_k &= (h_{m,k} \cdots h_{2,k} h_{1,k}) \cdot x_{k,0} \\ &= (h_{m,k} \cdots h_{3,k} h_{2,k}) \cdot x_{k,1} \cdot h_{1,k-1} \\ &= (h_{m,k} \cdots h_{4,k} h_{3,k}) \cdot x_{k,2} \cdot (h_{2,k-1} h_{1,k-1}) = \ldots \\ &= x_{k,m} \cdot (h_{m,k-1} \cdots h_{2,k-1} h_{1,k-1}) = y_k \cdot g_{k-1}. \end{aligned}$$

Or, in a more compact notation:

(3.1) $\qquad h_{i,k} \cdot x_{k,i-1} = x_{k,i} \cdot h_{i,k-1}, \qquad x_{k,0} = x_k, \qquad x_{k,m} = y_k.$

Hence we get the following sequence of elements of $\mathsf{X}^{-\omega} \cdot G$:

$$\begin{aligned} w_0 &= \ldots x_{2,0} x_{1,0} \cdot 1 = \ldots x_2 x_1 \cdot 1, \\ w_1 &= \ldots x_{2,1} x_{1,1} \cdot h_{1,0}, \\ w_2 &= \ldots x_{2,2} x_{1,2} \cdot h_{2,0} h_{1,0} \\ &\vdots \\ w_m &= \ldots x_{2,m} x_{1,m} \cdot h_{m,0} h_{m-1,0} \cdots h_{1,0} = \ldots y_2 y_1 \cdot h. \end{aligned}$$

But then the pairs

$$(w_0, w_1),$$
$$\left(w_1 \cdot h_{1,0}^{-1}, w_2 \cdot h_{1,0}^{-1}\right),$$
$$\left(w_2 \cdot (h_{2,0} h_{1,0})^{-1}, w_3 \cdot (h_{2,0} h_{1,0})^{-1}\right)$$
$$\vdots$$
$$\left(w_{m-1} \cdot (h_{m-1,0} \cdots h_{1,0})^{-1}, w_m \cdot (h_{m-1,0} \cdots h_{1,0})^{-1}\right)$$

belong to D, which proves that $(\ldots x_2 x_1 \cdot 1, \ldots y_2 y_1 \cdot h)$ belongs to the G-invariant equivalence relation generated by D. □

3.2.4. Basic properties of $\mathcal{X}_G$.

PROPOSITION 3.2.8. *The limit space $\mathcal{X}_G$ is metrizable and has topological dimension less than the size of the nucleus of the action.*

PROOF. It follows from Proposition 3.2.6 that every asymptotic equivalence class on $\mathsf{X}^{-\omega} \cdot G$ has not more than $|\mathcal{N}|$ elements, where $\mathcal{N}$ is the nucleus.

Now by [**Eng77**] Theorem 4.2.13, the quotient space $\mathcal{X}_G$ is metrizable, since it is a quotient of a locally compact separable metrizable space $\mathsf{X}^{-\omega} \cdot G$ by a closed equivalence relation with compact equivalence classes. The assertion about the dimension follows from the fact that the space $\mathsf{X}^{-\omega} \cdot G$ is 0-dimensional and that every equivalence class is of cardinality $\leq |\mathcal{N}|$, due to the Hurewicz formula (see [**Kur61**], page 52). □

The next two properties of $\mathcal{X}_G = \mathfrak{M}^{\otimes -\omega}$ follow directly from the definition of the asymptotic equivalence relation. The proof consists of just showing that the asymptotic equivalence is invariant under the respective transformations.

PROPOSITION 3.2.9. *The map*
$$\ldots a_3 a_2 a_1 \mapsto \ldots (a_{2n} \ldots a_{n+1})(a_n \ldots a_1)$$
from $\Omega(\mathfrak{M})$ *to* $\Omega(\mathfrak{M}^{\otimes n})$ *induces a G-equivariant homeomorphism*
$$\mathfrak{M}^{\otimes -\omega} \longrightarrow (\mathfrak{M}^{\otimes n})^{\otimes -\omega}.$$

This is an analog of the fact that the action of G on $(\mathsf{X}^n)^\omega$ is topologically conjugate with the action on X^ω.

PROPOSITION 3.2.10. *The map*
$$\ldots a_2 a_1 \mapsto \ldots a_2 a_1 \otimes v$$
induces for every $n \geq 0$ *and* $v \in \mathfrak{M}^{\otimes n}$ *a continuous map* $\zeta \mapsto \zeta \otimes v$ *of* $\mathfrak{M}^{\otimes -\omega}$.

Note that for the case $n = 0$ (when $\mathfrak{M}^{\otimes n} = G$) Proposition 3.2.10 defines the left action of G on $\mathfrak{M}^{\otimes -\omega}$. The map $\zeta \mapsto \zeta \otimes v$ in general is not a homeomorphism for $n > 0$.

EXAMPLE. In the case of the binary adding machine action over the alphabet $\mathcal{X} = \{0, 1\}$, we have
$$\xi \otimes 0 = \xi/2, \quad \xi \otimes 1 = (\xi + 1)/2,$$
where $\xi \in \mathbb{R}$ is a point of the limit space $\mathcal{X}_{\mathbb{Z}} = \mathbb{R}$. These relations are easy to check using the definition of the respective bimodule and the encoding of the points of $\mathbb{R}$ by the elements of $\mathsf{X}^{-\omega} \cdot \mathbb{Z}$.

3.3. Digit tiles

3.3.1. Let us fix some hyperbolic G-bimodule $\mathfrak{M}$ and a basis X of $\mathfrak{M}$. Let $\mathcal{N}$ be the nucleus of the action (G, X).

DEFINITION 3.3.1. *The (digit) tile* $\mathcal{T} = \mathcal{T}(\mathsf{X}) = \mathcal{T}(\mathfrak{M}, \mathsf{X})$ *is the image of* $\mathsf{X}^{-\omega} \cdot 1$ *in* $\mathcal{X}_G$, *i.e., the set of the points of* $\mathcal{X}_G$ *which can be represented in the form* $\ldots x_2 x_1$ *for* $x_i \in \mathsf{X}$.

The following is a direct corollary of Proposition 3.2.6.

PROPOSITION 3.3.2. *Two sequences* $\ldots x_2 x_1, \ldots y_2 y_1 \in \mathsf{X}^{-\omega}$ *represent the same point of the tile* $\mathcal{T}(\mathsf{X})$ *if and only if there exists a path* $\ldots e_2 e_1$ *in the Moore diagram of the nucleus* $\mathcal{N}$ *such that the arrow* e_1 *ends in the trivial state and every arrow* e_i *is labeled by* (x_i, y_i).

The tile $\mathcal{T}(\mathsf{X})$ *is homeomorphic to the quotient of the space* $\mathsf{X}^{-\omega}$ *by the described equivalence relation.* □

The second statement of the proposition follows from compactness of $\mathcal{T}$ and $\mathsf{X}^{-\omega}$.

We also get immediately that

(3.2) $$\mathcal{X}_G = \bigcup_{g \in G} \mathcal{T} \cdot g = \bigcup_{v \in \mathfrak{M}^{\otimes n}} \mathcal{T} \otimes v$$

and that

(3.3) $$\mathcal{T} = \bigcup_{v \in \mathsf{X}^n} \mathcal{T} \otimes v$$

for every $n \in \mathbb{N}$.

The sets $\mathcal{T} \otimes v$ for $v \in \mathfrak{M}^{\otimes n}$ are called the *tiles of the nth level*. It follows from Proposition 3.3.2 that the map $\xi \mapsto \xi \otimes v$ is an injective continuous map from $\mathcal{T}$ to $\mathcal{T} \otimes v$. It is thus a homeomorphism, since the tiles are compact.

Relation (3.2) shows that the tiles of a fixed level cover the space $\mathcal{X}_G$. The term "*tile*" originates from this fact and from a relation with the so-called "digit tilings" of the Euclidean space $\mathbb{R}^n$ (see Section 6.2).

EXAMPLE. We know that in the case of the adding machine action of the cyclic group $\langle a \rangle$ the limit space $\mathcal{X}_G$ is the real line $\mathbb{R}$, where a sequence $\ldots x_2 x_1 \cdot a^n$ encodes the point $n + \sum_{k=1}^{+\infty} x_k 2^{-k}$. Consequently, the digit tile is the set of the sums $\sum_{k=1}^{+\infty} x_k 2^{-k}$, where $x_k \in \{0, 1\}$. Hence, $\mathcal{T}$ is equal in this case to the closed segment $[0, 1]$. The tiles of the nth level are the segments of the form

$$\frac{1}{2^n}\left([0,1] + m\right) = \left[\frac{m}{2^n}, \frac{m+1}{2^n}\right],$$

where $m \in \mathbb{Z}$.

Our aim is to study the topology of the tiles and to show how combinatorial data related to the tiles determine the topology of $\mathcal{X}_G$.

The first step is the following lemma.

LEMMA 3.3.3. *A subset $C \subset \mathcal{X}_G$ is closed if and only if the set $C \cap \mathcal{T} \cdot g$ is closed for every $g \in G$.*

PROOF. Suppose that $C \cap \mathcal{T} \cdot g$ is closed for every $g \in G$. Denote by C_g the preimage of $C \cap \mathcal{T} \cdot g$ in $\mathsf{X}^{-\omega} \cdot g$. The sets C_g are then closed and the full preimage $\bigcup_{g \in G} C_g$ of C in $\mathsf{X}^{-\omega} \cdot G$ is therefore closed in $\mathsf{X}^{-\omega} \cdot G$, as G is discrete. This implies that C is closed due to Proposition 3.2.6. □

COROLLARY 3.3.4. *A subset $U \subset \mathcal{X}_G$ is open if and only if the set $U \cap \mathcal{T} \cdot g$ is relatively open in $\mathcal{T} \cdot g$ for every $g \in G$.* □

3.3.2. Adjacency of tiles.

PROPOSITION 3.3.5. *Two tiles $\mathcal{T} \otimes v_1$ and $\mathcal{T} \otimes v_2$ of the nth level intersect if and only if there exists $h \in \mathcal{N}$ such that $h \cdot v_1 = v_2$.*

PROOF. If the tiles $\mathcal{T} \otimes v$ and $\mathcal{T} \otimes u$ intersect, then there exist two asymptotically equivalent sequences of the form $\ldots x_2 x_1 \otimes v$ and $\ldots y_2 y_1 \otimes u$. It follows from Proposition 3.2.6 that there exists an element h of the nucleus such that $h \cdot v = u$.

Suppose now that there exists $h \in \mathcal{N}$ such that $h \cdot v = u$. The element h is a restriction of some element of the nucleus. Therefore there exists a letter $x_1 \in \mathsf{X}$ and an element $h_1 \in \mathcal{N}$ such that $h_1|_{x_1} = h$. Then $h_1 \cdot x_1 v = y_1 u$ for some $y_1 \in \mathsf{X}$. Similarly, there exists a letter $x_2 \in \mathsf{X}$ and an element $h_2 \in \mathcal{N}$ such that $h_2 \cdot x_2 x_1 v = y_2 y_1 u$ for some $y_2 \in \mathsf{X}$. Thus, inductively we prove that there exist infinite sequences $\ldots x_2 x_1 v, \ldots y_2 y_1 u \in \mathsf{X}^{-\omega} \cdot G$ and an infinite sequence $h_1, h_2, \ldots$ of elements of the nucleus such that $h_n \cdot x_n \ldots x_2 x_1 v = y_n \ldots y_2 y_1 u$ for all $n \in \mathbb{N}$. Therefore the sets $\mathsf{X}^{-\omega} v$ and $\mathsf{X}^{-\omega} u$ have two asymptotically equivalent elements, and the tiles $\mathcal{T} \otimes v$ and $\mathcal{T} \otimes u$ intersect. □

In particular, two tiles $\mathcal{T} \cdot g_1$ and $\mathcal{T} \cdot g_2$ of the 0th level intersect if and only if $g_1 g_2^{-1} \in \mathcal{N}$. So, if G is finitely generated and the action is recurrent, then the adjacency graph of the tiles of the 0th level coincides with the Calley graph of G with respect to the generating set $\mathcal{N}$ (see Proposition 2.11.3).

3.3.3. Boundary of tiles.

DEFINITION 3.3.6. We say that a contracting action of a group G satisfies the *open set condition* if for any element g of the nucleus there exists a finite word $v \in \mathsf{X}^*$ such that $g|_v = 1$.

The following is a complete answer to the question when two tiles have disjoint interiors.

PROPOSITION 3.3.7. *If the action satisfies the open set condition, then the set*
$$D = \mathcal{T} \cap \bigcup_{g \in G, g \neq 1} \mathcal{T} \cdot g$$
is equal to the boundary of $\mathcal{T}$, the set $\mathcal{T}$ is the closure of its interior and any two tiles of one level have disjoint interiors.

If the action does not satisfy the open set condition, then $D = \mathcal{T}$ and every tile is covered by the other tiles of the same level.

PROOF. Suppose that the action satisfies the open set condition. We are going to prove at first that $\mathcal{T} \setminus D$ is dense in $\mathcal{T}$.

Let $\mathcal{N} = \{h_1, h_2, \ldots, h_m\}$. Let the word $w_1 \in \mathsf{X}^*$ be such that $h_1|_{w_1} = 1$. We can find inductively for every h_i a word $w_i \in \mathsf{X}^*$ for which $h_i|_{w_1 w_2 \ldots w_i} = 1$. Then restriction of every element of the nucleus in the word $w = w_1 w_2 \ldots w_m$ will be trivial.

Let $\xi \in \mathcal{T}$ be an arbitrary point. Let $U \ni \xi$ be its neighborhood. The set U contains the image of a cylindrical set $\mathsf{X}^{-\omega} u$ for some u, i.e., the tile $\mathcal{T} \otimes u$. Consider the tile $\mathcal{T} \otimes wu \subset \mathcal{T} \otimes u \subset U$. Let $\zeta \in \mathcal{T} \otimes wu$ be an arbitrary point. It is represented by a sequence $\ldots x_2 x_1 wu \in \mathsf{X}^{-\omega}$. Suppose that $\ldots y_2 y_1 w'u' \cdot g$ is another sequence representing the same point ζ, where $|w'| = |w|, |u'| = |u|$ and $g \in G$. Then there exists a sequence $\{g_n\}_{n \geq 0}$ of elements of the nucleus such that $g_i \cdot x_i = y_i \cdot g_{i-1}$ for all $i \geq 1$ and $g_0 \cdot wu = w'u' \cdot g$. But $g_0|_w = 1$, so $g = 1$. Consequently, the point ζ does not belong to any tile $\mathcal{T} \cdot g$ with $g \neq 1$; i.e., it does not belong to D. We have proved that any neighborhood U of the point ξ contains an element of the set $\mathcal{T} \setminus D$.

The set $\bigcup_{g \in G, g \neq 1} \mathcal{T} \cdot g$ is closed by Lemma 3.3.3. Consequently, $\mathcal{T} \setminus D$ is an open dense subset of $\mathcal{T}$. In particular, D contains the boundary of $\mathcal{T}$. But if $\zeta \in D$ is an arbitrary point, then it also belongs to some other tile $\mathcal{T} \cdot g$ and thus every neighborhood contains a point of $\mathcal{T} \cdot g \setminus D \cdot g$, i.e., a point which does not belong to $\mathcal{T}$. Therefore every point of D is a boundary point and D coincides with the boundary of $\mathcal{T}$.

Suppose now that the action does not satisfy the open set condition, i.e., that there exists an element $h \in \mathcal{N}$ having no trivial restrictions. Then we can find a sub-automaton $\mathcal{N}_1$ of the nucleus which contains only states implementing non-trivial transformations (just take all the restrictions of h).

Let $\xi \in \mathcal{T}$ be an arbitrary point and let U be its neighborhood. Then U contains the tile $\mathcal{T} \otimes u$ for some $u \in \mathsf{X}^*$. Since the sub-automaton $\mathcal{N}_1$ is finite, its Moore diagram has a left-infinite path $\ldots e_2 e_1$. Let $\ldots x_2 x_1$ be the sequence of the letters which are read on the left parts of its labels. Then there exists a path γ such that on the left parts of its labels the sequence $\ldots x_2 x_1 u$ is read. It will end in a non-trivial state g and $\ldots x_2 x_1 u = \ldots y_2 y_1 \cdot g$ in $\mathcal{X}_G$ for some $\ldots y_2 y_1 \in \mathsf{X}^{-\omega}$; thus the point $\ldots x_2 x_1 u$ belongs to D. So, every neighborhood of the point ξ intersects with D; i.e., D is dense. But D is also closed; thus $\mathcal{T} = D$. □

COROLLARY 3.3.8. *Suppose that a contracting action of a group G on X^* satisfies the open set condition. Then a sequence $\ldots x_2 x_1 \in \mathsf{X}^{-\omega}$ represents a point of the boundary of $\mathcal{T}$ if and only if there exists a left-infinite path in the Moore diagram of the nucleus $\mathcal{N}$ which ends in a non-trivial state and is labeled by $\ldots (x_2, y_2)(x_1, y_1)$ for some $\ldots y_2 y_1 \in \mathsf{X}^{-\omega}$.*

PROOF. A direct corollary of Propositions 3.3.7 and 3.3.5. □

One can also describe the boundary of the tile using only the automaton generating the action without computing the nucleus.

PROPOSITION 3.3.9. *Suppose that the automaton (A, X) generates a contracting group action (G, X). Then:*

(1) *the action satisfies the open set condition if and only if for every state $g \in \mathsf{A}$ there exists a word $v \in \mathsf{X}^*$ such that $g|_v$ is trivial;*
(2) *if the action satisfies the open set condition, then a point $\xi \in \mathcal{T}$ belongs to the boundary of the tile if and only if there exists a path $\ldots e_2 e_1$ in the Moore diagram of A ending in a non-trivial state and labeled by $\ldots (x_2, y_2)(x_1, y_1)$, where $\ldots x_2 x_1$ or $\ldots y_2 y_1$ represents the point ξ.*

PROOF. If the action satisfies the open set condition, then for every $g \in G$ there exists $v \in \mathsf{X}^*$ such that $g|_v$ is trivial. One just has to find a word $v_1 \in \mathsf{X}^*$ such that $g|_{v_1}$ belongs to the nucleus and then a word $v_2 \in \mathsf{X}^*$ such that $g|_{v_1 v_2} = g|_{v_1}|_{v_2}$ is trivial. On the other hand, if for every $g \in \mathsf{A}$ there exists $v \in \mathsf{X}^*$ such that $g|_v$ is trivial, then for every $g^{-1} \in \mathsf{A}^{-1}$ there exists $g(v) \in \mathsf{X}^*$ such that $g^{-1}|_{g(v)} = g|_v^{-1} = 1$. We can find then for every product $g = g_n g_{n-1} \cdots g_1$ of states of $\mathsf{A} \cup \mathsf{A}^{-1}$ a word v_n such that $g|_{v_n} = 1$. We do it inductively: find a word v_1 such that $g_1|_{v_1} = 1$ and then for every $i = 2, \ldots, n$ find a word $v_i = u_i v_{i-1}$ such that $g_i|_{g_{i-1} \cdots g_1(u_i)}|_{u_i}$ is trivial. Then

$$g_i \cdots g_1|_{v_i} = g_i \cdots g_1|_{u_i v_{i-1}}$$
$$= g_i|_{g_{i-1} \cdots g_1(u_i)}|_{g_{i-1} \cdots g_1(v_i)} (g_{i-1} \cdots g_1)|_{v_{i-1}}$$
$$= 1|_{g_{i-1} \cdots g_1(v_i)} \cdot 1 = 1.$$

Thus the first claim of the proposition is proved.

Suppose that $\xi \in \mathcal{T}$ belongs to the boundary of the tile. Then it belongs also to a tile $\mathcal{T} \cdot g$ by Proposition 3.3.7. So it can be represented by two asymptotically equivalent sequences $\ldots x_2 x_1 \cdot 1$ and $\ldots y_2 y_1 \cdot g$. Proposition 3.2.7 implies then that there exists a path $\ldots e_2 e_1$ in the Moore diagram of A labeled either by $\ldots (\widetilde{x}_2, z_2)(\widetilde{x}_1, z_1)$ or by $\ldots (z_2, \widetilde{x}_2)(z_1, \widetilde{x}_1)$ for some $\ldots z_2 z_1 \in \mathsf{X}^{-\omega}$ and ending in a non-trivial state, where $\ldots \widetilde{x}_2 \widetilde{x}_1$ represents the point $\zeta = \ldots x_2 x_1$. □

3.3.4. Connectedness of the tiles. The following is a joint result with E. Bondarenko.

PROPOSITION 3.3.10. *Let T_n be the graph with the set of vertices X^n in which two vertices v_1, v_2 are connected by an edge if and only if there exists $h \in \mathcal{N}$ such that $h \cdot v_1 = v_2 \cdot 1$. Then the following conditions are equivalent:*

(1) *The tile $\mathcal{T}$ is connected.*

(2) *The graphs T_n are connected for all $n \geq 1$.*
(3) *The graph T_1 is connected.*

PROOF. Implication (1) $\Rightarrow$ (2) follows directly from Proposition 3.3.5.

Implication (2) $\Rightarrow$ (3) is trivial. Let us prove that (2) implies (1). Suppose that the graphs T_n are connected but the tile $\mathcal{T}$ is not. Then there exists a closed non-empty set $A \subset \mathcal{T}$ with non-empty closed complement $\mathcal{T} \setminus A$. Let $A_\omega \subset \mathsf{X}^{-\omega}$ be the preimage of A under the canonical projection $\mathsf{X}^{-\omega} \longrightarrow \mathcal{T}$. Then the set A_ω is also closed and has a non-empty closed complement.

For every $n \in \mathbb{N}$, let $A_n \subset \mathsf{X}^n$ be the set of all possible endings of length n of the infinite words belonging to A_ω. Since the set A_ω is closed, a sequence $\ldots x_2 x_1$ represents an element of A if and only if $x_n x_{n-1} \ldots x_1 \in A_n$ for every $n \in \mathbb{N}$.

There exists n_0 such that for all $n \geq n_0$, the sets A_n are not equal to X^n. Since the graph T_n is connected, there exists a word $v_n \in A_n$ and an element $s_n \in \mathcal{N}$ such that $s_n \cdot v_n \in \mathsf{X}^n \setminus A_n$. It follows from compactness that there exists an increasing sequence n_k such that both sequences v_{n_k} and $s_{n_k} \cdot v_{n_k}$ converge to certain elements $\xi = \ldots x_2 x_1$ and $\zeta = \ldots y_2 y_1$ of $\mathsf{X}^{-\omega}$ respectively. Then $\xi \in A_\omega$ and $\zeta \in \mathsf{X}^{-\omega} \setminus A_\omega$, since both sets A_ω and $\mathsf{X}^{-\omega} \setminus A_\omega$ are closed. For every $n \in \mathbb{N}$ the element $x_n x_{n-1} \ldots x_1$ is an ending of v_{n_k} and $y_n y_{n-1} \ldots y_1$ is an ending of $s_{n_k} \cdot v_{n_k}$, for all sufficiently big k. Let $g_n = s_{n_k}|_u$, where $v_{n_k} = u x_n \ldots x_1$. Then $g_n \in \mathcal{N}$ and $g_n \cdot x_n x_{n-1} \ldots x_1 = y_n y_{n-1} \ldots y_1$. Therefore, ξ and ζ are asymptotically equivalent and represent equal points of $\mathcal{T}$, which contradicts the choice of the set A.

It is sufficient now to prove that (3) implies (2). We argue by induction on n. Suppose that T_1 and T_{n-1} are connected. If $\{v_1, v_2\}$ is an edge of T_{n-1}, then for every letter $x \in \mathsf{X}$ the pair $\{v_1 x, v_2 x\}$ is an edge of T_n. Hence, $T_{n-1} x$ is a connected subgraph of T_n. Let $\{x, y\}$ be an arbitrary edge of the graph T_1. There exists an element $g \in \mathcal{N}$ such that $g \cdot x = y \cdot 1$. It follows from the definition of the nucleus that there exists a pair of words $v, u \in \mathsf{X}^n$ and an element $h \in \mathcal{N}$ such that $h \cdot v = u \cdot g$. Then we get an edge $\{vx, uy\}$ of the graph T_n, and thus the components $T_{n-1} x$ and $T_{n-1} y$ are connected by an edge in T_n if x, y are connected by an edge in T_1. Hence, connectivity of T_1 and T_{n-1} implies connectivity of T_n. $\square$

3.4. Axiomatic description of $\mathcal{X}_G$

3.4.1. A base of the topology on $\mathcal{X}_G$. We continue to study in this section the topology of $\mathcal{X}_G$ using the digit tiles.

PROPOSITION 3.4.1. *Let $U_n(\zeta)$ for $\zeta \in \mathcal{X}_G$ and $n \in \mathbb{N}$ denote the union of the tiles of the nth level to which ζ belongs. Then $\{U_n(\zeta)\}_{n \in \mathbb{N}}$ is a fundamental system of neighborhoods of ζ.*

PROOF. We have to prove that $U_n(\zeta)$ are neighborhoods of ζ and that every neighborhood of ζ contains $U_n(\zeta)$ for some $n \in \mathbb{N}$.

The first claim follows from the fact that the set

$$U_n(\zeta) \setminus \bigcup_{v \in \mathfrak{M}^{\otimes n}, \zeta \notin \mathcal{T} \otimes v} \mathcal{T} \otimes v = \mathcal{X}_G \setminus \bigcup_{v \in \mathfrak{M}^{\otimes n}, \zeta \notin \mathcal{T} \otimes v} \mathcal{T} \otimes v$$

contains ζ and is open by Lemma 3.3.3.

Let us prove the second claim. Let $U \ni \zeta$ be any neighborhood of ζ. Then the preimage $\hat{U}$ of U in $\mathsf{X}^{-\omega} \cdot G$ is a neighborhood of every preimage $\ldots x_2 x_1 \cdot g$ of ζ. Therefore $\hat{U}$ contains sets of the form $\mathsf{X}^{-\omega} x_n \ldots x_1 \cdot g$. We can find a common n

for all the preimages of ζ (since there is only a finite number of them). But then $U_n(\zeta) \subset U$. □

Note that Proposition 3.4.1, Proposition 3.3.5 and (3.3) on page 78 is a complete description of the topological space $\mathcal{X}_G$ together with the action of G.

We will use the following proposition in construction of the orbispace structure on $\mathcal{J}_G$.

LEMMA 3.4.2. *The map $T_x : \mathcal{X}_G \longrightarrow \mathcal{X}_G : \zeta \mapsto \zeta \otimes x$ is open for every $x \in \mathfrak{M}$.*

The map T_x was defined in Proposition 3.2.10.

PROOF. Let us fix some associated self-similar action (G, X) and let $x \in \mathsf{X}$. Suppose that $U \subset \mathcal{X}_G$ is open and let $\zeta \in U$ be an arbitrary point. There exists $n \in \mathbb{N}$ such that $U_{n-1}(\zeta) \subseteq U$.

Let us show that $U_n(\zeta \otimes x) \subseteq U_{n-1}(\zeta) \otimes x$, which will prove that $\zeta \otimes x = T_x(\zeta)$ is an internal point of $T_x(U)$, i.e., that T_x is open. Suppose that $\zeta \otimes x$ belongs to a tile $\mathcal{T} \otimes x_n \ldots x_1 \cdot g$, where $x_n \ldots x_1 \in \mathsf{X}^n$ and $g \in G$. Then it can be represented both by a sequence $\ldots x_{n+1} x_n \ldots x_1 \cdot g \in \mathsf{X}^{-\omega} \cdot G$ and by the sequence $\ldots y_3 y_2 h(x) \cdot h|_x$, where $\ldots y_3 y_2 \cdot h$ represents ζ.

There exists a path $\ldots e_2 e_1$ in the Moore diagram of the nucleus labeled by $\ldots (x_3, y_3)(x_2, y_2)(x_1, h(x))$ and ending in $h|_x g^{-1}$. Let f be the end of the path $\ldots e_3 e_2$. Then $\ldots x_3 x_2 = \ldots y_3 y_2 \cdot f$ in $\mathcal{X}_G$; hence $\zeta = \ldots x_3 x_2 \cdot f^{-1} h$. We have
$$f \cdot x_1 = h(x) \cdot h|_x g^{-1};$$
hence
$$f^{-1} h \cdot x = f^{-1} \cdot h(x) \cdot h|_x = x_1 \cdot \left(h|_x g^{-1}\right)^{-1} h|_x = x_1 \cdot g,$$
which implies that the image of the tile $\mathcal{T} \otimes x_n \ldots x_2 \cdot f^{-1} h \subseteq U_{n-1}(\zeta)$ under T_x is equal to $\mathcal{T} \otimes x_n \ldots x_1 \cdot g$, i.e., that $\mathcal{T} \otimes x_n \ldots x_1 \cdot g \subseteq U_{n-1}(\zeta) \otimes x$.

We have proved that every tile of the nth level containing $\zeta \otimes x$ is contained in $U_{n-1}(\zeta) \otimes x$, i.e., that $U_n(\zeta \otimes x) \subseteq U_{n-1}(\zeta) \otimes x$. □

3.4.2. The uniformity on $\mathcal{X}_G$. Another aspect of Proposition 3.4.1 is that it shows that the space $\mathcal{X}_G$ has a natural uniform structure.

Let us remember, following [**Bou71**] (Chapter II), the definition of a uniform structure.

We use the following notation. If $R_1, R_2 \subset A \times A$ are two relations on a set A, then $R_1 + R_2$ is the relation
$$(x, y) \in R_1 + R_2 \quad \Leftrightarrow \quad \exists z \in A \,:\, (x, z) \in R_1, (z, y) \in R_2.$$
The relations nR_i are defined for $n \in \mathbb{N}$ in the natural way.

We write sometimes $d(x, y) \leq R_i$ if $(x, y) \in R_i$, stressing in this way that uniformities are generalized metric spaces.

DEFINITION 3.4.3. A *uniformity* on a set $\mathcal{X}$ is a collection $\mathcal{U}$ of *entourages*, i.e., relations $R \subset \mathcal{X} \times \mathcal{X}$ containing the diagonal $\Delta_{\mathcal{X}} = \{(x, x) \,:\, x \in \mathcal{X}\}$, such that
  (1) If $R_2 \supset R_1$ and $R_1 \in \mathcal{U}$, then $R_2 \in \mathcal{U}$.
  (2) If $R_1, R_2 \in \mathcal{U}$, then $R_1 \cap R_2 \in \mathcal{U}$.
  (3) If $R \in \mathcal{U}$, then the *transposed* entourage $R^- = \{(x, y) \,:\, (y, x) \in R\}$ also belongs to $\mathcal{U}$.
  (4) For every $R \in \mathcal{U}$ there exists $R' \in \mathcal{U}$ such that $2R' \subseteq R$.

If $\mathcal{U}$ is a uniformity on $\mathcal{X}$, then a subset $A \subset \mathcal{X}$ is *open* if for every $x \in A$ there exists an entourage $R \in \mathcal{U}$ such that $y \in A$ for all $y \in \mathcal{X}$ such that $d(x, y) \leq R$. It is easy to see that every uniformity defines in this way a topology on $\mathcal{X}$ (which is the topology *induced* by the uniformity or *compatible* with it).

We consider only Hausdorff uniform spaces. A uniform space is Hausdorff if and only if the intersection $\bigcap_{R \in \mathcal{U}} R$ is equal to the diagonal $\Delta_{\mathcal{X}}$.

A collection of reflexive relations $\mathcal{B}$ on $\mathcal{X}$ is a *fundamental system of entourages* of a uniformity $\mathcal{U}$ if $\mathcal{U}$ consists of all the relations R for which there is a relation $R_0 \in \mathcal{B}$ such that $R \supseteq R_0$.

It is known that a set $\mathcal{B}$ of symmetric reflexive relations is a base of a uniformity if and only if

(1) for every pair $U, V \in \mathcal{B}$ there exists $W \in \mathcal{B}$ such that $W \subseteq U \cap V$, and
(2) for every $V \in \mathcal{B}$ there exists $U \in \mathcal{B}$ such that $2W \subseteq V$.

DEFINITION 3.4.4. Let Δ_n for $n \in \mathbb{N}$ be the set of the pairs $(\zeta_1, \zeta_2) \in \mathcal{X}_G \times \mathcal{X}_G$ of points which belong to two tiles $\mathcal{T}_1, \mathcal{T}_2$ of the nth level such that $\mathcal{T}_1 \cap \mathcal{T}_2 \neq \emptyset$. In other words, Δ_n is the union of the sets $(\mathcal{T} \otimes v_1) \times (\mathcal{T} \otimes v_2)$, for $v_1, v_2 \in \mathfrak{M}^{\otimes n}$, which have non-empty intersection with the diagonal of $\mathcal{X}_G \times \mathcal{X}_G$.

PROPOSITION 3.4.5. *The set $\{\Delta_n\}_{n \geq 0}$ is a fundamental system of entourages of a uniform structure on $\mathcal{X}_G$ compatible with the topology on it.*

PROOF. It is easy to see that Δ_n is symmetric. By Proposition 3.3.5

$$(3.4) \qquad \Delta_n = \bigcup_{v \in \mathfrak{M}^{\otimes n}, g \in \mathcal{N}} (\mathcal{T} \otimes v) \times (\mathcal{T} \otimes g \cdot v),$$

where $\mathcal{N}$ is the nucleus. Therefore,

$$2\Delta_n = \bigcup_{v \in \mathfrak{M}^{\otimes n}, g \in \mathcal{N}^2} (\mathcal{T} \otimes v) \times (\mathcal{T} \otimes g \cdot v).$$

There exists n_0 such that $\mathcal{N}^2|_{\mathsf{X}^{n_0}} \subset \mathcal{N}$, by definition of the nucleus. Let $u \in \mathsf{X}^{n_0}$, $v \in \mathfrak{M}^{\otimes n}$ and $g \in \mathcal{N}^2$ be arbitrary. Then

$$(\mathcal{T} \otimes u \otimes v) \times (\mathcal{T} \otimes g \cdot u \otimes v) = (\mathcal{T} \otimes u \otimes v) \times (\mathcal{T} \otimes u' \otimes g|_u \cdot v)$$
$$\subset (\mathcal{T} \otimes v) \times (\mathcal{T} \otimes g|_u \cdot v),$$

where $u' = g(u) \in \mathsf{X}^{n_0}$ and $g|_u \in \mathcal{N}$. Thus,

$$2\Delta_{n+n_0} \subset \Delta_n.$$

Consequently, $\{\Delta_n\}_{n \geq 0}$ is a fundamental system of entourages of a uniform structure on $\mathcal{X}_G$.

Let $U_n(\zeta)$ be as in Proposition 3.4.1 and denote $\Delta_n(\zeta) = \{\xi \,:\, (\xi, \zeta) \in \Delta_n\}$. Then we have

$$\Delta_{n+n_0}(\zeta) \subseteq U_n(\zeta) \subseteq \Delta_n(\zeta),$$

where n_0 is as above. Hence, $\Delta_n(\zeta)$ is a base of neighborhoods of ζ. This means that the uniform structure is compatible with the topology on $\mathcal{X}_G$. $\square$

PROPOSITION 3.4.6. *The action of G on $\mathcal{X}_G$ is uniformly equicontinuous; i.e., for every entourage U the intersection $\bigcap_{g \in G} U \cdot g$ is an entourage.*

Here G acts on $\mathcal{X}_G \times \mathcal{X}_G$ by the diagonal action $(\xi_1, \xi_2) \cdot g = (\xi_1 \cdot g, \xi_2 \cdot g)$.

PROOF. A direct corollary of the definition of the uniformity on $\mathcal{X}_G$. $\square$

3.4.3. Uniformities on co-compact G-spaces.
This is an auxiliary subsection whose aim is to show that proper co-compact G-spaces have a unique uniformity with respect to which the action is uniformly equicontinuous. It can be omitted during the first reading.

Recall that an action of a group G on a topological space $\mathcal{X}$ is called *co-compact* if there exists a compact subset $K \subset \mathcal{X}$ intersecting every G-orbit. The action is said to be *proper* if for every compact set $K \subset \mathcal{X}$ the set of the elements $g \in G$ such that $K \cdot g \cap K \neq \emptyset$ is finite.

An action of a group G on a uniform space $\mathcal{X}$ is said to be *uniformly equicontinuous* if for every entourage $U \in \mathcal{U}$ there exists an entourage $V \in \mathcal{U}$ such that $V \cdot g \subset U$ for all $g \in G$. Here G acts on $\mathcal{X} \times \mathcal{X}$ by the diagonal action: $(x,y) \cdot g = (x \cdot g, y \cdot g)$.

PROPOSITION 3.4.7. *Suppose that G acts on a locally compact Hausdorff space $\mathcal{X}$ by a proper co-compact action. Then there exists a unique uniformity on $\mathcal{X}$ such that the action of G is uniformly equicontinuous with respect to it. It is the uniformity defined by the fundamental system of entourages consisting of all G-invariant open neighborhoods of the diagonal in $\mathcal{X} \times \mathcal{X}$. This uniformity is complete.*

PROOF. We will use right actions here. Let us prove at first that the set of the G-invariant open neighborhoods of the diagonal is a fundamental system of neighborhoods of a complete uniform structure on $\mathcal{X}$ compatible with the topology on $\mathcal{X}$.

Let $K_0 \subset \mathcal{X}$ be a compact set intersecting every G-orbit. Let $K \supset K_0$ be its compact neighborhood. It also intersects every G-orbit.

Since K_0 intersects every G-orbit, the set $\bigcup_{g \in G} K_0 \cdot g \times K_0 \cdot g$ contains the diagonal of $\mathcal{X} \times \mathcal{X}$. Consequently, the set $\mathcal{K} = \bigcup_{g \in G} K \cdot g \times K \cdot g$ is a neighborhood of the diagonal. It is obviously G-invariant.

LEMMA 3.4.8. *If $U \subset \mathcal{K}$ is a G-invariant neighborhood of the diagonal $\Delta_\mathcal{X}$, then $V = U \cap (K \times K)$ is a neighborhood of the diagonal Δ_K of $K \times K$ and $\bigcup_{g \in G} V \cdot g$ is a G-invariant neighborhood of $\Delta_\mathcal{X}$.*

For every neighborhood $V \subset K \times K$ of the diagonal Δ_K there exists a G-invariant neighborhood U of the diagonal $\Delta_\mathcal{X}$ of $\mathcal{X} \times \mathcal{X}$ such that $U \cap (K \times K) \subset V$.

PROOF. The first statement of the lemma is obvious.

Let U be the set of the pairs $(x, y) \in \mathcal{X} \times \mathcal{X}$ such that $(x \cdot g, y \cdot g) \in V$ for all $g \in G$ such that $(x \cdot g, y \cdot g) \in K \times K$.

It is easy to see that the set U is G-invariant and that $U \cap (K \times K) \subset V$. Let us prove that U is a neighborhood of the diagonal $\Delta_\mathcal{X}$. Let (x, x) be a point of the diagonal. Since the action of G on $\mathcal{X}$ is proper, the set B of elements $g \in G$ such that $(x, x) \in K \cdot g \times K \cdot g$ is finite. For every $g_1 \in B$ the set of the elements $g_2 \notin B$ such that $V \cdot g_1 \cap K \cdot g_2 \times K \cdot g_2 \neq \emptyset$ is finite. Hence, the set $\bigcap_{g \in B} V \cdot g \setminus \bigcup_{g \notin B} K \cdot g \times K \cdot g$ is then a neighborhood of (x, x) contained in U. $\square$

It is known that there exists only one uniformity on the compact space K and that this uniformity consists of all the neighborhoods of the diagonal $\Delta_K \subset K \times K$. Moreover, it is known that this uniformity is complete (see [**Bou71**], Theorem 1, in Chapter II, §4).

In particular, the intersection of the neighborhoods of the diagonal Δ_K in $K \times K$ is equal to Δ_K. This implies together with Lemma 3.4.8 that the intersection of the open G-invariant neighborhoods of the diagonal $\Delta_\mathcal{X} \subset \mathcal{X} \times \mathcal{X}$ is equal to $\Delta_\mathcal{X}$.

Let us show that for every G-invariant entourage $U \subset \mathcal{X} \times \mathcal{X}$ there exists a G-invariant entourage U' such that $2U' \subset U$. Let $K_1 \subset \mathcal{X}$ be the union of the sets $K \cdot g$ for all $g \in G$ such that $K \cdot g \cap K \neq \emptyset$. The set K_1 is compact, since the action is proper.

The set of all the neighborhoods of the diagonal $\Delta_{K_1} \subset K_1 \times K_1$ is a uniformity on K_1; thus there exists an entourage $V \subset (K_1 \times K_1)$ such that $2V \subset U \cap (K_1 \times K_1)$ (where $2V$ is computed inside K_1). Then, by Lemma 3.4.8 (which also can be applied to K_1) there exists a G-invariant neighborhood $U' \subset (\mathcal{X} \times \mathcal{X})$ of the diagonal $\Delta_{\mathcal{X}}$ such that $U' \cap (K_1 \times K_1) \subset V$. We may assume that $U' \subset \mathcal{K} = \bigcup_{g \in G} K \cdot g \times K \cdot g$.

Let us show that $2U' \subset U$. Suppose that $(x, y), (y, z) \in U'$. Then $(x, y) = (x_1 \cdot g_1, y_1 \cdot g_1)$ for some $g_1 \in G$ and $x_1, y_1 \in K$ and $(y, z) = (y_2 \cdot g_2, z_2 \cdot g_2)$ for some $g_2 \in G$ and $y_2, z_2 \in K$. The entourage U' is G-invariant; hence $(x_1, y_1) \in U'$ and $(y_2, z_2) \in U'$. Using G-invariance of U', we get that $(y_2 \cdot g_2 g_1^{-1}, z_2 \cdot g_2 g_1^{-1}) \in U'$. But $y_2 \cdot g_2 g_1^{-1} = y_1$ and therefore $K \cdot g_2 g_1^{-1}$ has a non-empty intersection with K. Consequently, all the points $x_1, y_1 = y_2 \cdot g_2 g_1^{-1}$ and $z_2 \cdot g_2 g_1^{-1}$ belong to K_1. Thus $(x_1, y_1) \in V$ and $(y_2 \cdot g_2 g_1^{-1}, z_2 \cdot g_2 g_1^{-1}) \in V$, so that $(x_1, z_2 \cdot g_2 g_1^{-1}) \in 2V \subset U$. Using G-invariance of U, we get $(x, z) = (x_1 \cdot g_1, z_2 \cdot g_2) = (x_1, z_2 \cdot g_2 g_1^{-1}) \cdot g_1 \in U$. Thus, we have proved that $2U' \subset U$ and have finished proving that the set of G-invariant neighborhoods of $\Delta_{\mathcal{X}}$ is a base of a uniformity on $\mathcal{X}$.

Lemma 3.4.8 implies that restriction of the topology defined by this uniformity onto K coincides with the original topology on K. This, together with the G-invariance of the uniformity, implies that the uniformity agrees with the topology on $\mathcal{X}$.

Let us prove that this uniformity is complete. Let $x_n \in \mathcal{X}$ be a Cauchy sequence. There exists n_0 such that $d(x_k, x_m) \in \mathcal{K}$ for all $k, m \geq n_0$. Then there exists $g_{k,m} \in G$ such that $x_k, x_m \in K \cdot g_{k,m}$. In particular, $x_{n_0} \in K \cdot g_{n_0, n_0}$ and $x_{n_0}, x_m \in K \cdot g_{n_0, m}$ for all $m \geq n_0$. This implies that $K \cap K \cdot g_{n_0, m} g_{n_0, n_0}^{-1}$ contains $x_{n_0} \cdot g_{n_0, n_0}^{-1}$; hence the set of possible values of $g_{n_0, m} g_{n_0, n_0}^{-1}$ and the set of possible values of $g_{n_0, m}$ are finite. Similarly, $K \cap \left(K \cdot g_{m, k} g_{n_0, m}^{-1}\right)$ is non-empty, and therefore the set of possible $g_{m,k}$ is finite.

Consequently the set $K_2 = \bigcup_{m, k \geq n_0} K \cdot g_{m,k}$ is compact. The sequence x_n stays in K_2 for $n \geq n_0$. It is a Cauchy sequence with respect to the unique uniformity on K_2 (since this uniformity is the restriction of the G-invariant uniformity on $\mathcal{X}$). The uniformity on K_2 is complete; therefore the sequence x_n is convergent and the uniformity on $\mathcal{X}$ is complete.

Suppose now that we have a uniformity $\mathcal{U}$ on $\mathcal{X}$ such that the action of G on $\mathcal{X}$ is uniformly equicontinuous. This means that for every entourage U there exists an entourage V such that $d(x, y) \leq V$ implies $d(x \cdot g, y \cdot g) \leq U$ for all $g \in G$. But then $U' = \bigcap_{g \in G} U \cdot g \supseteq V$ and hence $U' \subseteq U$ is a G-invariant neighborhood of the diagonal belonging to $\mathcal{U}$.

It remains to prove that every G-invariant open neighborhood $U \subset \mathcal{X} \times \mathcal{X}$ of the diagonal belongs to $\mathcal{U}$. There exists an open G-invariant neighborhood $U' \subset \mathcal{X} \times \mathcal{X}$ of the diagonal such that $2U' \subseteq U$. We may assume that U' is symmetric.

Let $K \subset \mathcal{X}$ be a compact set intersecting every G-orbit. The restriction of the uniformity $\mathcal{U}$ onto K is the unique uniformity consisting of all the neighborhoods of the diagonal in $K \times K$. Hence there exists a G-invariant entourage $V \in \mathcal{U}$ such that $V \cap K \times K \subset U$. Then, using G-invariance of U, we get that $V \subset U$. $\square$

3.4.4. Coarse structure on a co-compact G-space.
We say that a relation $R \subset \mathcal{X} \times \mathcal{X}$ is *bounded* if there exists a compact set $C \subset \mathcal{X} \times \mathcal{X}$ such that $R \subset \bigcup_{g \in G} C \cdot g$. Here G acts on the direct square $\mathcal{X} \times \mathcal{X}$ by the diagonal action. It is easy to see that if R_1 and R_2 are bounded relations, then the relation $R_1 + R_2$ is bounded.

LEMMA 3.4.9. *If the group G is finitely generated, then there exists a bounded relation V such that $\bigcup_{n \geq 1} nV = \mathcal{X} \times \mathcal{X}$.*

PROOF. Let $K \subset \mathcal{X}$ be a compact set such that $\bigcup_{g \in G} K \cdot g = \mathcal{X}$ and let $S = S^{-1} \ni 1$ be a finite generating set of G. Let the relation $V \subset \mathcal{X} \times \mathcal{X}$ be the set of pairs of the form $(\xi_1 \cdot g, \xi_2 \cdot sg)$, where $\xi_1, \xi_2 \in K$, $g \in G$ and $s \in S$. It is obvious that V is bounded.

Any two points $\zeta_1, \zeta_2 \in \mathcal{X}$ can be written in the form $\zeta_1 = \xi_1 \cdot g$, $\zeta_2 = \xi_2 \cdot s_n \cdots s_2 s_1 g$ for $\xi_1, \xi_2 \in K$, $g \in G$ and $s_i \in S$. It is easy to see then that $(\zeta_1, \zeta_2) \in nV$. □

The set of all bounded relations on $\mathcal{X}$ is a structure of a *coarse space* (also called *asymptotic topology* or *uniformly bounded space*; see [**Roe03, Dra00, Nek99**]). Coarse structure, like the notion of a uniformity, also generalizes the notion of a metric space. However, the theory of coarse spaces studies the "large scale" properties of the space.

A coarse space $\mathcal{X}$ is given by a collection $\mathcal{A}$ of reflexive relations on $\mathcal{X}$ such that

(1) If $R \in \mathcal{A}$ and $R' \subset R$, then $R' \in \mathcal{A}$.
(2) If $R_1, R_2 \in \mathcal{A}$, then $R_1 + R_2 \in \mathcal{A}$.
(3) If $R \in \mathcal{A}$, then $R^- \in \mathcal{A}$.

Note that the second condition implies that $R_1 \cup R_2 \in \mathcal{A}$ for all $R_1, R_2 \in \mathcal{A}$.

3.4.5. Axiomatic description of $\mathcal{X}_G$.
Let $\mathfrak{M}$ be a hyperbolic bimodule over a group G. We allow the associated self-similar action to be non-faithful.

Let $\mathcal{X}$ be a locally compact Hausdorff topological space with a proper co-compact right G-action.

The tensor product $\mathcal{X} \otimes_G \mathfrak{M}$ is defined as the quotient of the topological space $\mathcal{X} \times \mathfrak{M}$ (where $\mathfrak{M}$ has the discrete topology) by the equivalence relation

$$\xi \otimes g \cdot a \sim \xi \cdot g \otimes a.$$

Then $\xi \otimes a \mapsto \xi \otimes a \cdot g$ is a well defined action by homeomorphisms of G on $\mathcal{X} \otimes \mathfrak{M}$.

DEFINITION 3.4.10. The G-space $\mathcal{X}$ is said to be $\mathfrak{M}$-*self-similar* (or just *self-similar* if $\mathfrak{M}$ is fixed) if the dynamical systems $(\mathcal{X}, G)$ and $(\mathcal{X} \otimes \mathfrak{M}, G)$ are topologically conjugate, i.e., if there exists a homeomorphism $\Phi : \mathcal{X} \otimes \mathfrak{M} \longrightarrow \mathcal{X}$ such that $\Phi(\xi \otimes a \cdot g) = \Phi(\xi \otimes a) \cdot g$ for all $\xi \otimes a \in \mathcal{X} \otimes \mathfrak{M}$ and $g \in G$. The homeomorphism Φ is called the *self-similarity structure* on $\mathcal{X}$.

Now let $\mathcal{X}$ be self-similar. We will write just $\xi \otimes a$ instead of $\Phi(\xi \otimes a)$, identifying $\mathcal{X} \otimes \mathfrak{M}$ with $\mathcal{X}$ by the homeomorphism Φ. If $v \in \mathfrak{M}^{\otimes n}$ and $\xi \in \mathcal{X}$, then $\xi \otimes v$ is defined inductively by

$$\xi \otimes (u \otimes a) = (\xi \otimes u) \otimes a.$$

If R is a relation on $\mathcal{X}$ and $v \in \mathfrak{M}^{\otimes n}$, then we denote by $R \otimes v$ the relation
$$\{(\xi \otimes v, \zeta \otimes v) : (\xi, \zeta) \in R\}.$$

DEFINITION 3.4.11. We say that the self-similarity structure on $\mathcal{X}$ is *contracting* if for every bounded relation $V \subset \mathcal{X} \times \mathcal{X}$ and every entourage U there exists $n_1 \in \mathbb{N}$ such that $V \otimes v \subseteq U$ for all $v \in \mathfrak{M}^{\otimes n}$, $n \geq n_1$.

Recall that here (in view of Proposition 3.4.7) a relation $U \subset \mathcal{X} \times \mathcal{X}$ is an entourage if and only if the set $\bigcap_{g \in G} U \cdot g$ is a neighborhood of the diagonal $\Delta_\mathcal{X} \subset \mathcal{X} \times \mathcal{X}$.

Readers who have omitted the previous two subsections should keep in mind the situation when G acts on a (proper) metric space $\mathcal{X}$ by isometries. Then Definition 3.4.11 is equivalent to the condition that for every $\varepsilon > 0$ and $R > 0$ there exists n_1 such that $d(x \otimes v, y \otimes v) < \varepsilon$ for all $x, y \in \mathcal{X}$ such that $d(x, y) < R$ and all $v \in \mathfrak{M}^{\otimes n}$, $n \geq n_1$.

LEMMA 3.4.12. *Suppose that the group G is finitely generated and that the self-similarity structure on $\mathcal{X}$ is contracting. Let $\mathsf{Y} \subset \mathfrak{M}$ be a finite set. Then for every sequence $\ldots x_2 x_1 \in \mathsf{Y}^{-\omega}$ and for every $\xi \in \mathcal{X}$ the sequence $\xi \otimes x_n \ldots x_2 x_1$ is convergent and the limit $\Psi(\ldots x_2 x_1)$ does not depend on ξ.*

Moreover, the convergence is uniform on compact sets over ξ and uniform on $\mathsf{Y}^{-\omega}$ over $\ldots x_2 x_1$; i.e., for every compact set $B \subset \mathcal{X}$ and every entourage $U \in \mathcal{U}$ there exists n_0 such that

$$(3.5) \qquad d(\zeta \otimes x_n \ldots x_2 x_1, \Psi(\ldots x_2 x_1)) \leq U$$

for all $\zeta \in B$, $n \geq n_0$ and $\ldots x_2 x_1 \in \mathsf{Y}^{-\omega}$.

PROOF. Let V be a bounded relation such that $\bigcup_{n \geq 1} nV = \mathcal{X} \times \mathcal{X}$. We may assume that V is an entourage. There exists n_0 such that $2V \otimes v \subseteq V$ for all $v \in \mathfrak{M}^{\otimes n}$, where $n \geq n_0$; hence $(2kV) \otimes v \subseteq kV$ for all $k \in \mathbb{N}$.

Take any $\xi \in \mathcal{X}$. There exists $k \in \mathbb{N}$ such that $d(\xi \otimes y_n \ldots y_2 y_1, \xi) \leq kV$ for all $n \leq n_0$ and all $y_i \in \mathsf{Y}$.

Let us prove that $d(\xi \otimes y_n \ldots y_2 y_1, \xi) \leq 2kV$ for all $n \geq 1$. It is true by the choice of k for all $n \leq n_0$. Let us prove it by induction on n. Suppose that it is true for all $n < m$. Then

$$d(\xi \otimes y_m \ldots y_2 y_1, \xi)$$
$$\leq d(\xi \otimes y_m \ldots y_2 y_1, \xi \otimes y_{n_0} \ldots y_2 y_1) + d(\xi \otimes y_{n_0} \ldots y_2 y_1, \xi)$$
$$\leq 2kV \otimes (y_{n_0} \ldots y_2 y_1) + kV \subseteq kV + kV = 2kV,$$

which finishes the inductive argument.

For every entourage U there exists $m_0 \in \mathbb{N}$ such that $2kV \otimes v \subseteq U$ for all $v \in \mathfrak{M}^{\otimes n}$, $n \geq m_0$. Then $d(\xi \otimes x_{n_1} \ldots x_2 x_1, \xi \otimes x_{n_2} \ldots x_2 x_1) \leq U$ for every sequence $\ldots x_2 x_1 \in \mathsf{Y}^{-\omega}$ and every pair of indices $n_1 \geq n_2 \geq m_0$; i.e., the sequence $\{\xi \otimes x_n \ldots x_2 x_1\}$ is Cauchy and thus is convergent. Note that the estimates do not depend on $\ldots x_2 x_1$.

If $\zeta \in \mathcal{X}$ is another point, then there exists $p \in \mathbb{N}$ such that $d(\xi, \zeta) \leq pV$. For every entourage U there exists $n_1 \in \mathbb{N}$ such that $pV \otimes x_n \ldots x_2 x_1 \subset U$ for all $n \geq n_1$. Then

$$d(\xi \otimes x_n \ldots x_2 x_1, \zeta \otimes x_n \ldots x_2 x_1) \leq U$$

for all $n \geq n_1$, which implies that the limit of the sequence $\{\xi \otimes x_n \ldots x_2 x_1\}$ does not depend on ξ.

Let us prove that the convergence is uniform. Let $B \subset \mathcal{X}$ be a compact set and let U be an entourage. Then there exists a number r_0 such that the diameter of B is less than $r_0 V$. Fix some point $\xi \in B$. There exists an entourage U' such that $2U' \leq U$. There exists a number n_0 such that $d(\xi \otimes x_n \ldots x_2 x_1, \Psi(\ldots x_2 x_1)) \leq U'$ and $r_0 V \otimes x_n \ldots x_2 x_1 \subseteq U'$ for all $n \geq n_0$ and $\ldots x_2 x_1 \in \mathsf{Y}^\omega$. The first inequality follows from the fact that the estimates proving that $\xi \otimes x_n \ldots x_2 x_1$ is a Cauchy sequence did not depend on $\ldots x_2 x_1 \in \mathsf{Y}^{-\omega}$.

Then

$$d(\zeta \otimes x_n \ldots x_2 x_1, \Psi(\ldots x_2 x_1))$$
$$\leq d(\zeta \otimes x_n \ldots x_2 x_1, \xi \otimes x_n \ldots x_2 x_1) + d(\xi \otimes x_n \ldots x_2 x_1, \Psi(\ldots x_2 x_1))$$
$$\leq U' + U' \leq U$$

for all $\zeta \in B, \ldots x_2 x_1 \in \mathsf{Y}^{-\omega}$ and $n \geq n_0$. □

THEOREM 3.4.13. *Let $\mathfrak{M}$ be a hyperbolic bimodule over a finitely generated group G and let $\mathcal{X}$ be a locally compact Hausdorff right G-space such that*
  (1) *the action of G on $\mathcal{X}$ is proper and co-compact, and*
  (2) *the G-space $\mathcal{X}$ is $\mathfrak{M}$-self-similar with a contracting self-similarity.*
Then there exists a uniformly continuous homeomorphism $\Psi : \mathcal{X}_G \longrightarrow \mathcal{X}$ such that

$$\Psi(\xi \cdot g) = \Psi(\xi) \cdot g \quad \text{and} \quad \Psi(\xi \otimes x) = \Psi(\xi) \otimes x$$

for all $\xi \in \mathcal{X}_G$, $g \in G$ and $x \in \mathfrak{M}$.

PROOF. Let us define the map Ψ, using Lemma 3.4.12, by the formula

$$\Psi(\ldots x_2 x_1) = \lim_{n \to \infty} \xi \otimes x_n \ldots x_2 x_1,$$

where $\ldots x_2 x_1 \in \Omega(\mathfrak{M})$, and let us prove that it satisfies the necessary conditions.

1. Ψ *is well defined.* Suppose that bounded sequences $\ldots y_2 y_1$ and $\ldots x_2 x_1$ define one point of $\mathcal{X}_G$, i.e., that they are asymptotically equivalent. Then there exists a bounded sequence $\{g_k\}$ of elements of the group G such that $g_n \cdot y_n \ldots y_2 y_1 = x_n \ldots x_2 x_1$. Let us find a constant sub-sequence $g = g_{n_k}$. Then

$$\Psi(\ldots x_2 x_1) = \lim_{k \to \infty} \xi \otimes x_{n_k} \ldots x_2 x_1$$
$$= \lim_{k \to \infty} \xi \otimes g \cdot y_{n_k} \ldots y_2 y_1 = \lim_{k \to \infty} \xi \cdot g \otimes y_{n_k} \ldots y_2 y_1 = \Psi(\ldots y_2 y_1),$$

since the limit in Lemma 3.4.12 does not depend on ξ.

2. *Equivariance.* Equalities

$$\Psi(\xi \cdot g) = \Psi(\xi) \cdot g \quad \text{and} \quad \Psi(\xi \otimes x) = \Psi(\xi) \otimes x$$

follow directly from the definitions.

3. Ψ *is uniformly continuous.* Choose a basis X of the bimodule $\mathfrak{M}$. Let $\mathcal{N}$ be the nucleus of the action (G, X). Let V be as in Lemma 3.4.9. We assume that V is an entourage. Choose a point $\xi \in \mathcal{X}$. We know that there exists $k \in \mathbb{N}$ such that $d(\xi \otimes x_n \ldots x_2 x_1 \cdot g, \xi) \leq 2kV$ for all n, $x_i \in \mathsf{X}$ and $g \in \mathcal{N}$ (see the proof of Lemma 3.4.12). This implies that $d(\Psi(\ldots x_2 x_1 \cdot g), \xi) \leq 3kV$. (One can find n such that $d(\Psi(\ldots x_2 x_1 \cdot g), \xi \otimes x_n \ldots x_2 x_1 \cdot g) \leq kV$.) Consequently, the set $\Psi(\mathcal{T} \cdot \mathcal{N})$ has diameter not greater than $6kV$, where $\mathcal{T}$ is the digit tile of the action (G, X).

For every entourage U there exists $n_0 \in \mathbb{N}$ such that $6kV \otimes v \subset U$ for all $v \in \mathfrak{M}^{\otimes n}$, $n \geq n_0$, due to the definition of a contracting self-similarity. Then

$$d(\Psi(\ldots x_n x_{n+1} \otimes g \cdot a_n \ldots a_2 a_1), \Psi(\ldots y_n y_{n+1} \otimes a_n \ldots a_2 a_1))$$
$$= d(\Psi(\ldots x_n x_{n+1} \cdot g) \otimes a_n \ldots a_2 a_1, \Psi(\ldots y_n y_{n+1}) \otimes a_n \ldots a_2 a_1)$$
$$\leq 6kV \otimes a_n \ldots a_2 a_1 \subseteq U,$$

for all $x_i, y_i, a_i \in \mathsf{X}$, $g \in \mathcal{N}$ and $n \geq n_0$. This proves that Ψ is uniformly continuous (see the definition of the uniformity on $\mathcal{X}_G$ and (3.4) on page 84).

4. Ψ *is surjective.* There exists a compact set $B \subset \mathcal{X}$ such that $\bigcup_{g \in G} B \cdot g = \mathcal{X}$. Since $\mathcal{X} \otimes \mathfrak{M}^{\otimes n} = \mathcal{X}$, this also implies that

(3.6)
$$\bigcup_{v \in \mathfrak{M}^{\otimes n}} B \otimes v = \mathcal{X}$$

for every $n \geq 0$.

Let us prove now that closure of the set $B' = \bigcup_{v \in \mathsf{X}^*} B \otimes v$ is compact. The set $\Psi(\mathcal{T})$ is compact as a continuous image of a compact set. Let W be a compact neighborhood of $\Psi(\mathcal{T})$. By (3.5), there exists n_0 such that $B \otimes v \subset W$ for all $v \in \mathfrak{M}^{\otimes n}$, $n \geq n_0$. Hence B' is a subset of the compact set $W \cup \bigcup_{v \in \mathsf{X}^*, |v| < n_0} B \otimes v$ and therefore it has a compact closure.

The action of G on $\mathcal{X}$ is proper; therefore there exists a finite set $C \subset G$ such that the condition $B \otimes v_1 \cdot g \cap B \otimes v_2 \neq \emptyset$ for $v_1, v_2 \in \mathsf{X}^*$ and $g \in G$ implies that $g \in C$.

Now take an arbitrary point $\xi \in \mathcal{X}$. For every $n \in \mathbb{N}$ there exists $v_n \in \mathsf{X}^n$ and $g_n \in G$ such that $\xi \in B \otimes v_n \cdot g_n$ (see (3.6)). Then $B \otimes v_1 \cap B \otimes v_n \cdot g_n g_1^{-1} \ni \xi \cdot g_1^{-1}$; hence $g_n g_1^{-1} \in C$ and the sequence $\{g_n\}$ is bounded. Therefore, there exists a sequence n_k such that the sequence v_{n_k} converges to a sequence $\ldots x_2 x_1 \in \mathsf{X}^{-\omega} \sqcup \mathsf{X}^*$ and the sequence $g_{n_k} = g$ is constant.

For every n there exists k_0 such that v_{n_k} ends by $x_n \ldots x_1$ for all $k \geq k_0$. Then

$$\xi \in (B \otimes x_{n_k} x_{n_k - 1} \ldots x_{n+1}) \otimes x_n \ldots x_2 x_1 \cdot g \subset B' \otimes x_n \ldots x_2 x_1 \cdot g,$$

and (3.5) applied to the precompact set B' implies that $\Psi(\ldots x_2 x_1 \cdot g) = \xi$.

5. Ψ *is injective.* Suppose that we have $\Psi(\ldots x_2 x_1 \cdot g) = \Psi(\ldots y_2 y_1 \cdot h)$ for some $\ldots x_2 x_1, \ldots y_2 y_1 \in \mathsf{X}^{-\omega}$ and $g, h \in G$, where X is a fixed basis of $\mathfrak{M}$. We have then for every n

$$\Psi(\ldots x_{n+2} x_{n+1}) \otimes x_n \ldots x_2 x_1 \cdot g = \Psi(\ldots y_{n+2} y_{n+1}) \otimes y_n \ldots y_2 y_1 \cdot h.$$

Thus, there exist $\xi, \zeta \in \Psi(\mathcal{T})$ such that $\xi \otimes x_n x_{n-1} \ldots x_1 \cdot g = \zeta \otimes y_n y_{n-1} \ldots y_1 \cdot h$. By definition of the tensor product, there exists $g_n \in G$ such that $\xi = \zeta \cdot g_n$ and $g_n \cdot x_n x_{n-1} \ldots x_1 \cdot g = y_n y_{n-1} \ldots y_1 \cdot h$. The set $\Psi(\mathcal{T})$ is compact; thus the first equality implies that the set of possible g_n is finite. Then the second equality implies that $\ldots x_2 x_1 \cdot g$ and $\ldots y_2 y_1 \cdot h$ are asymptotically equivalent, so that Ψ is injective.

This finishes the proof of the theorem, since every continuous bijection between locally compact Hausdorff spaces is a homeomorphism (see Theorem 10 on page 139 of [**Eng68**]). □

3.5. Connectedness of $\mathcal{X}_G$

THEOREM 3.5.1. *Let G be a finitely generated group with a contracting recurrent action (G, X). Then the limit G-space $\mathcal{X}_G$ is connected and locally connected.*

Let us prove the following technical lemma.

LEMMA 3.5.2. *Suppose that an action (G, X) of a finitely generated group G is recurrent and contracting. Let $\mathfrak{M} = \mathsf{X} \cdot G$ be the self-similarity bimodule and let $\mathcal{N}$ be the nucleus. Then there exists a finite set $B \subset G$ such that for any pair of words $v, u \in \mathsf{X}^*$ of equal lengths there exists a sequence $h_1, h_2, \ldots, h_m \in \mathcal{N}$ such that*
$$(h_m h_{m-1} \cdots h_1) \cdot v = u$$
in $\mathfrak{M}^{\otimes n}$ and $(h_k h_{k-1} \cdots h_1)|_v \in B$ for all $1 \le k \le m$.

PROOF. The group G is generated by $\mathcal{N}$ (see Proposition 2.11.3). For every pair $x, y \in \mathsf{X}$ there exists $g \in G$ such that $g \cdot x = y$ in $\mathfrak{M}$. Let us write g as a product $g_m \cdots g_1$ of the elements of $\mathcal{N}$ and let M be the maximal value of m for all pairs $x, y \in \mathsf{X}$. Denote by $A = \mathcal{N}^M$ the set of all elements of G which can be represented as products of at most M elements of the nucleus.

There exists by Proposition 2.11.5 (applied to the set $\mathsf{Y} = A \cdot \mathsf{X}$) a finite set $B \subset G$ such that $A \subseteq B$ and $(BA)|_\mathsf{X} \subset B$. Let us prove by induction on the length n of the words v and u that the set B satisfies the conditions of the lemma. The statement of the lemma is true for $n = 1$ by the choice of A. Suppose that it is true for n and let us prove it for $n+1$.

For every $h \in \mathcal{N}$ denote by $v_h, u_h \in \mathsf{X}^n$ and $h' \in \mathcal{N}$ the elements such that $h' \cdot v_h = u_h \cdot h$. Let vx_0 and uy_0 be arbitrary words of length $n+1$, where $v, u \in \mathsf{X}^n$ and $x_0, y_0 \in \mathsf{X}$. There exists an element $g \in G$ which can be written as a product $g = g_M \cdots g_1$ of the elements of the nucleus such that $g \cdot x_0 = y_0$.

There exists by the induction hypothesis a sequence $h_{m_1,1}, h_{m_1-1,1}, \ldots, h_{1,1} \in \mathcal{N}$ such that
$$(h_{m_1,1} h_{m_1-1,1} \cdots h_{1,1}) \cdot v = v_{g_1}$$
and $(h_{k,1} \cdots h_{1,1})|_v \in B$ for all $1 \le k \le m_1$.

We apply now g'_1 and get $g'_1 \cdot v_{g_1} x_0 = u_{g_1} \cdot g_1 \cdot x_0$. There exists, by the induction hypothesis, a sequence $h_{m_2,2}, h_{m_2-1,2}, \ldots, h_{1,2} \in \mathcal{N}$ such that
$$(h_{m_2,2} h_{m_2-1,2} \cdots h_{1,2}) \cdot u_{g_1} = v_{g_2}$$
and $(h_{k,2} \cdots h_{1,2})|_{u_{h_1}} \in B$ for all $1 \le k \le m_2$. Then we can apply g'_2 and get $g'_2 \cdot v_{g_2} \cdot g_1 \cdot x_0 = u_{g_2} \cdot g_2 g_1 \cdot x_0$.

We continue the process further and finally get a sequence
$$(3.7) \qquad g'_M, \ldots, h_{m_3,3}, \ldots, h_{1,3}, g'_2, h_{m_2,2}, \ldots, h_{1,2}, g'_1, h_{m_1,1} \ldots, h_{1,1}$$
such that
$$(h_{m_i,i} \cdots h_{1,i}) \cdot u_{g_{i-1}} = v_{g_i},$$
$(h_{k,i} \cdots h_{1,i})|_{u_{g_{i-1}}} \in B$ for all $1 \le k \le m_i$ and $i \ge 2$ and $g'_i \cdot v_{g_i} = u_{g_i} \cdot g_i$ for all $i \ge 1$. Note then that
$$(h_{k,i} \cdots h_{1,i} \cdots h_{1,1})|_v \in B \cdot g_{i-1} \cdots g_1$$
for $1 \le k < m_i$ and
$$(h_{m_i,i} \cdots h_{1,i} \cdots h_{1,1})|_v = g_{i-1} \cdots g_1.$$

There exists also a sequence $h_r, \ldots, h_1 \in \mathcal{N}$ such that
$$(h_r \cdots h_1) \cdot u_{g_M} = u$$
and $(h_k \cdots h_1)|_{u_{g_M}} \in B$ for all $1 \le k \le r$. Appending this sequence to the beginning of the sequence (3.7) we get a sequence $f_N, f_{N-1}, \ldots, f_1 \in \mathcal{N}$ such that
$$(f_N \cdots f_1) \cdot vx_0 = uy_0$$
and $(f_k \cdots f_1)|_v \in B \cdot A$ for all $1 \le k \le N$. Then
$$(f_k \cdots f_1)|_{vx_0} \in B \cdot A|_{x_0} \subset B$$
and thus the sequence $f_N, \ldots, f_1$ satisfies the conditions of the lemma. $\square$

PROOF OF THEOREM 3.5.1. Let B be as in Lemma 3.5.2. For every $n \in \mathbb{N}$ let Γ_n be the graph with the set of vertices $\mathsf{X}^n \cdot B$ where two vertices $v_1 \cdot g_1$ and $v_2 \cdot g_2$ are connected by an edge if and only if there exists an element $h \in \mathcal{N}$ such that $h \cdot v_1 \cdot g_1 = v_2 \cdot g_2$. We have proved that the set $\mathsf{X}^n \cdot 1$ belongs to one connected component of the graph Γ_n. Repeating the arguments from the proof of Proposition 3.3.10, we get that the tile $\mathcal{T}$ belongs to one connected component $\mathcal{C}$ of $\mathcal{T} \cdot B$.

If $g_1, g_2 \in G$ are such that $g_1 g_2^{-1} \in \mathcal{N}$, then $\mathcal{C} \cdot g_1 \cap \mathcal{C} \cdot g_1 \supset \mathcal{T} \cdot g_1 \cap \mathcal{T} \cdot g_2 \ne \emptyset$ due to Proposition 3.3.5. The set $\mathcal{N}$ generates the group G; hence we obtain that the space $\mathcal{X}_G$ is connected.

For a point $\xi \in \mathcal{X}_G$ and $n \in \mathbb{N}$ denote by U_n the union of the sets of the form $\mathcal{C} \cdot v$ containing ξ, where $v \in \mathfrak{M}^{\otimes n}$. Then U_n is a base of connected neighborhoods of ξ by Proposition 3.4.1. $\square$

Arguments similar to those of the proof of Lemma 3.5.2 where used at first by K. Pilgrim and P. Haissinsky in their proof that the space $\mathcal{J}_G = \mathcal{X}_G/G$ is locally connected for recurrent actions (private communication).

COROLLARY 3.5.3. *If an action (G, X) of a finitely generated group is contracting and recurrent, then the limit space $\mathcal{X}_G$ is path-connected and locally path-connected.*

PROOF. Every locally compact metrizable connected and locally connected space is path-connected and locally path-connected by a theorem of R. L. Moore and S. Mazurkiewicz (see Problem J on page 260 of [**Eng68**] and references therein). $\square$

3.6. The limit space $\mathcal{J}_G$

3.6.1. Definition and basic properties. The *limit space* $\mathcal{J}_G$ is the space of orbits $\mathcal{X}_G/G$ of the right action of G on $\mathcal{X}_G$.

Since the action of G on $\mathcal{X}_G$ is described in terms of the space $\mathsf{X}^{-\omega} \cdot G$ as multiplication:
$$(\ldots x_2 x_1 \cdot h) \cdot g = \ldots x_2 x_1 \cdot hg,$$
we can encode the points of $\mathcal{J}_G$ by the left-infinite sequences $\ldots x_2 x_1 \in \mathsf{X}^{-\omega}$.

The corresponding equivalence relation is described in the following way.

DEFINITION 3.6.1. Two sequences $\ldots x_2 x_1, \ldots y_2 y_1 \in \mathsf{X}^{-\omega}$ are asymptotically equivalent if there exists a bounded sequence $\{g_n\}$ of elements of G such that
$$g_n(x_n \ldots x_1) = y_n \ldots y_1$$
for all $n \ge 1$.

The proof of the following proposition is straightforward.

PROPOSITION 3.6.2. *The quotient of the space $\mathsf{X}^{-\omega}$ by the asymptotic equivalence relation is homeomorphic to the limit space $\mathcal{J}_G$. The homeomorphism sends the equivalence class of the sequence $\ldots x_2 x_1 \in \mathsf{X}^{-\omega}$ to the orbit of the image of $\ldots x_2 x_1 \cdot 1$ in $\mathcal{X}_G$.* □

Let us list topological properties of the space $\mathcal{J}_G$. The proofs are identical to the proofs of the similar properties of $\mathcal{X}_G$ (or follow directly from them). The statement about connectedness is proved similarly to the proof of Proposition 3.3.10.

THEOREM 3.6.3. *Two sequences $\ldots x_2 x_1, \ldots y_2 y_1$ are asymptotically equivalent if and only if there exists a left-infinite path $\ldots e_2 e_1$ in the Moore diagram of the nucleus such that the edge e_i is labeled by (x_i, y_i).*

The topological space $\mathcal{J}_G$ is compact, metrizable and has topological dimension less than the size of the nucleus. It is connected if the group G is finitely generated and level-transitive. It is locally connected if the group G is finitely generated and recurrent.

The following proposition also follows directly from its analog for $\mathcal{X}_G$ (Proposition 3.2.7).

PROPOSITION 3.6.4. *Let (G, X) be a contracting action of a finitely generated group. Let (A, X) be a finite automaton generating the action. Denote by $D \subset \mathsf{X}^{-\omega} \times \mathsf{X}^{-\omega}$ the set of the pairs $(\ldots x_2 x_1, \ldots y_2 y_1)$ such that there exists a path $\ldots e_2 e_1$ in the Moore diagram of A such that (x_i, y_i) is the label of e_i. Then the equivalence relation on $\mathsf{X}^{-\omega}$ generated by D coincides with the asymptotic equivalence relation.*

Note that the equivalence generated by D will be automatically closed.

3.6.2. The limit dynamical system and its Markov partition. A special property of the limit space $\mathcal{J}_G$ is existence of the *shift map*. It is easy to see that the asymptotic equivalence relation on $\mathsf{X}^{-\omega}$ is invariant under the shift

$$\sigma : \ldots x_2 x_1 \mapsto \ldots x_3 x_2;$$

therefore σ induces a continuous map $\mathsf{s} : \mathcal{J}_G \longrightarrow \mathcal{J}_G$. It is surjective, and every point $\zeta \in \mathcal{J}_G$ has at most $d = |\mathsf{X}|$ preimages under s.

The dynamical system $(\mathcal{J}_G, \mathsf{s})$ is called the *limit dynamical system* of the self-similar action.

The images of the tiles $\mathcal{T} \otimes v$ for $v \in \mathsf{X}^n$ are called the *tiles of the nth level* of $\mathcal{J}_G$ and are denoted $\mathcal{T}_v$. In particular, $\mathcal{T}_\varnothing = \mathcal{J}_G$.

We see that $\mathcal{T}_v = \bigcup_{x \in \mathsf{X}} \mathcal{T}_{vx}$ and that $\mathsf{s}(\mathcal{T}_{vx}) = \mathcal{T}_v$ for every $v \in \mathsf{X}^*$ and $x \in \mathsf{X}$.

One can prove, in the same way as for the tiles of $\mathcal{X}_G$, that the following proposition holds.

PROPOSITION 3.6.5. *If the action satisfies the open set condition, then every tile $\mathcal{T}_v$ is equal to the closure of its interior, any two different tiles of one level have disjoint interiors and the boundary of $\mathcal{T}_v$ for $v \in \mathsf{X}^n$ is equal to $\mathcal{T}_v \cap \bigcup_{u \in \mathsf{X}^n, u \neq v} \mathcal{T}_u$.*

Consequently, if the action satisfies the open set condition, then every collection $\{\mathcal{T}_v\}_{v \in \mathsf{X}^n}$ is a Markov partition of the limit dynamical system $(\mathcal{J}_G, \mathsf{s})$.

3.6.3. Schreier graphs as approximations of $\mathcal{J}_G$.

Let G be a group generated by a finite set S and acting on a set M. Then the corresponding *Schreier graph* $\Gamma(S, M)$ is the graph with the set of vertices M and the set of arrows $S \times M$, where the arrow (s, v) starts in v and ends in $s(v)$. The *simplicial* Schreier graph $\overline{\Gamma}(S, M)$ remembers only the vertex adjacency: its set of vertices is M and two vertices are adjacent if and only if one is the image of the other under the action of a generator $s \in S$.

If (G, X) is a self-similar action and G is generated by a finite set S, then we get the sequence $\Gamma_n = \Gamma(S, \mathsf{X}^n)$ of the Schreier graphs of the action of G on the levels X^n of the tree X^* (and the sequence $\overline{\Gamma}_n = \overline{\Gamma}(S, \mathsf{X}^n)$ of the respective simplicial Schreier graphs).

If (G, X) is generated by a finite automaton (A, X), then the graphs $\Gamma(\mathsf{A}, \mathsf{X}^n)$ coincide with the dual Moore diagrams of the automata $(\mathsf{A}, \mathsf{X}^n)$ (see the end of Subsection 1.3.6).

Note first of all that Definition 3.6.1 can be formulated in the following way.

PROPOSITION 3.6.6. *Let (G, X) be a contracting self-similar action of a group generated by a finite set S. Then sequences $\ldots x_2 x_1, \ldots y_2 y_1 \in \mathsf{X}^{-\omega}$ are asymptotically equivalent with respect to the action if and only if there exists a number C such that the distance between $x_n \ldots x_2 x_1$ and $y_n \ldots y_2 y_1$ in $\overline{\Gamma}(S, \mathsf{X}^n)$ is less than C.*

PROOF. If the vertices $x_n \ldots x_2 x_1$ and $y_n \ldots y_2 y_1$ are a distance less than C in $\overline{\Gamma}(S, \mathsf{X})$, then there exists an element $g_n \in G$ such that $g_n(x_n \ldots x_2 x_1) = y_n \ldots y_2 y_1$ and g_n is a product of less than C elements of $S \cup S^{-1}$. The set of such elements g_n is finite; i.e., the sequence $\{g_n\}$ is bounded.

On the other hand, if $\{g_n\}$ is a bounded sequence, then there exists C such that every g_n is a product of less than C elements of $S \cup S^{-1}$. $\square$

COROLLARY 3.6.7. *Let (G_1, X) and (G_2, X) be contracting self-similar actions and let S_1 and S_2 be finite generating sets of G_1 and G_2, respectively. Suppose that for every $n \in \mathbb{N}$ the identical map on X^n is an isomorphism of the simplicial Schreier graphs $\overline{\Gamma}(S_1, \mathsf{X}^n)$ and $\overline{\Gamma}(S_2, \mathsf{X}^n)$. Then the limit dynamical systems $(\mathcal{J}_{G_1}, \mathsf{s})$ and $(\mathcal{J}_{G_2}, \mathsf{s})$ are topologically conjugate. In particular, the limit spaces $\mathcal{J}_{G_1}$ and $\mathcal{J}_{G_2}$ are homeomorphic.* $\square$

EXAMPLE. The simplicial Schreier graphs $\overline{\Gamma}(S, \mathsf{X}^n)$ of the action of the Grigorchuk group with respect to the standard generating set $S = \{a, b, c, d\}$ coincides with the simplicial Schreier graphs of the dihedral group generated by the transformations

$$a = \sigma, \qquad B = (a, B).$$

This follows from the fact that for every $v \in \mathsf{X}^*$ and $g \in \{b, c, d\}$ either $g(v) = B(v)$ or $g(v) = v$, which is easily checked looking at the wreath recursions or the portraits defining the automorphisms b, c, d and B.

Consequently, the limit dynamical system of the Grigorchuk group coincides with that of the dihedral group. We will see later (Subsection 6.3.1) that the limit space of the dihedral group is the segment $[0, 1]$ on which the shift s acts as the *tent map* $x \mapsto |2x - 1|$.

This example shows that the limit dynamical system $(\mathcal{J}_G, \mathsf{s})$ carries less information then the group action. This is the reason why it is important to consider the limit space $\mathcal{J}_G$ as the *orbispace* of the action of G on $\mathcal{X}_G$, i.e., to preserve the information about the stabilizers (*isotropy groups*) of the action. We will do this in Section 4.6.

The next proposition is proved in the same way as Proposition 3.3.5.

PROPOSITION 3.6.8. *Two tiles $\mathcal{T}_{v_1}$ and $\mathcal{T}_{v_2}$ of the nth level intersect if and only if there exists an element $g \in \mathcal{N}$ such that $g(v_1) = v_2$.* □

Proposition 3.6.8 shows that two tiles $\mathcal{T}_v, \mathcal{T}_u$ for $v, u \in \mathsf{X}^n$ are adjacent if and only if the vertices v and u are adjacent in the graph $\overline{\Gamma}(\mathcal{N}, \mathsf{X})$.

This (together with Proposition 3.4.1) shows that the Schreier graphs $\overline{\Gamma}(S, \mathsf{X}^n)$ are good approximations of the limit space $\mathcal{J}_G$. A more precise statement is the following theorem. (Another interpretation will be given in Section 3.8.)

THEOREM 3.6.9. *A compact Hausdorff space $\mathcal{X}$ is homeomorphic to the limit space $\mathcal{J}_G$ if and only if there exists a collection $\mathfrak{T} = \{T_v : v \in \mathsf{X}^*\}$ of closed subsets of $\mathcal{X}$ such that the following conditions hold.*
(1) *$T_\varnothing = \mathcal{X}$ and $T_v = \bigcup_{x \in \mathsf{X}} T_{xv}$ for every $v \in \mathsf{X}^*$.*
(2) *The set $\bigcap_{n=1}^\infty T_{x_n x_{n-1} \ldots x_1}$ contains only one point for every word $\ldots x_2 x_1 \in \mathsf{X}^{-\omega}$.*
(3) *The intersection $T_v \cap T_u$ for $u, v \in \mathsf{X}^n$ is non-empty if and only if there exists an element s of the nucleus such that $s(v) = u$.*

If $\mathcal{X}$ is a metric space, then (2) is equivalent to the condition
$$\lim_{n \to \infty} \max_{v \in \mathsf{X}^n} \operatorname{diam}(T_v) = 0.$$

PROOF. The limit space $\mathcal{J}_G$ satisfies the conditions of the theorem for the sets $T_v = \mathcal{T}_v$.

Suppose now that a topological space $\mathcal{X}$ satisfies the conditions of the theorem for a collection $\mathfrak{T} = \{T_v\}_{v \in \mathsf{X}^*}$. Let us prove that the map
$$\Pi : \ldots x_2 x_1 \mapsto \bigcap_{n=1}^\infty T_{x_n x_{n-1} \ldots x_1}$$
is a continuous surjection from $\mathsf{X}^{-\omega}$ to $\mathcal{X}$. Let $A \subseteq \mathcal{X}$ be a closed subset. Denote by A_n the set of all the words $v \in \mathsf{X}^n$ for which the set T_v has a non-empty intersection with A. Then obviously, $A \subseteq \bigcap_{n=1}^\infty \bigcup_{v \in A_n} T_v$. On the other hand, if a point a does not belong to A, then the set of the words $v \in \bigcup_{n \geq 1} A_n$ such that $a \in T_v$ is finite. Otherwise there would exist an infinite word $\ldots x_2 x_1 \in \mathsf{X}^{-\omega}$ such that the intersection $\bigcap_{n \geq 1} T_{x_n \ldots x_1}$ contains a and a point of A, which contradicts condition (2). Hence
$$A = \bigcap_{n=1}^\infty \bigcup_{v \in A_n} T_v.$$
Denote
$$A^* = \bigcap_{n=1}^\infty \bigcup_{v \in A_n} \mathsf{X}^{-\omega} v.$$
If a sequence $\xi = \ldots x_2 x_1$ belongs to A^*, then for any $n \geq 1$ the word $x_n x_{n-1} \ldots x_1$ belongs to A_n. But then $\Pi(\xi) \in A$. On the other hand, if $\Pi(\xi)$ belongs to A, then for any $n \in \mathbb{N}$ the word $x_n x_{n-1} \ldots x_1$ belongs to A_n, so $\xi \in A^*$. Thus, A^* is equal to the preimage of A under Π. The set A^* is closed, so the preimage of every closed

set under the map Π is closed and the map is continuous. The fact that it is onto follows directly from condition (1) of the theorem.

The spaces $\mathsf{X}^{-\omega}$ and $\mathcal{X}$ are compact and the surjection Π is closed, so it is a quotient map. So it is sufficient to prove that $\Pi(\xi) = \Pi(\zeta)$ if and only if ξ and ζ are asymptotically equivalent.

Suppose that $\Pi(\ldots x_2 x_1) = \Pi(\ldots y_2 y_1)$. Then for every n the sets $T_{x_n x_{n-1} \ldots x_1}$ and $T_{y_n y_{n-1} \ldots y_1}$ intersect; thus $x_n x_{n-1} \ldots x_1 = s_n(y_n y_{n-1} \ldots y_1)$ for some element s_n of the nucleus. Hence ξ and ζ are asymptotically equivalent.

On the other hand, if $\xi = \ldots x_2 x_1$ and $\zeta = \ldots y_2 y_1$ are asymptotically equivalent, then for every $n \in \mathbb{N}$ there exists an element s_n of the nucleus such that $x_n x_{n-1} \ldots x_1 = s_n(y_n y_{n-1} \ldots y_1)$, so the sets $T_{x_n x_{n-1} \ldots x_1}$ and $T_{y_n y_{n-1} \ldots y_1}$ intersect for every n; hence $\Pi(\xi) = \Pi(\zeta)$. $\square$

3.7. Limit spaces of self-similar subgroups

Let ϕ be a virtual endomorphism of a group G. Recall that a subgroup $H \leq G$ is said to be ϕ-semi-invariant if

$$\phi(H \cap \operatorname{Dom} \phi) \leq H.$$

Recall that a subgroup H is ϕ-semi-invariant if and only if it is *self-similar*, i.e., if there exists a basis X of the bimodule $\phi(G)G$ and a subset $\mathsf{Y} \subset \mathsf{X}$ such that $\phi(1)1 \in \mathsf{Y}$ and the equality $g \cdot x = y \cdot h$ for $g \in H$ and $x \in \mathsf{Y}$ implies that $y \in \mathsf{Y}$ and $h \in H$ (see 2.7.1, in particular Lemma 2.7.2 and the comment after the proof).

The set Y is a basis of the H-bimodule $\phi_H(H)H \subset \phi(G)G$, where ϕ_H is the restriction of ϕ onto H.

THEOREM 3.7.1. *Suppose that (G, X) is a contracting self-similar action and let $H \leq G$ be a self-similar subgroup with the respective H-invariant alphabet $\mathsf{Y} \subset \mathsf{X}$. Then the action (H, Y) is also contracting, and there exists an H-equivariant continuous map $F : \mathcal{X}_H \longrightarrow \mathcal{X}_G$. Let $f : \mathcal{J}_H \longrightarrow \mathcal{J}_G$ be the induced map of the orbit spaces.*

(1) *If H is transitive on the first level, then f is surjective.*
(2) *If H has finite index in G, the actions (G, X), (H, Y) are recurrent and $\phi : G \dashrightarrow G$ is injective, then F is a homeomorphism.*

The map $f : \mathcal{J}_H \longrightarrow \mathcal{J}_G$ agrees with the shift maps on the limit spaces (i.e., f is a semi-conjugacy of the limit dynamical systems).

PROOF. The fact that (H, Y) is contracting is straightforward.

It follows directly from the definition of a self-similar subgroup that if two sequences $\ldots a_2 a_1 \cdot g$ and $\ldots b_2 b_1 \cdot h$, where $a_i, b_i \in \mathsf{Y}$ and $g, h \in H$ are asymptotically equivalent with respect to the action (H, Y), then they are asymptotically equivalent with respect to the action (G, X).

This means that the natural embedding $\mathsf{Y}^{-\omega} \cdot H \hookrightarrow \mathsf{X}^{-\omega} \cdot G$ induces a continuous H-equivariant map $F : \mathcal{X}_H \longrightarrow \mathcal{X}_G$.

It is easy to see that the induced map $f : \mathcal{J}_H \longrightarrow \mathcal{J}_G$ is well defined and agrees with the shifts.

If the subgroup H is transitive on the first level, then $\mathsf{Y} = \mathsf{X}$ and thus the induced map $f : \mathcal{J}_H \longrightarrow \mathcal{J}_G$ is surjective.

Let us prove that if H has finite index, ϕ is injective and the actions (G, Y) and (H, Y) are recurrent, then F is a homeomorphism. We shall use Theorem 3.4.13.

The space $\mathcal{X}_G$ is a proper right H-space, since H is a subgroup of G. It is also co-compact, since the action of G on $\mathcal{X}_G$ is co-compact and H is a subgroup of finite index.

The bimodule $\mathfrak{M}_H = \phi_H(H)H = \mathsf{Y} \cdot H$ is a sub-bimodule of the H-bimodule $\mathfrak{M}_G = \mathsf{X} \cdot G$. Let us define an $\mathfrak{M}_H$-self-similarity structure on $\mathcal{X}_G$ just restricting the $\mathfrak{M}_G$-self-similarity structure. We have to prove that it is really a self-similarity.

Let us prove that $\xi_1 \otimes_G m_1 = \xi_2 \otimes_G m_2$ is equivalent to $\xi_1 \otimes_H m_1 = \xi_2 \otimes_H m_2$ for all $\xi_1, \xi_2 \in \mathcal{X}_G$ and $m_1, m_2 \in \mathfrak{M}_H$. It is obvious that $\xi_1 \otimes_H m_1 = \xi_2 \otimes_H m_2$ implies $\xi_1 \otimes_G m_1 = \xi_2 \otimes_G m_2$. Let us prove the implication in the other direction. Fix some $y \in \mathsf{Y}$. Then $\xi_i \otimes_H m_i = \xi_i \otimes_H h_i \cdot y = \xi_i \cdot h_i \otimes_H y$ for some $h_i \in H$. The equality $\xi_1 \cdot h_1 \otimes_G y = \xi_2 \cdot h_2 \otimes_G y$ means that there exists $g \in G$ such that $\xi_1 \cdot h_1 = \xi_2 \cdot h_2 g$ and $g \cdot y = y$. But the virtual endomorphism ϕ is injective, so the second equality implies $g = 1$; hence $\xi_1 \cdot h_1 = \xi_2 \cdot h_2$ and therefore $\xi_1 \otimes_H m_1 = \xi_2 \otimes_H m_2$.

It remains to prove that $\mathcal{X}_G \otimes \mathfrak{M}_H = \mathcal{X}_G$. Every element of $\mathfrak{M}_H$ is equal to $h \cdot y$ for some $h \in H$. Consequently,

$$\mathcal{X}_G \otimes \mathfrak{M}_H = \bigcup_{m \in \mathfrak{M}_H} \mathcal{X}_G \otimes m = \bigcup_{h \in H} \mathcal{X}_G \otimes h \cdot y = \bigcup_{h \in H} \mathcal{X}_G \cdot h \otimes y = \mathcal{X}_G \otimes y = \mathcal{X}_G.$$

The defined $\mathfrak{M}_H$-self-similarity on $\mathcal{X}_G$ is obviously contracting, and thus Theorem 3.4.13 provides a homeomorphism $\Psi : \mathcal{X}_H \longrightarrow \mathcal{X}_G$ whose definition agrees with the definition of the map F. $\square$

3.8. The limit space $\mathcal{J}_G$ as a hyperbolic boundary

3.8.1. Self-similarity graph.

DEFINITION 3.8.1. Let $\mathfrak{M}$ be a d-fold covering bimodule over a finitely generated group G. For a given finite generating set S of G and a basis X of $\mathfrak{M}$ we define the *self-similarity graph* $\Sigma(G, S, \mathsf{X})$ as the graph with the set of vertices X^*, where two vertices $v_1, v_2 \in \mathsf{X}^*$ belong to a common edge if and only if either $v_i = xv_j$ for some $x \in \mathsf{X}$ (the *vertical* edges) or $s(v_i) = v_j$ for some $s \in S$ (the *horizontal* edges), for $\{i, j\} = \{1, 2\}$.

As an example, see a part of the self-similarity graph of the adding machine in Figure 3.1.

If all the restrictions of the elements of the generating set S also belong to S, then the self-similarity graph $\Sigma(G, S, \mathsf{X})$ is an *augmented tree* in the sense of V. Kaimanovich (see [**Kai03**]).

The definition of the self-similarity graph depends on the choice of the generating set S and the basis X. We will use the classical notion of quasi-isometry in order to make it more canonical (see [**Gro93, GH90**]).

DEFINITION 3.8.2. Two metric spaces $(\mathfrak{X}, d_{\mathfrak{X}})$ and $(\mathfrak{Y}, d_{\mathfrak{Y}})$ are said to be *quasi-isometric* if there exists a map (which is then called *quasi-isometry*) $f : \mathfrak{X} \longrightarrow \mathfrak{Y}$ and constants $L > 1$, $C > 0$ such that

(i)
$$L^{-1} d_{\mathfrak{X}}(x_1, x_2) - C < d_{\mathfrak{Y}}(f(x_1), f(x_2)) < L d_{\mathfrak{X}}(x_1, x_2) + C,$$
for all $x_1, x_2 \in \mathfrak{X}$, and

(ii) for every $y \in \mathfrak{Y}$ there exists $x \in \mathfrak{X}$ such that $d_{\mathfrak{Y}}(y, f(x)) < C$.

We will also use the following equivalent definition.

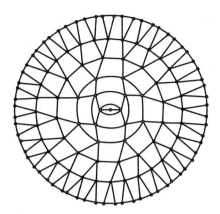

FIGURE 3.1. The self-similarity graph of the adding machine

DEFINITION 3.8.3. (1) Two maps $f_1, f_2 : \mathfrak{X} \longrightarrow \mathfrak{Y}$ are *shift-equivalent* if
$$\sup_{x \in \mathfrak{X}} d_{\mathfrak{Y}}(f_1(x), f_2(x)) < \infty.$$

(2) A map $f : \mathfrak{X} \longrightarrow \mathfrak{Y}$ is *quasi-Lipschitz* if there exist $C_1, C_2 > 0$ such that
$$d_{\mathfrak{Y}}(f(x_1), f(x_2)) \leq C_1 d_{\mathfrak{X}}(x_1, x_2) + C_2$$
for all $x_1, x_2 \in \mathfrak{X}$.

(3) Two maps $f_1 : \mathfrak{X} \longrightarrow \mathfrak{Y}$ and $f_2 : \mathfrak{Y} \longrightarrow \mathfrak{X}$ are a *pair of inverse quasi-isometries* if they are quasi-Lipschitz and $f_1 \circ f_2$ and $f_2 \circ f_1$ are shift-equivalent to the identical maps.

It is easy to prove that f is a quasi-isometry in the sense of Definition 3.8.2 if and only if it belongs to a pair of inverse quasi-isometries in the sense of Definition 3.8.3. Moreover, the inverse of a quasi-isometry is defined uniquely up to a shift-equivalence.

Let us show that the self-similarity graph $\Sigma(G, S, \mathsf{X})$ depends, up to a quasi-isometry, only on the group G and the bimodule $\mathfrak{M}$.

LEMMA 3.8.4. *Let G be a finitely generated group with a self-similar action, and let $\mathfrak{M}$ be the self-similarity bimodule.*
  (1) *The self-similarity graphs $\Sigma(G, S_1, \mathsf{X})$ and $\Sigma(G, S_2, \mathsf{X})$, where S_1, S_2 are two finite generating sets of G, are quasi-isometric.*
  (2) *If X and Y are bases of $\mathfrak{M}$, then $\Sigma(G, S, \mathsf{X})$ and $\Sigma(G, S, \mathsf{Y})$ are quasi-isometric.*
  (3) *The self-similarity graph of the nth tensor power of the self-similar action is quasi-isometric to the self-similarity graph of the original action.*

PROOF. 1) The identical map $\Sigma(G, S_1, \mathsf{X}) \longrightarrow \Sigma(G, S_2, \mathsf{X})$ of the sets of vertices is a quasi-isometry. The constant L is any number such that the length of every element of one of the generating sets has length less than L with respect to the other generating set. The constant C can be any positive number.

2) Let $\alpha : \mathsf{X}^* \longrightarrow \mathsf{Y}^*$ be the isomorphism conjugating the actions (G, X) and (G, Y) (see Proposition 2.3.4). Let us prove that the map α is a quasi-isometry

of the self-similarity graphs $\Sigma(G, S, \mathsf{X})$ and $\Sigma(G, S, \mathsf{Y})$. The map α preserves the horizontal edges, since it conjugates the actions.

If (v, xv) is a vertical edge, then $\alpha(xv) = \alpha(x)\alpha|_x(v) = \alpha(x)h_x\alpha(v)$, where h_x is such that $x = \alpha(x) \cdot h_x$ in $\mathfrak{M}$ (see Proposition 2.3.4). Let L be the maximal length of the elements $h_x \in G$ for all $x \in \mathsf{X}$. Then the distance between $\alpha(v)$ and $h_x\alpha(v)$ in $\Sigma(G, S, \mathsf{Y})$ is not greater than L; therefore

$$d(\alpha(v), \alpha(xv)) \leq d(\alpha(v), h_x\alpha(v)) + d(h_x\alpha(v), \alpha(xv)) \leq L + 1.$$

Hence the map $\alpha : \mathsf{X}^* \longrightarrow \mathsf{Y}^*$ is a quasi-isometry.

3) The set of the vertices of $\Sigma(G, S, \mathsf{X}^n)$ is equal to $\{\varnothing\} \cup \mathsf{X}^n \cup \mathsf{X}^{2n} \cup \mathsf{X}^{3n} \cup \ldots$. Let $F : \Sigma(G, S, \mathsf{X}^n) \longrightarrow \Sigma(G, S, \mathsf{X})$ be the natural inclusion.

It is easy to see that $d(F(u), F(v)) \leq n \cdot d(u, v)$ and $d(F(u), F(v)) \geq d(u, v)$ for all $u, v \in \Sigma(G, S, \mathsf{X})$.

For every vertex $v = x_1 x_2 \ldots x_m \in \mathsf{X}^*$ of the graph $\Sigma(G, S, \mathsf{X})$ there exists a vertex $x_r x_{r+1} \ldots x_m$ belonging to the vertex set of the graph $\Sigma(G, S, \mathsf{X}^n)$ which is at distance less than n from v (one must take r to be the minimal number, such that $m - r + 1$ is divisible by n). So the map F satisfies both conditions of Definition 3.8.2. $\square$

3.8.2. Hyperbolicity of the self-similarity graph. Let us recall the definition of Gromov-hyperbolic metric spaces [**Gro87**].

Let $(\mathfrak{X}, d)$ be a metric space. The *Gromov product* of two points $x, y \in \mathfrak{X}$ with respect to the basepoint $x_0 \in \mathfrak{X}$ is the number

$$\langle x \cdot y \rangle = \langle x \cdot y \rangle_{x_0} = \frac{1}{2}\left(d(x, x_0) + d(y, x_0) - d(x, y)\right).$$

DEFINITION 3.8.5. A metric space $\mathfrak{X}$ is said to be *Gromov-hyperbolic* if there exists $\delta > 0$ such that the inequality

(3.8) $$\langle x \cdot y \rangle \geq \min\left(\langle x \cdot z \rangle, \langle y \cdot z \rangle\right) - \delta$$

holds for all $x, y, z \in \mathfrak{X}$.

The standard definition requires that inequality (3.8) hold for any choice of the basepoint. However, we can fix the basepoint, and these two versions of the definition will be equivalent (see, for example, Proposition 1.2 in [**CDP90**]).

If a proper geodesic metric space (for instance a graph) is quasi-isometric to a hyperbolic space, then it is also hyperbolic. For the proofs of the mentioned facts and for other properties of hyperbolic spaces and groups, look at one of these books: [**Gro87, CDP90, GH90**].

THEOREM 3.8.6. *If the action of a finitely generated group G is contracting, then the self-similarity graph $\Sigma(G, S, \mathsf{X})$ is a Gromov-hyperbolic space.*

PROOF. It is sufficient to prove that some quasi-isometric graph is hyperbolic. Therefore, we can change by statement (1) of Lemma 3.8.4 the set of generators S so that $S = S^{-1}$ and it will contain all the restrictions of its elements. Then the length of any restriction of an element $g \in G$ is not greater than the length of g.

We may also assume that the nucleus of the action is a subset of S. Then there exists $N \in \mathbb{N}$ such that for every element $g \in G$ of length ≤ 4 and any word $x_1 x_2 \ldots x_N \in \mathsf{X}^*$, the restriction $g|_{x_1 x_2 \ldots x_N}$ belongs to S. After passing to the Nth power of the action (using Lemma 3.8.4), we may assume that $g|_x \in S$ for every $g \in G$ of length ≤ 4 and $x \in \mathsf{X}$.

Let us prove the following lemma.

LEMMA 3.8.7. *Any two vertices w_1, w_2 of the graph $\Sigma(G, S, \mathsf{X})$ can be written in the form $w_1 = a_1 a_2 \ldots a_n w, w_2 = b_1 b_2 \ldots b_m g(w)$, where $a_i, b_i \in \mathsf{X}, w \in \mathsf{X}^*$, $g \in G$, $l(g) \le 4$ and $d(w_1, w_2) = n + m + l(g)$.*

Then the Gromov product $\langle w_1 \cdot w_2 \rangle$ with respect to the basepoint $\varnothing$ is equal to $|w| - l(g)/2$.

Here $l(g)$ denotes the length of the element g with respect to the generating set S.

PROOF. Let $v_1 = w_1, v_2, \ldots v_k = w_2$ be the consecutive vertices of the shortest path connecting w_1 and w_2. Then every v_{i+1} is obtained from v_i by application of one of the following procedures:

(1) deletion of the first letter $a \in \mathsf{X}$ in v_i (*descending edges*),
(2) appending a letter $a \in \mathsf{X}$ to the beginning of v_i (*ascending edges*),
(3) application of an element of S to v_i (*horizontal edges*).

If the path has three consecutive vertices v_i, v_{i+1}, v_{i+2} such that $v_{i+1} = av_i$, $a \in \mathsf{X}$ and $v_{i+2} = s(v_{i+1})$ for $s \in S$, then $v_{i+2} = bs'(v_i)$, where $b = s(a) \in \mathsf{X}$ and $s' = s|_a \in S$. We replace the segment $\{v_i, v_{i+1}, v_{i+2}\}$ of the path by the segment $\{v_i, s'(v_i), bs'(v_i) = v_{i+2}\}$.

If the path has three consecutive vertices v_i, v_{i+1}, v_{i+2} such that $v_{i+1} = s(v_i)$ for $s \in S$ and $v_{i+1} = av_{i+2}$, then $v_i = s^{-1}(av_{i+2}) = bs'(v_{i+2})$, where $b = s^{-1}(a) \in \mathsf{X}$ and $s' = s^{-1}|_a \in S$. Then we replace the segment $\{v_i, v_{i+1}, v_{i+2}\}$ of the path by the segment $\{v_i = bs'(v_{i+2}), s'(v_{i+2}), v_{i+2}\}$.

Let us perform these replacements as many times as possible. Then we will not change the length of the path, so each time we will get a geodesic path connecting the vertices w_1, w_2. Note that a geodesic path cannot have a descending edge after an ascending one. Therefore, eventually after a finite number of replacements we will get a geodesic path in which we have at first only descending, then horizontal and then only ascending edges. Then $w_1 = a_1 a_2 \ldots a_n w, w_2 = b_1 b_2 \ldots b_m g(w)$, with $a_i, b_i \in \mathsf{X}, w \in \mathsf{X}^*, g \in G$, and $d(w_1, w_2) = n + m + l(g)$.

Suppose that $l(g) > 4$. Let $w = aw'$, $a \in \mathsf{X}$ and denote $b = g(a)$ and $h = g|_a$. Then we have $l(h) \le l(g) - 3$. Since $w_1 = a_1 a_2 \ldots a_n aw'$ and $w_2 = b_1 b_2 \ldots b_m bh(w')$, we have $d(w_1, w_2) \le n + 1 + m + 1 + l(h) \le n + m + l(g) - 1$, which contradicts the fact that the original path was the shortest one.

We have
$$\langle w_1 \cdot w_2 \rangle = \frac{1}{2}\left(n + |w| + m + |w| - (n + m + l(g))\right) = |w| - \frac{l(g)}{2}.$$
□

Let us take three points w_1, w_2, w_3. We can write them by Lemma 3.8.7 as
$$w_1 = a_1 a_2 \ldots a_n w, \qquad w_2 = b_1 b_2 \ldots b_m g_1(w)$$
and
$$w_2 = b_1 b_2 \ldots b_p u, \qquad w_3 = c_1 c_2 \ldots c_q g_2(u),$$
where $a_i, b_i, c_i \in \mathsf{X}$; $g_1, g_2 \in G$; $l(g_1), l(g_2) \le 4$; and
$$\langle w_1 \cdot w_2 \rangle = |w| - l(g_1)/2, \qquad \langle w_2 \cdot w_3 \rangle = |u| - l(g_2)/2.$$

We can assume that $p \leq m$. Then $|u| \leq |w| = |g_1(w)|$, so we can write $u = vg_1(w)$ for some $v \in \mathsf{X}^*$. Then $w_3 = c_1 c_2 \ldots c_q g_2(v) h g_1(w)$, where $h = g_2|_v$. We have $l(h) \leq l(g_2) \leq 4$ and $d(w_1, w_3) \leq n + l(h) + l(g_1) + q + |v|$; hence

$$\langle w_1 \cdot w_3 \rangle = \frac{1}{2}(n + |w| + q + |v| + |w| - d(w_1, w_3)) \geq |w| - (l(h) + l(g_1))/2 \geq |w| - 4.$$

Finally, $\min(\langle w_1 \cdot w_2 \rangle, \langle w_2 \cdot w_3 \rangle) \leq \langle w_1 \cdot w_2 \rangle \leq |w|$, so

$$\langle w_1 \cdot w_3 \rangle \geq \min(\langle w_1 \cdot w_2 \rangle, \langle w_2 \cdot w_3 \rangle) - 4,$$

and the graph $\Sigma(G, S, \mathsf{X})$ is 4-hyperbolic. $\square$

3.8.3. The space $\mathcal{J}_G$ as a hyperbolic boundary. Let $\mathfrak{X}$ be a hyperbolic space. We say that a sequence $\{x_n\}$ of points of $\mathfrak{X}$ *converges to infinity* if the Gromov product $\langle x_n \cdot x_m \rangle$ goes to infinity when $m, n \to \infty$. This definition does not depend on the choice of the basepoint. We say that two sequences $\{x_n\}$ and $\{y_n\}$, convergent to infinity, are *equivalent* if $\lim_{n,m \to \infty} \langle x_n \cdot y_m \rangle = \infty$.

The set of the equivalence classes of the sequences convergent to infinity in the space $\mathfrak{X}$ is called the *hyperbolic boundary* of the space $\mathfrak{X}$ and is denoted $\partial \mathfrak{X}$. If a sequence $\{x_n\}$ converges to infinity, then its *limit* is the equivalence class $a \in \partial \mathfrak{X}$, to which belongs $\{x_n\}$, and we say that $\{x_n\}$ *converges to a*.

If $a, b \in \partial \mathfrak{X}$ are two points of the boundary, then their *Gromov product* is defined as

$$\langle a \cdot b \rangle = \sup_{\{x_n\} \in a, \{y_m\} \in b} \liminf_{m, n \to \infty} \langle x_n \cdot y_m \rangle.$$

For every $r > 0$ define

(3.9) $$V_r = \{(a, b) \in \partial \mathfrak{X} \times \partial \mathfrak{X} : \langle a \cdot b \rangle \geq r\}.$$

Then $\{V_r : r \geq 0\}$ is a fundamental system of entourages of a uniformity on $\partial \mathfrak{X}$ (see [**GH90**] for proofs). We introduce on the boundary $\partial \mathfrak{X}$ the topology defined by this uniform structure.

THEOREM 3.8.8. *The limit space $\mathcal{J}_G$ of a contracting action (G, X) of a finitely generated group G is homeomorphic to the hyperbolic boundary $\partial \Sigma(G, S, \mathsf{X})$ of the self-similarity graph $\Sigma(G, S, \mathsf{X})$. Moreover, there exists a homeomorphism $F : \mathcal{J}_G \longrightarrow \partial \Sigma(G, S, \mathsf{X})$, such that $D = F \circ \pi$, where $\pi : \mathsf{X}^{-\omega} \longrightarrow \mathcal{J}_G$ is the canonical projection and $D : \mathsf{X}^{-\omega} \longrightarrow \partial \Sigma(G, S, \mathsf{X})$ carries every sequence $\ldots x_2 x_1 \in \mathsf{X}^{-\omega}$ to its limit*

$$\lim_{n \to \infty} x_n x_{n-1} \ldots x_1 \in \partial \Sigma(G, S, \mathsf{X}).$$

PROOF. We will need the following well known result (see, for example, Theorem 2.2 from [**CDP90**]).

LEMMA 3.8.9. *Let $\mathfrak{X}_1, \mathfrak{X}_2$ be proper geodesic hyperbolic spaces and let $f_1 : \mathfrak{X}_1 \longrightarrow \mathfrak{X}_2$ be a quasi-isometry. Then a sequence $\{x_n\}$ of points of $\mathfrak{X}_1$ converges to infinity if and only if the sequence $\{f_1(x_n)\}$ does. The map $\partial f_1 : \{x_n\} \mapsto \{f_1(x_n)\}$ defines a homeomorphism $\partial f_1 : \partial \mathfrak{X}_1 \longrightarrow \partial \mathfrak{X}_2$ of the boundaries.* $\square$

We pass, using Lemmata 3.8.9 and 3.8.4, to an Nth power of the self-similar action and change S, if necessary, in the same way as in the proof of Theorem 3.8.6, so that we assume that for every $g \in G$ such that $l(g) \leq 4$ and for every $a \in \mathsf{X}$, the restriction $g|_a$ belongs to the generating set. The nucleus of the action is contained then in the generating set S.

Suppose that the sequence $\{w_n\}$ converges to infinity in $\Sigma(G, S, \mathsf{X})$. Choose its convergent subsequence in $\mathsf{X}^{-\omega} \sqcup \mathsf{X}^*$. Suppose that the limit is $\ldots x_2 x_1 \in \mathsf{X}^{-\omega}$. The Gromov product $\langle w_i \cdot w_j \rangle$ with respect to the basepoint $\varnothing$ is equal to $|w| - l(g)/2 \leq |w|$, where w and g are as in Lemma 3.8.7. It follows that the length of w goes to infinity as $i, j \to \infty$. Consequently, if $\ldots y_2 y_1 \in \mathsf{X}^{-\omega}$ is another accumulation point of $\{w_n\}$, then for every n there exists $g_n \in G$ such that $l(g_n) \leq 4$ and $g_n(x_n \ldots x_1) = y_n \ldots y_1$. Thus, all the accumulation points of $\{w_n\}$ in $\mathsf{X}^{-\omega}$ are asymptotically equivalent to $\ldots x_2 x_1$.

If, on the other hand, $\{w_n\}$ is a sequence convergent to $\ldots x_2 x_1$ in $\mathsf{X}^{-\omega}$, then for every $n \in \mathbb{N}$, if w_i and w_j have a common ending of length $\geq n$, then $\langle w_i \cdot w_j \rangle \geq n$. Hence, every sequence $\{w_n\} \subset \mathsf{X}^*$ convergent in $\mathsf{X}^{-\omega} \sqcup \mathsf{X}^*$ is convergent to infinity in $\Sigma(G, S, \mathsf{X})$, and if two sequences converge to one point of $\mathsf{X}^{-\omega}$, then they converge to one point of the hyperbolic boundary.

Thus, the map $D : \mathsf{X}^{-\omega} \longrightarrow \partial \Sigma(G, S, \mathsf{X})$ is surjective and the map $F : \mathcal{J}_G \longrightarrow \partial \Sigma(G, S, \mathsf{X})$, satisfying the conditions of the theorem, is uniquely defined.

Let $A = \{g \in G : l(g) \leq 4\}$ and for every $n \in \mathbb{N}$ define
$$U_n = \{(w_1 v, w_2 s(v)) : w_1, w_2 \in \mathsf{X}^{-\omega}, v \in \mathsf{X}^n, s \in A\} \subset \mathsf{X}^{-\omega} \times \mathsf{X}^{-\omega}.$$

By $\tilde{U}_n$ we denote the image of U_n in $\mathcal{J}_G \times \mathcal{J}_G$. We know that $g|_x$ belongs to the nucleus for all $x \in \mathsf{X}$ whenever $l(g) \leq 4$. Therefore
$$\Delta_n \subseteq \tilde{U}_n \subseteq \Delta_{n-1},$$
where Δ_n are the images in $\mathcal{J}_G \times \mathcal{J}_G$ of the entourages given in Definition 3.4.4 (see the first equality in the proof of Proposition 3.4.5 on page 84). Consequently, the fundamental system of entourages $\tilde{U}_n$ defines the topology of $\mathcal{J}_G$.

On the other hand, Lemma 3.8.7 implies
$$V_n \subseteq D \times D(U_n) \subseteq V_{n-2},$$
where V_n are as in (3.9). Hence the map F is a homeomorphism. $\square$

As an example, consider the adding machine action. It is not hard to prove that its self-similarity graph (shown in Figure 3.1) is quasi-isometric to the hyperbolic plane $\mathbb{H}$, whose boundary is homeomorphic to the circle.

3.9. Groups of bounded automata

3.9.1. Tiles with finite boundary and bounded automata.
Ideas of this section are close to the paper [**Sid00**] by S. Sidki. The central idea of [**Sid00**] is to stratify the group of finite automata using the cyclic structure of the automata. We will try to do this using the topological dimension of the limit space. The groups with zero dimensional limit space are exactly the subgroups of the group of finitary automorphisms of X^* (see below). Hence, the finitary groups are "rank 0" in our classification.

The next step is the one-dimensional limit space. It seems, however, that more important is the dimension of the boundary of the tiles. In this line, the next "rank 1" step should be the case when the boundary is finite. This condition does not correspond exactly to the case when the limit space has dimension 1 (there are one-dimensional limit spaces with infinite boundary of the tile), but it has many important group-theoretical and dynamical implications.

Note that self-similar spaces (fractals) having finite boundaries of "tiles" (i.e., of the parts similar to the whole fractal) were studied by different authors from the point of view of harmonic analysis and Brownian motion on fractals. Such classes of fractals where called *post-critically finite self-similar sets*, *nested fractals* or *finitely ramified fractals*. See the papers [**Kig01, Kig92, Lin90, Sab97**] for properties of such self-similar sets.

We will see that the tiles of a contracting self-similar group are finite; i.e., the limit space is "post-critically finite" if and only if the group is generated by *bounded automata* in the sense of S. Sidki. This shows an interesting relation between two notions which appeared totally independently in different parts of Mathematics: Analysis on Fractals and Automata Groups.

Similar to the finite boundary of tiles being an important condition, making possible, for example, the study of the Brownian motion on the fractal, the class of groups generated by bounded automata is the most studied and most convenient class of self-similar groups.

Most of the examples mentioned in Chapter 1 belong to this class. In particular are the Grigorchuk group [**Gri80**], the adding machine action of $\mathbb{Z}$ and all the examples of Section 1.8. These particular examples where generalized to different classes of groups acting on rooted trees: branch groups [**Gri00**], GGS-groups [**Bau93**], AT groups [**Mer83, Roz96**], and spinal groups [**BGŠ03**]. All groups belonging to these classes have finite boundary of the tiles (if they are finite-state and self-similar). Also the groups constructed by V. Sushchansky [**Sus79**] are generated by bounded automata, though they are not self-similar.

3.9.2. Activity growth of automata. We remember here some results of Said Sidki and show their relation to the properties of the limit spaces.

The activity growth of automata is only one of the possible notions of growth of automata. For another notion see, for instance, the paper [**Gri88**].

Let us denote by $\alpha(k,q)$ the number of the words $v \in \mathsf{X}^k$ such that $q|_v \neq 1$, where q is an automorphism of the tree X^*.

S. Sidki used in [**Sid00**] the function $\theta(k,q)$ equal to the number of the words $v \in \mathsf{X}^k$ such that $q|_v$ is *active*, i.e., acts non-trivially on X^1, but our approach is equivalent to his.

Suppose that (A, X) is a finite automaton. Denote by A' the set of non-trivial states of A. Consider the vector space $\mathbb{R}^{\mathsf{A}'}$. Then the *adjacency matrix* of A' is the matrix of the linear operator A given by

$$A(q) = \sum_{x \in \mathsf{X}} \pi(q,x),$$

where $\pi(q,x)$ is equal to $q|_x$ if $q|_x \neq 1$ and zero otherwise. Let the linear functional $I : \mathbb{R}^{\mathsf{A}'} \longrightarrow \mathbb{R}$ be given on the basis A' by $I(q) = 1$; i.e., $I(v)$ is the sum of the coordinates of the vector $v \in \mathbb{R}^{\mathsf{A}'}$.

It follows that

$$\alpha(k,q) = I\left(A^k(q)\right).$$

Hence, the generating function of the sequence $\alpha(k,q)$ is the rational function

(3.10) $$B_q(t) = \sum_{k=0}^{\infty} \alpha(k,q) t^k = I\left(\sum_{k=0}^{\infty} t^k A^k(q)\right) = I\left((1-tA)^{-1}(q)\right).$$

The general facts about rational generating functions imply that the limit
$$\alpha(q) = \lim_{k\to\infty} \sqrt[k]{\alpha(k,q)}$$
exists and is equal to one of the positive eigenvalues of A (note that $\alpha(k,q)$ is monotone on k).

If $\alpha(q) = 1$, then $\alpha(k,q)$ has a polynomial growth of some degree $n(q) \in \mathbb{N}$. If $n(q) = 0$, then $\alpha(k,q)$ is periodic and thus bounded.

It is straightforward that

(3.11) $\qquad \alpha(k, q_1 q_2) \leq \alpha(k, q_1) + \alpha(k, q_2), \quad \text{and} \quad \alpha(k, q^{-1}) = \alpha(k, q).$

It follows that the set $\mathcal{B}_n(\mathsf{X}) = \mathcal{B}_n$ of finite-state automorphisms q of X^* such that $\alpha(k,q)$ is bounded by a polynomial of degree $\leq n$ is a group.

Similarly, for every $a > 1$ the set of finite-state automorphisms $q \in \operatorname{Aut} \mathsf{X}^*$ such that $\alpha(q) \leq a$ (or $\alpha(q) < a$) is a group, since (3.11) implies that

$$\alpha(q_1 q_2) \leq \max(\alpha(q_1), \alpha(q_2)), \quad \text{and} \quad \alpha(q^{-1}) = \alpha(q).$$

DEFINITION 3.9.1. The group $\mathcal{B}_0 = \mathcal{B}_0(\mathsf{X})$ is the *group of bounded automata*.

A finite-state automorphism q of the tree X^* is bounded if and only if the number of the words $v \in \mathsf{X}^n$ such that $q|_v \neq 1$ is uniformly bounded. In this case the activity of q is concentrated around a finite number of "directions" of the tree X^*.

An automatic transformation q of X^* is said to be *finitary* (see [**GNS00**]) if there exists $n \in \mathbb{N}$ such that $q|_v = 1$ for all $v \in \mathsf{X}^n$ (i.e., if q changes at most the first n letters of every word). The minimal number n is called the *depth* of q.

In other words, q is finitary if and only if $\alpha(k,q)$ is equal to zero for all k big enough. It follows from (3.11) that the set of all finitary automatic transformations of X^* is a locally finite group.

3.9.3. Growth of the nucleus. Consider now a contracting group $G \leq \operatorname{Aut} \mathsf{X}^*$. Let $\mathcal{N}$ be its nucleus and let $\mathcal{N}' = \mathcal{N} \setminus \{1\}$.

The growth of $\alpha(k,q)$ for $q \in \mathcal{N}$ determines the growth of $\alpha(k,g)$ for all $g \in G$.

LEMMA 3.9.2. *For every $g \in G$ there exist coefficients $c_s \in \mathbb{N}$, where $s \in \mathcal{N}'$, and a number $n_0 \in \mathbb{N}$ such that*

$$\alpha(k,g) = \sum_{s \in \mathcal{N}'} c_s \cdot \alpha(k - n_0, s)$$

for all $k \geq n_0$.

PROOF. Take $n_0 \in \mathbb{N}$ such that $g|_v \in \mathcal{N}$ for every $v \in \mathsf{X}^{n_0}$. Then let c_s be the number of the words $v \in \mathsf{X}^{n_0}$ such that $g|_v = s$. Then the equality from the lemma obviously holds. $\square$

COROLLARY 3.9.3. *(1) If $\mathcal{N} \subset \mathcal{B}_n$, then $G \leq \mathcal{B}_n$.*
(2) $\alpha(g) \leq \max_{s \in \mathcal{N}} \alpha(s)$ for every $g \in G$. $\square$

Inequality (2) of the corollary can also be written as equality

$$\max_{g \in G} \alpha(g) = \max_{s \in \mathcal{N}} \alpha(s).$$

Let us denote $\alpha(G) = \max_{g \in G} \alpha(g)$.

PROPOSITION 3.9.4. *Let A be the adjacency matrix of $\mathcal{N}' = \mathcal{N} \setminus \{1\}$. Then $\alpha(G)$ is equal to the principal eigenvalue of A.*

If $\alpha(G) = 1$ and n is the multiplicity of the eigenvalue 1 of the matrix A, then $G \leq \mathcal{B}_{n-1}$ and $G \not\leq \mathcal{B}_{n-2}$.

PROOF. We have
$$\alpha(A) = \lim_{k \to \infty} \sqrt[k]{\sum_{s \in \mathcal{N}'} \alpha(k, s)}$$
and, by (3.10):
$$\sum_{k=0}^{\infty} t^k \sum_{s \in \mathcal{N}'} \alpha(k, s) = I\left(\sum_{k=0}^{\infty} t^k A^k \left(\sum_{s \in \mathcal{N}'} s\right)\right) = I\left((1 - tA)^{-1} \left(\sum_{s \in \mathcal{N}'} s\right)\right),$$
where I is the linear functional mapping a vector of $\mathbb{R}^{\mathcal{N}'}$ to the sum of its coordinates. The statement of the proposition follows now from the Perron-Frobenius theorem (see, for instance [**Gan59**], Chapter XIII) and the theory of rational generating functions. □

If we pass to the nth power of the action, then the number $\alpha(G)$ is changed to $\alpha(G)^n$. Hence the number
$$h(G) = \frac{\log \alpha(G)}{\log |\mathsf{X}|}$$
does not change after passing to tensor powers of actions. Note also that $\alpha(k, g) \leq |\mathsf{X}|^k$ for all g and k; therefore $\alpha(G) \leq |\mathsf{X}|$, i.e., $h(G) \leq 1$.

PROPOSITION 3.9.5. *If the action satisfies the open set condition, then $h(G) < 1$.*

PROOF. We know that $\alpha(G) = \max_{s \in \mathcal{N}} \alpha(s)$. If the action satisfies the open set condition, then we can find a word $v \in \mathsf{X}^*$ such that $s|_v = 1$ for all $s \in \mathcal{N}$ (see the proof of Proposition 3.3.7). Then if $w \in \mathsf{X}^*$ is any word containing v as a subword, $s|_w = 1$ for all $s \in \mathcal{N}$. Hence, $\alpha(k, s)$ is not greater than the number $p(k, w)$ of words of length k which do not contain the word v. But it is well known (see, for example, Chapter XIII, §3 of [**Gan59**]) that this implies that $\lim_{n \to \infty} \sqrt[n]{p(k, s)} < |X|$. □

3.9.4. Theorem of S. Sidki. The following result shows an interesting connection between the activity growth of automata and properties of the groups that they generate.

THEOREM 3.9.6 (S. Sidki). *The group $\mathcal{B}_n(\mathsf{X})$ has no free subgroups for any finite alphabet X.*

For the proof (of a more general theorem also covering infinite alphabets) see [**Sid04a**].

For $n = 0$ the theorem of S. Sidki is a partial case of a theorem proved by L. Bartholdi, V. Kaimanovich, B. Virag and the author, which states that the group $\mathcal{B}_0$ is amenable (see Theorem 6.12.2). It is an interesting open question whether the groups $\mathcal{B}_n$ are also amenable.

3.9.5. Relation with the boundary of $\mathcal{T}$. The number $h(G)$ measures the size of the boundary of the tile, as the following proposition shows.

PROPOSITION 3.9.7. *Suppose that the action satisfies the open set condition. Let b_k be the number of words $v \in \mathsf{X}^k$ such that $\mathcal{T} \otimes v$ intersects the boundary of $\mathcal{T}$. Then*
$$\lim_{k \to \infty} \sqrt[k]{b_k} = \alpha(G),$$
and the sequence b_k is bounded if and only if $G \le \mathcal{B}_0$.

PROOF. Denote by $\tilde{b}_k$ the number of pairs $(v, q) \in \mathsf{X}^k \times \mathcal{N}'$ such that the tile $\mathcal{T} \otimes v$ intersects the tile $\mathcal{T} \cdot q$.

Proposition 3.3.5 implies that
$$\tilde{b}_k = \sum_{q \in \mathcal{N}'} \alpha(k, q). \tag{3.12}$$

Consequently, the generating function $\sum_{k=0}^{\infty} \tilde{b}_k t^k$ is rational and
$$\lim_{k \to \infty} \sqrt[k]{\tilde{b}_k} = \max_{q \in \mathcal{N}} \alpha(q).$$

It follows from Propositions 3.3.7 and 3.3.5 that
$$|\mathcal{N}|^{-1} \tilde{b}_k \le b_k \le \tilde{b}_k,$$
which finishes the proof. $\square$

COROLLARY 3.9.8. *Suppose that the action (G, X) is contracting and satisfies the open set condition. Then the tile $\mathcal{T}$ has finite boundary if and only if G is a subgroup of the group of bounded automata $\mathcal{B}_0$.*

PROOF. If the boundary of $\mathcal{T}$ has b points, then the sequence b_k in Proposition 3.9.7 is bounded by $b \cdot |\mathcal{N}|$, since every point belongs to not more than $|\mathcal{N}|$ tiles. Consequently, $\alpha(k, q)$ is also bounded for every $q \in \mathcal{N}$, due to (3.12). Thus $\mathcal{N} \subset \mathcal{B}_0$, which implies $G \le \mathcal{B}_0$, by Corollary 3.9.3.

If $G \le \mathcal{B}_0$, then by Proposition 3.9.7, the number of the tiles of the kth level, intersecting the boundary of $\mathcal{T}$ is uniformly bounded. But this is possible only when the boundary is finite. $\square$

3.9.6. Structure of bounded automata. We say that an automatic transformation q is *bounded* if it belongs to $\mathcal{B}_0$, i.e., if it is defined by a bounded automaton. The following is proved in [**Sid00**] (Corollary 14).

PROPOSITION 3.9.9. *An automatic transformation is bounded if and only if it is defined by a finite automaton in whose Moore diagram every two non-trivial cycles are disjoint and are not connected by a directed path.*

Here a cycle is called *trivial* if it has only one vertex, which is the trivial state. In particular, every finitary transformation is bounded, since it has no non-trivial cycles.

DEFINITION 3.9.10. A finitely automatic automorphism $g \in \operatorname{Aut} \mathsf{X}^*$ is said to be *directed* if there exists a periodic sequence $w = u^{\omega}$ such that $g|_v$ is not finitary if and only if the word v is a beginning of the sequence w.

3.9. GROUPS OF BOUNDED AUTOMATA

PROPOSITION 3.9.11. *An automorphism g of the tree X^* is bounded if and only if, after passing to a power X^n of the alphabet, it is written*

$$g = \pi(g_1, \ldots, g_{d^n}),$$

where each g_i is either finitary or directed. Moreover, we may assume that the directed restrictions $g_i = h$ satisfy a recursion

$$h = \sigma(h_1, \ldots, h_{d^n}),$$

where all h_i are finitary, except for one which is equal to h.

PROOF. The "if" statement is obvious.

Consider an arbitrary element $g \in \mathcal{B}_0$. If g is not finitary, then there exists an infinite word $x_1 x_2 \ldots$ such that all restrictions $g|_{x_1 \ldots x_n}$ are non-trivial. Then two of these restrictions, say $g|_{x_1 \ldots x_n}$ and $g|_{x_1 \ldots x_m}$, $m > n$, are equal. Thus the restriction $h = g|_{x_1 \ldots x_n}$ belongs to a cycle; i.e., there exists $v \in \mathsf{X}^*$ such that $h|_v = h$ (in our case $v = x_{n+1} \ldots x_m$).

This proves that for every element $g \in \mathcal{B}_0$ there exists $n \in \mathbb{N}$ such that every restriction $g|_v$ for every $v \in \mathsf{X}^n$ either is finitary or belongs to a cycle.

Suppose now that $g \in \mathcal{B}_0$ belongs to a cycle., i.e., that there exists a non-empty word $v = x_1 x_2 \ldots x_n \in \mathsf{X}^*$ such that $g|_v = g$. Then for every word $v_1 \in \mathsf{X}^n$ different from v, the restriction $g|_{v_1}$ is finitary. Otherwise we get either two intersecting cycles or two cycles connected by a directed path, which contradicts Proposition 3.9.9.

Hence, if we pass to the alphabet X^n, then g is given by a recursion

$$g = \pi(g_1, g_2, \ldots, g_{d^n}),$$

where $g_i = g$ for one of the indices and g_j is finitary for all $j \neq i$. □

The following is a joint result with E. Bondarenko (see [**BN03**]).

THEOREM 3.9.12. *Every self-similar finitely generated subgroup G of $\mathcal{B}_0$ is contracting. Its nucleus is equal to the set of restrictions of the elements belonging to the cycles of the Moore diagram of (G, X), i.e., to the set of elements $g \in G$ for which there exist $h \in G$ and $v, u \in \mathsf{X}^*, v \neq \varnothing$ such that $h|_v = h$ and $h|_u = g$.*

PROOF. Let S be a finite automaton generating G, i.e., a finite generating set such that $s|_x \in S$ for all $s \in S$ and $x \in \mathsf{X}$. Then the non-trivial cycles of S are disjoint. Let n_1 be a common multiple of their lengths and let $S_1 \subset S$ be their union. We can make n_1 sufficiently big, replacing it by its multiple, so that for every $s \in S$ and every $v \in \mathsf{X}^{n_1}$ the restriction $s|_v$ is either finitary or belongs to S_1. We may also assume that n_1 is bigger than the depth of every finitary element of S. Then for every finitary $s \in S$ and $v \in \mathsf{X}^{n_1}$ the restriction $s|_{v_1}$ is trivial.

If we choose n_1 so that it satisfies all the above conditions, then for every $s \in S$ and every $v \in \mathsf{X}^{n_1}$ the restriction $s|_v$ is either finitary or belongs to S_1. In the first case $s|_{vu} = 1$ for all $u \in \mathsf{X}^{n_1}$. In the second case $s|_{vu_1} = s|_v$ for a unique word $u_1 \in \mathsf{X}^{n_1}$ and $s|_{vu}$ is finitary for any $u \in \mathsf{X}^{n_1}, u \neq u_1$. Hence $s|_{vuw} = 1$ for all $u, w \in \mathsf{X}^{n_1}, u \neq u_1$.

Denote by S_0 the set of the finitary elements of S. Let $\mathcal{N}_1$ be the set of all the elements $h \in G \setminus 1$ such that there exists a unique word $u(h) \in \mathsf{X}^{n_1}$ such that $h|_{u(h)} = h$ and for all the words $u \in \mathsf{X}^{n_1}$ not equal to $u(h)$ the restriction $h|_u$ belongs to $\langle S_0 \rangle$. It is easy to see that the set $\mathcal{N}_1$ is finite (every element h of $\mathcal{N}_1$ is uniquely defined by the permutation it induces on X^{n_1} and by the restrictions $h|_u$ for $u \in \mathsf{X}^{n_1}$; note also that the group $\langle S_0 \rangle$ is finite).

Let us denote by $l_1(g)$ the minimal number of the elements of $S_1 \cup S_1^{-1}$ in a decomposition of g into a product of the elements of $S \cup S^{-1}$.

Let us prove that there exists for every $g \in G$ a number k such that for every $v \in \mathsf{X}^{n_1 k}$ the restriction $g|_v$ belongs to $\mathcal{N}_1 \cup \langle S_0 \rangle$. We will prove this by induction on $l_1(g)$.

If $l_1(g) = 1$, then $g = h_1 s h_2$, where $h_1, h_2 \in \langle S_0 \rangle$ and $s \in S_1$. The elements h_1, h_2 are finitary; thus there exists k such that for every $v \in \mathsf{X}^{n_1 k}$ the restriction $h_i|_v$ is trivial. Then we have $h_1 s h_2|_v = s|_{h_2(v)}$; thus $g|_v$ is either equal to $s \in \mathcal{N}_1$ or belongs to $S_0 \cup S_0^{-1}$. Thus the claim is proved for the case $l_1(g) = 1$.

Suppose that the claim is proved for all the elements $g \in G$ such that $l_1(g) < m$. Let $g = s_1 s_2 \ldots s_k$, where $s_i \in S \cup S^{-1}$. For every $u \in \mathsf{X}^{n_1}$ the restriction $s_i|_u$ is either equal to s_i or belongs to S_0. Consequently, either $g|_u = g$ for one u and $g|_v \in \langle S_0 \rangle$ for all $v \in \mathsf{X}^{n_1} \setminus \{u\}$ or $l_1(g|_u) < l_1(g)$ for every $u \in \mathsf{X}^{n_1}$. In the first case we have $g \in \mathcal{N}_1$, and in the second one we apply the induction hypothesis, and the claim is proved.

The set $\mathcal{N}_1$ obviously belongs to the nucleus. Consequently, the group G is contracting with the nucleus equal to $\{g|_v \; : \; g \in \mathcal{N}_1, v \in \mathsf{X}^*, |v| < n_1\}$, which is equal to the set of the restrictions of the states belonging to the cycles of (G, X). □

3.9.7. Connectedness of the tiles.

PROPOSITION 3.9.13. *Let G be a finitely generated recurrent subgroup of $\mathcal{B}_0(\mathsf{X})$. Then there exists $n \in \mathbb{N}$ and a basis Y of $\mathfrak{M}^{\otimes n}$ such that the self-similar action (G, Y) is a subgroup of $\mathcal{B}_0(\mathsf{Y})$ and the tile $\mathcal{T}(\mathsf{Y})$ is connected.*

PROOF. Let $\mathcal{N}$ be the nucleus of the action (G, X). Recall that it consists of all cycles of the complete automaton of (G, X) and their restrictions.

Let us take n such that it is a multiple of the length of every cycle of the nucleus $\mathcal{N}$ and greater than the depth of every finitary element of $\mathcal{N}$.

Let us pass to the nth tensor power of the action, i.e., to the action of (G, X^n). Then $\mathcal{N}$ is also the nucleus of the nth tensor power, and every element $g \in \mathcal{N}$ is either *rooted* (i.e., changes at most the first letter $x \in \mathsf{X}^n$ of a word xv) or there exists a unique $x \in \mathsf{X}^n$ such that $g|_x = g$, whereas for any $y \in \mathsf{X}^n, y \neq x$, the restriction $g|_y$ is rooted. Let $\mathcal{N}_0$ be the set of the rooted elements of $\mathcal{N}$ and let $\mathcal{N}_1 = \mathcal{N} \setminus \mathcal{N}_0$.

The tile $\mathcal{T}(\mathsf{X}^n)$ coincides with the tile $\mathcal{T}(\mathsf{X})$ and is connected if and only if the graph T_n is connected, where T_n is as in Proposition 3.3.10. Recall that T_n is the graph with the set of vertices X^n in which the vertices $v, u \in \mathsf{X}^n$ are connected by an edge in T_n if and only if there exists $g \in \mathcal{N}$ such that $g \cdot v = y \cdot 1$.

If $g \in \mathcal{N}_0$ is rooted, then $g \cdot v = g(v) \cdot 1$ for any $v \in \mathsf{X}^n$. Therefore, if $G_0 = \langle \mathcal{N}_0 \rangle$ is transitive on X^n, then the tile $\mathcal{T}(\mathsf{X}^n) = \mathcal{T}(\mathsf{X})$ is connected and there is nothing to prove.

Suppose that G_0 is not transitive on X^n. Let T_n^0 be the subgraph of T_n in which two vertices $v, u \in \mathsf{X}^n$ are connected by an edge if and only if there exists $s \in \mathcal{N}_0$ such that $s(v) = u$. Note that connected components of T_n^0 are exactly the G_0-orbits of X^n.

LEMMA 3.9.14. *Let R_n be the graph with the set of vertices X^n where two vertices $u, v \in \mathsf{X}^n$ are connected by an edge if and only if there exists $g \in \mathcal{N}$ such that $g \cdot u = v \cdot h$ with rooted h. Then R_n is connected.*

3.9. GROUPS OF BOUNDED AUTOMATA

PROOF. The action of $G = \langle \mathcal{N} \rangle$ is transitive on X^n; therefore the *Schreier graph* of the action is connected. The Schreier graph is the graph with the set of vertices X^n where two vertices $v, u \in \mathsf{X}^n$ are connected by an edge if and only if there exists $s \in \mathcal{N}$ such that $s(v) = u$. The graph R_n is obviously a subgraph of the Schreier graph.

For every $g \in \mathcal{N}$ there exists at most one word $v \in \mathsf{X}^n$ such that the restriction $g|_v$ is not rooted. Therefore, in every cycle of the action of g on X^n at most one respective edge of the Schreier graph is absent in R_n. Hence, R_n is also connected. $\square$

Choose a spanning forest T' of T_n^0. The graph T_n^0 and its spanning forest T' are subgraphs of R_n. Hence we can extend T' to a spanning tree R' of R_n. For every edge $(v, u) \in R' \setminus T'$ there exists $s(v, u) = s \in \mathcal{N}$ such that $s \cdot v = u \cdot h$ and the restriction h is rooted. Let us choose such an element $s(v, u)$ for every $(v, u) \in R' \setminus T'$ and denote $h(v, u) = h = s(v, u)|_v$. Note that if $s \cdot v = u \cdot h$, then $s^{-1} \cdot u = v \cdot h^{-1}$. We may thus assume that $s(v, u) = s(u, v)^{-1}$ and $h(v, u) = h(u, v)^{-1}$.

Choose an element $v_0 \in \mathsf{X}^n$. For every $v \in \mathsf{X}^n$ there exists a unique simple path $e_1 e_2 \ldots e_k$ in R' starting in v and ending in v_0. Denote $\gamma(v) = h(e_1) h(e_2) \cdots h(e_k)$, where $h(e_i)$ is defined above for $e_i \in R' \setminus T'$ and is the identity for $e_i \in T'$. Then $\gamma(v)$ is rooted for every $v \in \mathsf{X}^n$. It is also easy to see that $\gamma(v)$ is constant on components of T', i.e., on the G_0-orbits of X^n.

Let us change the basis X^n of $\mathfrak{M}^{\otimes n}$ to $\mathsf{Y} = \{v \cdot \gamma(v) : v \in \mathsf{X}^n\}$. Let $T_1(\mathsf{Y})$ be the graph with the set of vertices Y, where two vertices y_1, y_2 are connected by an edge if and only if there exists an element g of the nucleus of the action (G, Y) such that $g \cdot y_1 = y_2 \cdot 1$. We have to prove that the graph $T_1(\mathsf{Y})$ is connected.

If $g \in G_0$, $v \cdot \gamma(v) \in \mathsf{Y}$, then $g \cdot v = u \cdot 1$ for some $u \in \mathsf{X}^n$ and $\gamma(u) = \gamma(v)$, since u and v belong to the same G_0-orbit. Therefore,

$$g \cdot v \cdot \gamma(v) = u \cdot \gamma(v) = u \cdot \gamma(u);$$

i.e., every element $g \in \mathcal{N}_0$ acts as a rooted automorphism on Y^*. Moreover, the bijection $v \mapsto v \cdot \gamma(v)$ conjugates the action of g on X^n to the action of g on Y. In particular G_0-orbits are connected subgraphs of $T_1(\mathsf{Y})$.

If $g \in \mathcal{N}_1$, then there exists a unique $v_g \in \mathsf{X}^n$ such that $g \cdot v_g = u \cdot g$ for some $u \in \mathsf{X}^n$. If $v_g \neq v \in \mathsf{X}^n$, then $g \cdot v = w \cdot h$ for some $w \in \mathsf{X}^n$, where h is rooted.

Consider $g' = \gamma(u)^{-1} g \gamma(v_g)$ and denote $v'_g = \gamma(v_g)^{-1} \cdot v_g$. Then v'_g and v_g belong to the same G_0-orbit; hence $\gamma(v_g) = \gamma(v'_g)$. Let $u' = \gamma(u)^{-1} \cdot u$. Then $u' \in \mathsf{X}^n$ belongs to the same G_0-orbit with u; hence $\gamma(u') = \gamma(u)$. We have therefore

$$g' \cdot v'_g \cdot \gamma(v'_g) = \gamma(u)^{-1} g \cdot v_g \cdot \gamma(v_g) = \gamma(u)^{-1} \cdot u \cdot g\gamma(v_g)$$
$$= u' \cdot g\gamma(v_g) = u' \cdot \gamma(u') \cdot \gamma(u)^{-1} g\gamma(v_g) = u' \cdot \gamma(u') \cdot g'.$$

If $v' \in \mathsf{X}^n \setminus \{v'_g\}$, then the word $v = \gamma(v_g) \cdot v'$ is different from v_g and

$$g' \cdot v' \cdot \gamma(v') = \gamma(u)^{-1} g \gamma(v_g) \cdot v' \cdot \gamma(v') = \gamma(u)^{-1} g \cdot v \cdot \gamma(v')$$
$$= \gamma(u)^{-1} \cdot w \cdot h\gamma(v') = w' \cdot h\gamma(v') = w' \cdot \gamma(w') \cdot \gamma(w)^{-1} h\gamma(v),$$

where $g \cdot v = w \cdot h$, $w' = \gamma(u)^{-1} \cdot w$. We have $h, \gamma(w), \gamma(v) \in \mathcal{N}_0$; hence $\gamma(w)^{-1} h \gamma(v')$ is a rooted automorphism of Y^*.

We have proved that g' belongs to the nucleus of the action of G on Y^*.

Now take an edge in $e = (v, w) \in R' \setminus T'$. Let $s = s(e) \in \mathcal{N}_1$ and $h = h(e) = s|_v \in \mathcal{N}_0$ be the respective elements (then $s \cdot v = w \cdot h$) and consider the element $s' = \gamma(u)^{-1} s \gamma(v_s)$, where v_s and u are such that $s \cdot v_s = u \cdot s$. Then for the vertex $v' = \gamma(v_s)^{-1} \cdot v$ by the above calculations, we have

$$s' \cdot v' \cdot \gamma(v') = w' \cdot \gamma(w') \cdot \gamma(w)^{-1} h \gamma(v),$$

where $w' = \gamma(u)^{-1} \cdot w$. But $\gamma(w) = h\gamma(v)$, by definition of γ; hence $\gamma(w)^{-1} h \gamma(v) = 1$ and s' gives us an edge from v' to w' in $T_1(\mathsf{Y})$. Hence, the G_0-orbit $G_0(v') = G_0(v)$ is connected in $T_1(\mathsf{Y})$ to the G_0-orbit $G_0(w') = G_0(w)$. Now connectedness of R' implies connectedness of $T_1(\mathsf{Y})$. □

3.10. One-dimensional subdivision rules

3.10.1. Tile diagrams. We present here an iterative algorithm which can be used to produce approximations of the limit space of a group generated by a bounded automaton.

Let G be a self-similar finitely generated group of bounded automata and let $\mathcal{N}$ be its nucleus. Then the Moore diagram of the set of non-finitary elements of $\mathcal{N}$ is a disjoint union of simple cycles. We assume also that the tile $\mathcal{T}$ of the group is connected (see Proposition 3.9.13).

A sequence $\ldots x_2 x_1 \in \mathsf{X}^{-\omega}$ represents a point of $\partial \mathcal{T}$ if and only if there exists a left-infinite directed edge-path $\ldots e_2 e_1$ in the Moore diagram of $\mathcal{N}$ labeled by labels $\ldots (x_2, y_2)(x_1, y_1)$ and ending in a non-trivial state (see Corollary 3.3.8). Such a path is obviously (pre-)periodic, i.e., is of the form $(e_m \ldots e_{n+1})^{-\omega} e_n \ldots e_1$, where $e_m \ldots e_{n+1}$ is one of the cycles of the nucleus. Thus there is only a finite number of them, and it is easy to find all such paths.

Two sequences $\ldots x_2 x_1$ and $\ldots y_2 y_1$ represent the same point of $\mathcal{T}$ if and only if there exists a path $\ldots e_2 e_1$ in $\mathcal{N}$ labeled by $\ldots (x_2, y_2)(x_1, y_1)$ and ending in the trivial state (see Proposition 3.3.2).

Thus, we can effectively find all the points of the boundary $\partial \mathcal{T}$, finding all the sequences encoding them and finding out which sequences encode the same points.

DEFINITION 3.10.1. A *tile diagram* of an action (G, X) is a compact connected topological space Γ together with a bijective correspondence between $\partial \mathcal{T}$ and a set of *marked points* of Γ.

If Γ is a tile diagram, then its *inflation* $\Gamma \cdot \mathsf{X}$ is the tile diagram obtained by the following procedure:

(1) Take $|\mathsf{X}|$ copies $\Gamma \cdot x$ of Γ. Here $x \in \mathsf{X}$ is a label, and if v is a point of Γ, then $v \cdot x$ is the corresponding point of $\Gamma \cdot x$.
(2) Identify two points $v_1 \cdot x_1$ and $v_2 \cdot x_2$ if and only if v_1 and v_2 are marked in Γ and the corresponding points $\xi_1, \xi_2 \in \partial \mathcal{T}$ are such that $\xi_1 \otimes x_1 = \xi_2 \otimes x_2$ in $\mathcal{X}_G$.
(3) A point $v \cdot x$ is marked and corresponds to a point $\zeta \in \partial \mathcal{T}$ if and only if v is marked in Γ and $\zeta = \xi \otimes x \in \partial \mathcal{T}$, where $\xi \in \partial \mathcal{T}$ is the point corresponding to v.

The condition of connectedness of the tile $\mathcal{T}$ ensures that the inflation of a tile diagram is again connected.

The inflation $\Gamma \cdot \mathsf{X}$ can be easily found knowing the nucleus and Proposition 3.2.6.

3.10. ONE-DIMENSIONAL SUBDIVISION RULES

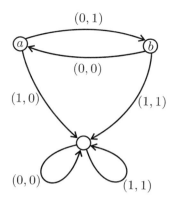

FIGURE 3.2. The automaton generating IMG $(z^2 - 1)$

We denote by $\Gamma \cdot \mathsf{X}^n$ the nth iteration of the inflation. The space $\Gamma \cdot \mathsf{X}^n$ consists of $|\mathsf{X}|^n$ pieces $\Gamma \cdot v$, $v \in \mathsf{X}^n$, glued together using the adjacency rule of the tiles, described above.

If we rescale the spaces $\Gamma \cdot \mathsf{X}^n$ so that the pieces $\Gamma \cdot v$ become small, then the space $\Gamma \cdot \mathsf{X}^n$ will be a good approximation of the tile $\mathcal{T}$. The original shape of Γ is irrelevant.

A trivial example of a tile diagram is the tile $\mathcal{T}$ itself with the identical bijection between the boundary and the set of the marked points. But the notion of inflation of a tile diagram is purely combinatorial (unlike the tile, which may have complicated topology).

We may, for example, consider tile diagrams, which are *graphs* such that the marked points are vertices. It is easy to see that inflation of a graph will again be a graph.

3.10.2. Examples.

Basilica graphs. Consider the group generated by the automaton shown in Figure 3.2. We will see later that it is the *iterated monodromy group* IMG $(z^2 - 1)$ of the polynomial $z^2 - 1$.

The nucleus of IMG $(z^2 - 1)$ is equal, by Theorem 3.9.12, to $\{1, a, b, a^{-1}, b^{-1}\}$. The left-infinite paths ending in a non-trivial state are the paths going along the loop between the states a and b or along the loop between the states a^{-1} and b^{-1} (see Figure 3.2).

Looking at the labels along these paths, we see that the boundary of the tile $\mathcal{T}$ is $\{0^{-\omega}, (01)^{-\omega}, (10)^{-\omega}\}$. All these points are identified with one point in $\mathcal{J}_G$ but are different points of $\mathcal{T}$. We also have the following identifications of the points of $\mathcal{T}$:

$$0^{-\omega}1 \sim (10)^{-\omega}0 \sim (01)^{-\omega}1.$$

Consequently, if A, B and C are the marked points of a tile diagram Γ corresponding to the points $0^{-\omega}, (01)^{-\omega}$ and $(10)^{-\omega}$, respectively, then in the inflated diagram $\Gamma \cdot \mathsf{X}$ the point $A \cdot 0$ becomes A; $B \cdot 0$ becomes C; $C \cdot 1$ becomes B; and the points $A \cdot 1$, $B \cdot 1$ and $C \cdot 0$ are glued together and are not marked.

Let us start from the graph Γ shown on the left-hand side of Figure 3.3. The right-hand side of the figure shows the graph $\Gamma \cdot \mathsf{X}^6$. If we now identify the points

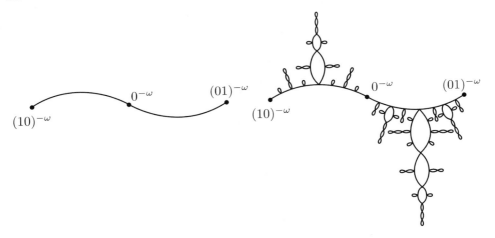

FIGURE 3.3. Basilica graphs

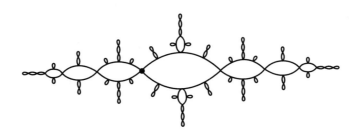

FIGURE 3.4. Approximation of $\mathcal{J}_{\mathrm{IMG}(z^2-1)}$

$(10)^{-\omega}$, $0^{-\omega}$ and $(01)^{-\omega}$, then we get the graph shown in Figure 3.4, which is an approximation of the limit space of the group $\mathrm{IMG}\left(z^{2}-1\right)$. This limit space is homeomorphic, as we will see later, to the Julia set of the polynomial $z^{2}-1$.

Sierpinski gasket. Let $\mathsf{X} = \{0, 1, 2\}$ and consider the group G generated by the automaton shown in Figure 3.5. The central vertex of the Moore diagram is the trivial state.

It is the group generated by the transformations b_i defined by the conditions $b_i(iw) = iw$ and $b_i(jw) = kw$, where $\{i, j, k\} = \{0, 1, 2\}$. This group coincides with $\mathcal{P}(\mathfrak{S}(3))$ (see 1.8.2).

A straightforward check (for example using Theorem 3.9.12) shows that the automaton shown in Figure 3.5 is the nucleus of the action. Looking at the Moore diagram of the nucleus, we see that the boundary of the tile $\mathcal{T}$ consists of three points represented by the words $0^{-\omega}, 1^{-\omega}$ and $2^{-\omega}$ and that we have identifications

$$0^{-\omega}1 \sim 0^{-\omega}2, \quad 1^{-\omega}0 \sim 1^{-\omega}2, \quad 2^{-\omega}0 \sim 2^{-\omega}1$$

in the tile. Take the tile diagram Γ, shown on the left-hand side of Figure 3.6. The right-hand side of Figure 3.6 shows then the diagram $\Gamma \cdot \mathsf{X}^5$. We see from the diagrams that the tile $\mathcal{T}$ and the limit space $\mathcal{J}_G$ are homeomorphic to the Sierpinski gasket.

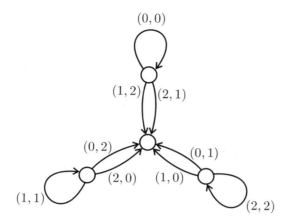

FIGURE 3.5. Automaton generating "Sierpinski group"

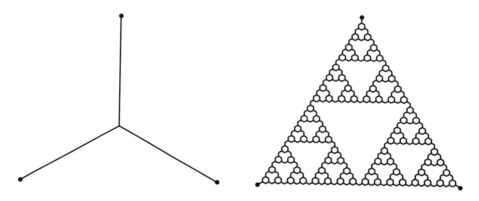

FIGURE 3.6. Approximation of the Sierpinski gasket

3.11. Uniqueness of the limit space

The aim of this section is to show that the limit space $\mathcal{X}_G$ and the action of G on it are uniquely determined by the algebraic structure of G for a wide class of self-similar groups.

3.11.1. Finite-state conjugator.

PROPOSITION 3.11.1. *Let G_1 and $G_2 = \alpha G_1 \alpha^{-1}$ be conjugated recurrent groups generated by finite bounded automata over the alphabet X and suppose that their centralizers in $\operatorname{Aut} \mathsf{X}^*$ are trivial. Then the conjugator α is finite-state.*

PROOF. We may assume, by Proposition 3.9.13 and Corollary 2.11.7, that the tiles of G_1 are connected.

Let $\mathcal{N}_1$ and $\mathcal{N}_2$ be the nuclei of the groups G_1 and G_2, respectively. They are also their generating sets. Take any $u \in \mathsf{X}^*$ and $a \in \mathcal{N}_1$ and let $k = |u|$. There exists an element $b \in \mathcal{N}_1$ and a word $w \in \mathsf{X}^k$ such that $b|_w = a$.

Let $T_k(G_i)$ be the graph with the set of vertices X^k, where $v_1, v_2 \in \mathsf{X}^k$ are connected by an edge if and only if there exists $g \in \mathcal{N}_i$ such that $g \cdot v_1 = v_2 \cdot 1$. The graphs $T_k(G_i)$ are connected, since we assume that the tiles of G_1 are connected (see Proposition 3.3.10).

Let us choose a simple path in $T_k(G_1)$ from u to w and a simple path from $b(w)$ to u. Let τ_1 and τ_2 be the consecutive products of the generators along the respective paths, so that $\tau_1 \cdot u = w \cdot 1$ and $\tau_2 \cdot b(w) = u \cdot 1$. Then $\tau_2 b \tau_1 \cdot u = u \cdot a$. It follows that

$$\alpha \tau_2 b \tau_1 \alpha^{-1} \cdot \alpha(u) \cdot \alpha|_u = \alpha \tau_2 b \tau_1 \alpha^{-1} \cdot \alpha \cdot u = \alpha \tau_2 b \tau_1 \cdot u = \alpha \cdot u \cdot a = \alpha(u) \cdot \alpha|_u a.$$

Consequently

$$(\alpha|_u) \, a \, (\alpha|_u)^{-1} = \left(\alpha \tau_2 b \tau_1 \alpha^{-1}\right)\big|_{\alpha(u)}$$
$$= \left(\alpha \tau_2 \alpha^{-1}\right)\big|_{\alpha b(w)} \cdot \left(\alpha b \alpha^{-1}\right)\big|_{\alpha(w)} \cdot \left(\alpha \tau_1 \alpha^{-1}\right)\big|_{\alpha(u)},$$

since $\alpha \tau_1 \alpha^{-1}(\alpha(u)) = \alpha(w)$, and $\alpha b \alpha^{-1}(\alpha(w)) = \alpha b(w)$.

The elements $\alpha g \alpha^{-1}$, for $g \in \mathcal{N}_1$, are generators of G_2. The elements $\alpha \tau_1 \alpha^{-1}$ and $\alpha \tau_2 \alpha^{-1}$ are products of the generators $\alpha g \alpha^{-1}$ along simple paths in the graph of the action of G_2 on X^k defined with respect to the generating set $\alpha \mathcal{N}_1 \alpha^{-1}$. Therefore, $\left(\alpha \tau_1 \alpha^{-1}\right)\big|_{\alpha(u)} = b_1|_{u_1} \cdot b_2|_{u_2} \cdots b_r|_{u_r}$, where b_s are equal to $\alpha g \alpha^{-1}$ for some $g \in \mathcal{N}_1$, and $u_1, u_2, \ldots, u_r$ are the vertices of the path. These vertices are pairwise different; thus the length of $\left(\alpha \tau_1 \alpha^{-1}\right)\big|_{\alpha(u)}$ with respect to the generating set $\mathcal{N}_2$ of G_2 is not greater than the sum $\sum_{g \in \mathcal{N}_1} \sum_{v \in \mathsf{X}^k} l_{\mathcal{N}_2}\left(\left(\alpha g \alpha^{-1}\right)|_v\right)$. The same is true about the length of the restriction $\left(\alpha \tau_2 \alpha^{-1}\right)\big|_{\alpha b(w)}$.

But the sum $\sum_{v \in \mathsf{X}^k} l_{\mathcal{N}_2}\left(\left(\alpha g \alpha^{-1}\right)\big|_w\right)$ is not greater than some constant C_g, not depending on k, by the definition of a bounded automaton. Therefore, the lengths of $\alpha \tau_1 \alpha^{-1}\big|_{\alpha(u)}$ and $\alpha \tau_2 \alpha^{-1}\big|_{\alpha a_j(w_i)}$ are not greater than $R = \sum_{g \in \mathcal{N}_1} C_g$.

So we get a uniform bound $3R$ on the length of the element $(\alpha|_u) \, a \, (\alpha|_u)^{-1}$, where $a \in \mathcal{N}_1$ and $u \in \mathsf{X}^*$ are arbitrary. But the centralizer of G_1 is trivial; thus $\alpha|_u$ is uniquely determined by the values of $(\alpha|_u) \, a \, (\alpha|_u)^{-1}$. Hence there is only a finite number of possibilities for $\alpha|_u$, and the automorphism α is finite-state. $\square$

Let us investigate possible structure of the conjugator α.

PROPOSITION 3.11.2. *Let G_1 and G_2 be recurrent groups acting on X^*. Suppose that they are conjugate by a finite-state automorphism of X^*. Let $\mathfrak{M}_i$ be the self-similarity G_i-bimodules for $i = 1, 2$. Then there exists $n \in \mathbb{N}$, a bijection $\Psi : \mathfrak{M}_1^{\otimes n} \longrightarrow \mathfrak{M}_2^{\otimes n}$ and a finite-state automorphism $\alpha \in \mathrm{Aut}\,\mathsf{X}^*$ such that $\alpha \cdot G_1 \cdot \alpha^{-1} = G_2$ and*

$$\Psi(g \cdot m \cdot h) = \alpha g \alpha^{-1} \cdot \Psi(m) \cdot \alpha h \alpha^{-1},$$

for all $g, h \in G_1$ and $m \in \mathfrak{M}_1^{\otimes n}$.

In other words, after we pass to the nth power X^n of the alphabet, the conjugator α can be chosen of the form

$$\alpha = \sigma \left(g_1 \cdot \alpha, g_2 \cdot \alpha, \ldots, g_{d^n} \cdot \alpha\right),$$

where $g_i \in G_2$, $\sigma \in \mathfrak{S}(\mathsf{X}^n)$.

PROOF. Equivalence of the two statements of the proposition follows from Proposition 2.3.4.

For every $u \in \mathsf{X}^*$ the image of the stabilizer G_u of the vertex u in G_1 under the restriction map $|_u$ is G_1 (definition of a recurrent action), and the same is true for G_2. This implies that for every $u \in \mathsf{X}^*$ we have $\alpha|_u G_1 \alpha|_u^{-1} = G_2$ (apply the restriction map $|_u$ to the equality $\alpha \cdot G_u \cdot \alpha^{-1} = G_{\alpha(u)}$).

Thus, every restriction $\alpha|_u$ of a (finite-state) conjugator is a (finite-state) conjugator.

Note that if $G_2 \alpha = G_2 \beta$, i.e., if $\alpha = h\beta$ for some $h \in G_2$, then $\alpha|_u = h|_{\beta(u)} \cdot \beta|_u$; hence $G_2 \alpha|_u = G_2 \beta|_u$. Thus, if we identify in the automaton defining α the states which belong to the same right coset of G_2 in $\operatorname{Aut} \mathsf{X}^*$, then we get a graph of a well defined automaton (without the output function), which we will denote A.

Let $u, v \in \mathsf{X}^k$ be arbitrary finite words of the same length. It follows from Proposition 2.8.2 that there exists $g \in G_1$ such that $g \cdot u = v \cdot 1$. Then
$$\left(\alpha g \alpha^{-1}\right)|_{\alpha(u)} = \alpha|_v \alpha|_u^{-1};$$
hence $\alpha|_v = h \cdot (\alpha|_u)$ for some $h \in G_2$, i.e., $G_2 \alpha|_v = G_2 \alpha|_u$.

This implies that the value of $\pi(q, x)$, where π is the transition function of A, depends only on the state q. The automaton A is finite; hence we can find a state $q_1 = G_2 \cdot \alpha|_u$ and a number n such that $\pi(q_1, x_1 \ldots x_n) = q_1$ for all $x_1 \ldots x_n \in \mathsf{X}^n$. This means that $\beta = \alpha|_u$ can be written with respect to the alphabet X^n in the form
$$\beta = \sigma \cdot (g_1 \cdot \beta, g_2 \cdot \beta, \ldots, g_{d^n} \cdot \beta),$$
for $g_i \in G_2$, which finishes the proof, since $\alpha|_u G_1 \alpha|_u^{-1} = G_2$. $\square$

3.11.2. Uniqueness of $\mathcal{X}_G$. See Definitions 1.2.4 and 2.10.6 for the notions of a weakly branch group and a saturated isomorphism.

THEOREM 3.11.3. *Let G_i, $i = 1, 2$, be recurrent weakly branch groups generated by bounded automata over the alphabet X and let $\mathfrak{M}_i$, $i = 1, 2$ be the self-similarity G_i-bimodule. Suppose that there exists a saturated isomorphism between G_1 and G_2. Then*

(1) *there exists $n \in \mathbb{N}$, a bijection $\Psi : \mathfrak{M}_1^{\otimes n} \longrightarrow \mathfrak{M}_2^{\otimes n}$ and an isomorphism $\psi : G_1 \longrightarrow G_2$, such that*
$$\Psi(g \cdot m \cdot h) = \psi(g) \cdot \Psi(m) \cdot \psi(h)$$
for all $g, h \in G$ and $m \in \mathfrak{M}_1^{\otimes n}$;

(2) *there exists a homeomorphism $F : \mathcal{X}_{G_1} \longrightarrow \mathcal{X}_{G_2}$ such that*
$$F(\zeta \cdot g) = F(\zeta) \cdot \psi(g), \qquad F(\zeta \otimes v) = F(\zeta) \otimes \Psi(v)$$
for all $\zeta \in \mathcal{X}_{G_1}$, $v \in \mathfrak{M}^{\otimes n}$ and $g \in G_1$.

PROOF. Since $(\mathfrak{M}^{\otimes n})^{\otimes -\omega} = \mathfrak{M}^{\otimes -\omega}$ for every hyperbolic bimodule $\mathfrak{M}$, (1) implies (2).

Statement (1) follows directly from Propositions 2.10.7 and 3.11.2. $\square$

EXAMPLE. Consider the following two groups. The group G_1 is generated by
$$a_1 = \sigma(1, a_3), \quad a_2 = (1, a_1), \quad a_3 = (a_2, 1);$$
and the group G_2 is generated by
$$a_1 = \sigma(1, a_3), \quad a_2 = (1, a_1), \quad a_3 = (1, a_2).$$

We will see later that these groups are the iterated monodromy groups of the polynomials $z^2 + c$, where c is either the real root $-1.7549\ldots$ of the polynomial

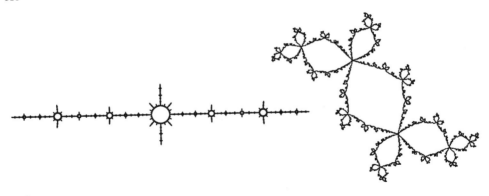

FIGURE 3.7. Airplane and Rabbit

$x^3 + 2x^2 + x + 1$ (for G_1) or one of the two complex roots $-0.1226\ldots \pm 0.7449\ldots i$ (for G_2).

It is not hard to prove that these groups are weakly branch. (In fact, we have the inclusion $G_i' > G_i' \times G_i'$.) Let $G_i^{2^0} = G_i$ and define inductively $G_i^{2^n}$ to be equal to the subgroups generated by the squares of the elements of $G_i^{2^{n-1}}$. Then the subgroups $G_i^{2^n}$ belong to the nth level stabilizer $\mathsf{St}(n)$ and act level-transitively on the subtrees with the roots on the nth level. This is proved by simple inductive arguments. It is also obvious that if $\psi : G_1 \longrightarrow G_2$ is an isomorphism, then $\psi\left(G_1^{2^n}\right) = G_2^{2^n}$; i.e., every isomorphism between G_1 and G_2 is saturated.

Consequently, if the groups G_1 and G_2 are isomorphic, then we can apply Theorem 3.11.3 and conclude, for example, that the limit spaces $\mathcal{J}_{G_1}$ and $\mathcal{J}_{G_2}$ are homeomorphic.

We will prove later that the limit spaces of the iterated monodromy groups G_1 and G_2 are homeomorphic to the Julia sets of the respective polynomials. These Julia sets are called in the literature the "Airplane" and the "Douady Rabbit" (see [**Mil99**]) and are shown in Figure 3.7.

It is more or less evident from Figure 3.7 that these Julia sets are not homeomorphic; hence the groups G_1 and G_2 are not isomorphic. A rigorous proof, for example, is to show that it is possible to cut the Rabbit into three connected components by deletion of a point, while the Airplane can be divided into no more than two components.

It seems, however, that it is hard to prove that the groups G_1 and G_2 are not isomorphic using "classical" group-theoretical invariants.

CHAPTER 4

Orbispaces

4.1. Pseudogroups and étale groupoids

We present here the main definitions and properties of pseudogroups of local homeomorphisms and étale groupoids. Our approach is similar to that of [**BH99**], where more details can be found.

4.1.1. Pseudogroups.

DEFINITION 4.1.1. Let $\mathcal{X}$ be a topological space. A *pseudogroup of local homeomorphisms* of $\mathcal{X}$ is a collection $\mathcal{H}$ of homeomorphisms $H : U_1 \longrightarrow U_2$ between open subsets of $\mathcal{X}$ (including the *empty*, or *zero* homeomorphism $0 : \emptyset \longrightarrow \emptyset$), which satisfies the following conditions.

(1) *Composition:* if $H_1 : U_1 \longrightarrow U_2$, $H_2 : U_3 \longrightarrow U_4$ belong to $\mathcal{H}$, then $H_1 \circ H_2 : H_2^{-1}(U_4 \cap U_1) \longrightarrow H_1(U_4 \cap U_1)$ also belongs to $\mathcal{H}$.
(2) *Inversion:* if $H : U_1 \longrightarrow U_2$ belongs to $\mathcal{H}$, then $H^{-1} : U_2 \longrightarrow U_1$ belongs to $\mathcal{H}$.
(3) *Restriction:* if $H : U_1 \longrightarrow U_2$ belongs to $\mathcal{H}$ and $U_1' \subset U_1$ is an open subset, then $H|_{U_1'} : U_1' \longrightarrow H(U_1')$ belongs to $\mathcal{H}$.
(4) *Union:* if $H : U \longrightarrow V$ is a homeomorphism between two open sets and there exists a cover $\{U_i\}$ of U by open subsets $U_i \subset U$ such that $H|_{U_i} : U_i \longrightarrow H(U_i)$ belongs to $\mathcal{H}$, then H also belongs to $\mathcal{H}$.

If G is a group acting on $\mathcal{X}$ by homeomorphisms, then it generates a pseudogroup of local homeomorphisms whose elements are unions of restrictions of the group elements onto open sets.

4.1.2. Etale groupoids.

A *groupoid* $(\mathcal{G}, \mathcal{X})$ is a small category of isomorphisms. A category is called small if the classes of morphisms and objects are sets. Here $\mathcal{G}$ is the set of morphisms and $\mathcal{X}$ is the set of objects of the category. We identify every object $x \in \mathcal{X}$ with the trivial automorphism id_x of x; thus $\mathcal{X}$ is the *set of units* of the groupoid $\mathcal{G}$. The set of units $\mathcal{X}$ is often denoted $\mathcal{G}^{(0)}$.

Every element g of a groupoid $\mathcal{G}$ is an isomorphism from its *source* $\mathbf{s}(g) = g^{-1}g$ to its *range* $\mathbf{r}(g) = gg^{-1}$. A product $g_1 g_2$ is defined if and only if $\mathbf{r}(g_2) = \mathbf{s}(g_1)$.

We denote by $\mathcal{G}^{(2)}$ the set of *composable pairs* $(g_1, g_2) \in \mathcal{G} \times \mathcal{G}$, i.e., such pairs that $g_1 g_2$ is defined. The groupoid structure is defined by the *multiplication* map

$$\mathcal{G}^{(2)} \longrightarrow \mathcal{G} : (g_1, g_2) \longrightarrow g_1 g_2,$$

and the *inversion map*

$$\mathcal{G} \longrightarrow \mathcal{G} : g \longrightarrow g^{-1}.$$

A *topological* groupoid is a groupoid $(\mathcal{G}, \mathcal{X})$ with topology on $\mathcal{G}$ (and induced topology on $\mathcal{X} = \mathcal{G}^{(0)} \subset \mathcal{G}$) for which these maps are continuous. It is called *étale* if the maps $\mathbf{s} : g \mapsto g^{-1}g$ and $\mathbf{r} : g \mapsto gg^{-1}$ are étale, i.e., are homeomorphisms in a neighborhood of each point $g \in \mathcal{G}$. An equivalent condition is that the set $\mathcal{G}^{(0)}$ and the maps $\mathbf{s}$ and $\mathbf{r}$ are open.

Two points $x, y \in \mathcal{X}$ belong to the same $\mathcal{G}$-*orbit* if there exists $g \in \mathcal{G}$ such that $x = \mathbf{s}(g)$ and $y = \mathbf{r}(g)$, i.e., if the objects x, y are isomorphic in the category.

The set of $\mathcal{G}$-orbits is denoted $\mathcal{G}\backslash\mathcal{X}$. If $\mathcal{G}$ is a topological groupoid, then the set of orbits is a topological space with the quotient topology.

The *isotropy group* of a point $x \in \mathcal{X}$ is the set $\mathcal{G}_x$ of the groupoid elements $g \in \mathcal{G}$ such that $\mathbf{s}(g) = \mathbf{r}(g) = x$, i.e., the automorphism group of the object x in the category $(\mathcal{G}, \mathcal{X})$. If $x, y \in \mathcal{X}$ belong to one $\mathcal{G}$-orbit, then their isotropy groups $\mathcal{G}_x$ and $\mathcal{G}_y$ are isomorphic. If $h \in \mathcal{G}$ is such that $\mathbf{s}(h) = x$ and $\mathbf{r}(h) = y$, then the map $g \mapsto h \cdot g \cdot h^{-1}$ is an isomorphism $\mathcal{G}_x \longrightarrow \mathcal{G}_y$.

We always assume that the space of units $\mathcal{X}$ and the spaces on which groups and pseudogroups act are locally compact and Hausdorff.

4.1.3. The groupoid of germs and the associated pseudogroup.

If $\mathcal{H}$ is a pseudogroup of local homeomorphisms of a topological space $\mathcal{X}$, then its *groupoid of germs* is the set of equivalence classes of pairs (H, x), where $H : U \longrightarrow V$ is an element of $\mathcal{H}$ and $x \in U$. Two pairs (H_1, x_1), (H_2, x_2) are identified if and only if $x_1 = x_2 = x$ and there exists a neighborhood U of x such that $H_1|_U = H_2|_U$. The *germ topology* is defined by the fundamental system of open sets of the form

$$\mathcal{U}_H = \{(H, x) \,:\, x \in U\},$$

where $H : U \longrightarrow V$ is an element of $\mathcal{H}$. Then the maps $(H, x) \mapsto x$ and $(H, x) \mapsto H(x)$ are homeomorphisms $\mathcal{U}_H \longrightarrow U$ and $\mathcal{U}_H \longrightarrow H(U)$, respectively.

It is easy to prove that the groupoid of germs is an étale groupoid with respect to the multiplication

$$(H_1, x)(H_2, y) = (H_1 H_2, y),$$

where the product is defined if and only if $H_2(y) = x$. The space of units is equal to the set of the germs (Id, x), $x \in \mathcal{X}$. The natural identification $(Id, x) \mapsto x$ is a homeomorphism of the space of units with $\mathcal{X}$. The source and the range maps are defined by

$$\mathbf{s}(H, x) = x, \qquad \mathbf{r}(H, x) = H(x).$$

If G is a discrete group acting on a topological space $\mathcal{X}$, then we can also define the *groupoid of the action* with the set of elements $G \times \mathcal{X}$ and the multiplication

$$(h_1, x)(h_2, y) = (h_1 h_2, y),$$

where the product is defined if and only if $h_2(y) = x$. The space of units is also naturally identified with $\mathcal{X}$ with the same formulae for the source and the range maps as in the case of the groupoid of germs.

The groupoid of germs is a quotient of the groupoid of the action. The groupoid of germs coincides with the groupoid of the action if and only if every non-trivial element of G acts non-trivially on every non-empty open subset of $\mathcal{X}$.

If $\mathcal{G}$ is the groupoid of germs of a pseudogroup $\mathcal{H}$ acting on $\mathcal{X}$, then $\mathcal{H}$ can be reconstructed as the *pseudogroup of open $\mathcal{G}$-sets*. A subset $H \subset \mathcal{G}$ is a $\mathcal{G}$-set if the maps

$$\mathbf{s}|_H, \mathbf{r}|_H : H \longrightarrow \mathcal{X}$$

are homeomorphisms. Then the map $\mathbf{s}(h) \mapsto \mathbf{r}(h)$, $h \in H$ is a well defined homeomorphism from $\mathbf{s}(H)$ to $\mathbf{r}(H)$, which we will also denote by H. If H_1 and H_2 are

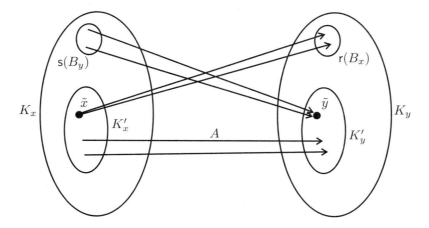

FIGURE 4.1

$\mathcal{G}$-sets, then $H_1 H_2 = \{h_1 h_2 \ : \ h_i \in H_i\}$ is also a $\mathcal{G}$-set defining the homeomorphism $H_1 \circ H_2$. It is also easy to prove that the set of all open $\mathcal{G}$-sets is a pseudogroup of local homeomorphisms of $\mathcal{X}$.

One can prove that the pseudogroup of $\mathcal{G}$-sets of the groupoid $\mathcal{G}$ of germs of a pseudogroup $\mathcal{H}$ coincides with $\mathcal{H}$.

4.2. Orbispaces

4.2.1. Proper groupoids. A groupoid $(\mathcal{G}, \mathcal{X})$ is said to be *proper* if the map

$$(\mathbf{s}, \mathbf{r}) : \mathcal{G} \longrightarrow \mathcal{X} \times \mathcal{X}$$

is proper, i.e., if for any compact subset $\mathcal{K} \subset \mathcal{X} \times \mathcal{X}$ the set $\{g \in \mathcal{G} \ : \ (\mathbf{s}(g), \mathbf{r}(g)) \in \mathcal{K}\}$ is compact.

A groupoid $\mathcal{G}$ is proper if and only if for any compact subset $\mathcal{K} \subset \mathcal{X}$ the set $\{g \in \mathcal{G} \ : \ \mathbf{s}(g), \mathbf{r}(g) \in \mathcal{K}\}$ is compact.

A pseudogroup of local homeomorphisms is said to be *proper* if its groupoid of germs is proper.

Recall that an action of a discrete group G on a topological space $\mathcal{X}$ is called proper if for every compact set $\mathcal{K} \subset \mathcal{X}$ the set of elements $g \in G$ such that $g(\mathcal{K}) \cap \mathcal{K} \neq \emptyset$ is finite. An action is proper if and only if the groupoid of the action is proper.

We have the following well known (at least for the group case) fact.

PROPOSITION 4.2.1. *If $(\mathcal{G}, \mathcal{X})$ is a proper groupoid, then the space of orbits $\mathcal{G}\backslash\mathcal{X}$ is Hausdorff.*

PROOF. Suppose that $x, y \in \mathcal{G}\backslash\mathcal{X}$ are two different points of the space of orbits and let $\tilde{x}, \tilde{y}$ be some of their preimages in $\mathcal{X}$.

Let K_x and K_y be compact neighborhoods of the points $\tilde{x}$ and $\tilde{y}$ respectively. The sets $B_x = \{g \in \mathcal{G} \ : \ \mathbf{s}(g) = \tilde{x}, \mathbf{r}(g) \in K_y\}$ and $B_y = \{g \in \mathcal{G} \ : \ \mathbf{s}(g) \in K_x, \mathbf{r}(g) = \tilde{y}\}$ are compact. The points $\tilde{x}$ and $\tilde{y}$ belong to different orbits; therefore $\tilde{x} \notin \mathbf{s}(B_y)$ and $\tilde{y} \notin \mathbf{r}(B_x)$. The sets $\mathbf{s}(B_y)$ and $\mathbf{r}(B_x)$ are compact as continuous images of compact sets.

The space $\mathcal{X}$ is Hausdorff and locally compact; therefore there exist compact neighborhoods $K'_x \subseteq K_x$ and $K'_y \subseteq K_y$ of the points $\tilde{x}$ and $\tilde{y}$ such that $K'_x \cap \mathbf{s}(B_y) = \emptyset$ and $K'_y \cap \mathbf{r}(B_x) = \emptyset$.

Let $A = \{g \in \mathcal{G} : \mathbf{s}(g) \in K'_x, \mathbf{r}(g) \in K'_y\}$. The set A is compact, $\tilde{x} \notin \mathbf{s}(A)$ and $\tilde{y} \notin \mathbf{r}(A)$. The sets $\mathbf{s}(A)$ and $\mathbf{r}(A)$ are compact; therefore the sets $U_x = K'_x \setminus \mathbf{s}(A)$ and $U_y = K'_y \setminus \mathbf{r}(A)$ are neighborhoods of the points $\tilde{x}$ and $\tilde{y}$. There are no elements $g \in \mathcal{G}$ such that $\mathbf{s}(g) \in U_x$ and $\mathbf{r}(g) \in U_y$; i.e., the images of U_x and U_y in $\mathcal{G}\backslash\mathcal{X}$ are disjoint neighborhoods of the points x and y. □

4.2.2. Equivalence of groupoids. A model example of an étale groupoid is the groupoid of the germs of changes of charts in an atlas of a manifold. Namely, if $\mathcal{M}$ is a manifold and $\mathfrak{A}$ is its atlas consisting of charts $q_i : \mathcal{X}_i \longrightarrow \mathcal{M}$, where $\mathcal{X}_i$ are open subsets of $\mathbb{R}^n$, then the respective *groupoid of changes of charts* is the groupoid of the germs of the maps

$$q_j^{-1} \circ q_i : q_i^{-1}(q_i(\mathcal{X}_i) \cap q_j(\mathcal{X}_j)) \longrightarrow q_j^{-1}(q_i(\mathcal{X}_i) \cap q_j(\mathcal{X}_j)),$$

which are considered to be local homeomorphisms of the disjoint union $\bigsqcup \mathcal{X}_i$.

Two atlases are equivalent if their union is again an atlas. Manifolds are then defined as the equivalence classes of the atlases. Note that the groupoid of changes of charts is always proper and free (a groupoid is *free* if all its isotropy groups are trivial).

The notion of an orbispace is a generalization of these notions. The only condition that we drop is the freeness of the groupoid. But we need to generalize the notion of equivalence of atlases, which is done here.

If $A \subset \mathcal{X}$, then *restriction* of $\mathcal{G}$ onto A is the groupoid $\mathcal{G}|_A = \{g \in \mathcal{G} : \mathbf{s}(g), \mathbf{r}(g) \in A\}$. It is the maximal sub-groupoid of $\mathcal{G}$ with the space of units A.

DEFINITION 4.2.2. An *equivalence* of two étale groupoids $(\mathcal{G}_1, \mathcal{X}_1)$ and $(\mathcal{G}_2, \mathcal{X}_2)$ is an étale groupoid $\mathcal{G}$ (denoted $\mathcal{G}_1 \vee \mathcal{G}_2$) with the space of units $\mathcal{X}_1 \sqcup \mathcal{X}_2$ ($\sqcup$ is the disjoint union of topological spaces) such that the restriction of $\mathcal{G}_1 \vee \mathcal{G}_2$ onto $\mathcal{X}_i$ coincides with $\mathcal{G}_i$ for $i = 1, 2$ and every $\mathcal{G}_1 \vee \mathcal{G}_2$-orbit is the union of a $\mathcal{G}_1$-orbit and a $\mathcal{G}_2$-orbit.

Two pseudogroups of local homeomorphisms are said to be equivalent if their groupoids of germs are equivalent. This can be formulated without use of groupoids in the following way.

DEFINITION 4.2.3. Let $(\mathcal{H}_1, \mathcal{X}_1)$ and $(\mathcal{H}_2, \mathcal{X}_2)$ be pseudogroups of local homeomorphisms. An *equivalence* $\mathcal{E} : \mathcal{H}_1 \longrightarrow \mathcal{H}_2$ is a collection of homeomorphisms $U_1 \longrightarrow U_2$, where $U_1 \subset \mathcal{X}_1$ and $U_2 \subset \mathcal{X}_2$ are open subsets, such that $\mathcal{H} = \mathcal{H}_1 \sqcup \mathcal{H}_2 \sqcup \mathcal{E} \sqcup \mathcal{E}^{-1}$ is a pseudogroup of local homeomorphisms of the space $\mathcal{X}_1 \sqcup \mathcal{X}_2$ and every $\mathcal{H}$-orbit is a union of an $\mathcal{H}_1$-orbit and an $\mathcal{H}_2$-orbit.

If $\mathcal{E}_1 : \mathcal{H}_1 \longrightarrow \mathcal{H}_2$ and $\mathcal{E}_2 : \mathcal{H}_2 \longrightarrow \mathcal{H}_3$ are equivalences, then $\mathcal{E}_1^{-1} : \mathcal{H}_2 \longrightarrow \mathcal{H}_1$ and $\mathcal{E}_2 \circ \mathcal{E}_1 : \mathcal{H}_1 \longrightarrow \mathcal{H}_3$ are equivalences.

If $\mathcal{E} : \mathcal{H}_1 \longrightarrow \mathcal{H}_2$ is an equivalence in the sense of Definition 4.2.3 and $\mathcal{G}_i$ is the groupoid of germs of $\mathcal{H}_i$, then the respective equivalence in the sense of Definition 4.2.2 is the groupoid $\mathcal{G}_1 \vee \mathcal{G}_2$ equal to the disjoint union of $\mathcal{G}_1$, $\mathcal{G}_2$, the space of germs of $\mathcal{E}$ and the space of germs of $\mathcal{E}^{-1}$. We will denote this equivalence of groupoids also by $\mathcal{E} : \mathcal{G}_1 \longrightarrow \mathcal{G}_2$.

4.2.3. Restrictions and localizations.
We will often use two important ways to define an equivalence of étale groupoids: *restrictions* and *localizations*.

Let $(\mathcal{G}, \mathcal{X})$ be an étale groupoid and let $\mathcal{X}' \subset \mathcal{X}$ be an open subset intersecting every $\mathcal{G}$-orbit. Then the *restriction* $\mathcal{G}'$ of the groupoid $\mathcal{G}$ onto $\mathcal{X}'$ is an étale groupoid $(\mathcal{G}', \mathcal{X}')$ equivalent to $(\mathcal{G}, \mathcal{X})$. The equivalence map is the collection of the local homeomorphisms between open subsets $U' \subset \mathcal{X}'$ and $U \subset \mathcal{X}$, which are defined by $\mathcal{G}$-sets.

One can prove that two groupoids are equivalent if and only if they are restrictions of a common groupoid.

If $\mathcal{U} = \{U_i\}_{i \in I}$ is a cover of $\mathcal{X}$ by open subsets indexed by a set I, then the *localization* of the groupoid $\mathcal{G}$ onto $\mathcal{U}$ (see [**Hae01**]) is the groupoid $(\mathcal{G}_\mathcal{U}, \mathcal{X}_\mathcal{U})$, where:

(1) the set of units $\mathcal{X}_\mathcal{U}$ is equal to the disjoint union $\bigsqcup_{i \in I}(U_i, i)$, where (U_i, i) is a copy of U_i;
(2) the set of elements $\mathcal{G}_\mathcal{U}$ is equal to the set of triples (i, g, j), where $g \in \mathcal{G}$ and $i, j \in I$ are such that $\mathbf{s}(g) \in U_j$ and $\mathbf{r}(g) \in U_i$;
(3) the groupoid structure is given by the equalities
$$\mathbf{s}(i, g, j) = (\mathbf{s}(g), j), \quad \mathbf{r}(i, g, j) = (\mathbf{r}(g), i)$$
and
$$(i, g_1, k) \cdot (k, g_2, j) = (i, g_1 g_2, j).$$
(4) The topology on $\mathcal{G}_\mathcal{U}$ is given by the fundamental system of open sets of the form
$$\{(i, g, j) \;:\; g \in H\},$$
where H is any open $\mathcal{G}$-set such that $\mathbf{s}(H) \subseteq U_j$ and $\mathbf{r}(H) \subseteq U_i$.

It is not hard to prove that the localization $(\mathcal{G}_\mathcal{U}, \mathcal{X}_\mathcal{U})$ is equivalent to $(\mathcal{G}, \mathcal{X})$. The equivalence groupoid $(\mathcal{G} \vee \mathcal{G}_\mathcal{U}, \mathcal{X} \sqcup \mathcal{X}_\mathcal{U})$ is the groupoid of germs of the pseudogroup generated by the union of the pseudogroup associated to $(\mathcal{G}, \mathcal{X})$ and the set of local homeomorphisms $I_i : (U_i, i) \longrightarrow U_i : (x, i) \mapsto x$.

One can also prove that two groupoids are equivalent if and only if they have a common localization. For more on equivalence of groupoids see [**BH99**] and [**Hae01**].

4.2.4. Orbispaces.

DEFINITION 4.2.4. An *orbispace* $\mathcal{O}$ is an equivalence class of proper pseudogroups. Every pseudogroup $(\mathcal{H}, \mathcal{X})$ belonging to the class is called the *pseudogroup of changes of charts* of the orbispace. The space $|\mathcal{O}|$ of the orbits of the pseudogroup of changes of charts is the *underlying space* of the orbispace. The canonical quotient map $q : \mathcal{X} \longrightarrow |\mathcal{O}|$ is the *uniformizing map*.

Every equivalence $\mathcal{E} : (\mathcal{G}_1, \mathcal{X}_1) \longrightarrow (\mathcal{G}_2, \mathcal{X}_2)$ induces a homeomorphism of the spaces of orbits $\mathcal{G}_1 \backslash \mathcal{X}_1$ and $\mathcal{G}_2 \backslash \mathcal{X}_2$. Therefore, the underlying space is defined uniquely up to a homeomorphism.

The pseudogroup of changes of charts $(\mathcal{H}, \mathcal{X})$ (and the respective groupoid of germs, which is called the *groupoid of changes of charts*), together with the uniformizing map $q : \mathcal{X} \longrightarrow |\mathcal{O}|$, is called the *atlas* of the orbispace $\mathcal{O}$.

We use pseudogroups and their groupoids of germs interchangeably, keeping in mind the relation between them described in 4.1.3.

We will usually denote the underlying space $|\mathcal{O}|$ just by $\mathcal{O}$ when it does not lead to confusion.

If $(\mathcal{G}_1, \mathcal{X}_1)$ and $(\mathcal{G}_2, \mathcal{X}_2)$ are two atlases of an orbispace $\mathcal{O}$, i.e., two equivalent étale groupoids, then we want to have a preferred equivalence groupoid $(\mathcal{G}_1 \vee \mathcal{G}_2, \mathcal{X}_1 \sqcup \mathcal{X}_2)$, called the *union* of the atlases. The union of the atlases is given by an equivalence $\mathcal{E} : \mathcal{G}_1 \longrightarrow \mathcal{G}_2$ (see Definition 4.2.3 and comments after it). Therefore, every time we introduce a new atlas $(\mathcal{G}', \mathcal{X}')$ of an orbispace $\mathcal{O}$, we fix an equivalence $\mathcal{E} : \mathcal{G}' \longrightarrow \mathcal{G}$ with some old atlas $(\mathcal{G}, \mathcal{X})$. If $\mathcal{E}_1 : \mathcal{G}_1 \longrightarrow \mathcal{G}$ and $\mathcal{E}_2 : \mathcal{G}_2 \longrightarrow \mathcal{G}$ are two such preferred equivalences, then the preferred equivalence between $\mathcal{G}_1$ and $\mathcal{G}_2$ is $\mathcal{E}_2^{-1} \circ \mathcal{E}_1$. The preferred equivalence is introduced in many cases implicitly (for example, if the new atlas is given as a restriction or a localization of an old one).

The *isotropy group* of a point $x \in \mathcal{O}$ is the isotropy group of any of its preimages in an atlas $(\mathcal{G}, \mathcal{X})$ of $\mathcal{O}$. We denote it by $\mathcal{G}_x$. The isotropy group is unique up to an isomorphism, since the isotropy groups of the points belonging to one orbit of a groupoid are isomorphic.

4.2.5. Rigid orbispaces. An orbispace is *rigid* if its atlas is a Hausdorff groupoid.

EXAMPLE. Consider a bouquet Y of three segments, i.e., the space $[0,1] \times \{a,b,c\}/\sim$, where $\sim$ is the equivalence relation $(0,a) \sim (0,b) \sim (0,c)$. Denote by O the equivalence class of $(0,a)$. Let the group $G = \mathfrak{S}(a,b,c)$ act on Y by permutations of the second coordinate.

The underlying space of the orbispace $G\backslash Y$ is homeomorphic to $[0,1]$. Let $\sigma = (a,b)$ be the transposition. Then every neighborhood of the germ (σ, O) contains germs of the trivial transformation; hence every two neighborhoods of (σ, O) and (id, O) intersect and the respective groupoid of germs is not Hausdorff.

4.2.6. Orbispaces with additional structure. If we have some local structure (like $C^{(k)}$-differentiable, analytic, piecewise linear, Riemannian, etc.), then orbispace with this structure is defined by pseudogroups of local homeomorphisms preserving the structure.

For example, a differentiable n-dimensional *orbifold* is an orbispace defined by an atlas $(\mathcal{H}, \mathcal{X})$, where $\mathcal{X}$ is a disjoint union of open subsets of $\mathbb{R}^n$ and $\mathcal{H}$ is a pseudogroup of local diffeomorphisms. An equivalence between atlases of n-dimensional orbifolds must also be given by local diffeomorphisms.

4.3. Open sub-orbispaces and coverings

4.3.1. Open mappings of orbispaces.

DEFINITION 4.3.1. An *open map* $f : \mathcal{M}_1 \longrightarrow \mathcal{M}_2$ between orbispaces is defined by an open continuous functor of groupoids $F : (\mathcal{G}_1, \mathcal{X}_1) \longrightarrow (\mathcal{G}_2, \mathcal{X}_2)$, where $(\mathcal{G}_i, \mathcal{X}_i)$ is an atlas of $\mathcal{M}_i$.

PROPOSITION 4.3.2. *An open functor F is uniquely determined by its restriction $F|_{\mathcal{X}_1} : \mathcal{X}_1 \longrightarrow \mathcal{X}_2$ onto the spaces of units.*

A continuous open map $F : \mathcal{X}_1 \longrightarrow \mathcal{X}_2$ defines an open map of orbispaces if and only if for every germ $g = (H, x) \in \mathcal{G}_1$ there exists a germ $F(g) = (H', F(x)) \in \mathcal{G}_2$ such that

(4.1) $$H' \circ F = F \circ H$$

on a neighborhood of x.

PROOF. It is sufficient to show that the germ $(H', F(x))$, satisfying (4.1), is unique.

Consider a small neighborhood U of the point x. The set $F(U)$ is a neighborhood of the point $F(x)$, since the map F is open. We have a commutative diagram of continuous maps

$$\begin{array}{ccc} U & \xrightarrow{H} & H(U) \\ \downarrow F & & \downarrow F \\ F(U) & \xrightarrow{H'} & F \circ H(U) \end{array}$$

with surjective vertical arrows. Then the map H' closing the diagram is unique. $\square$

One may interpret (4.1) as an explicit formulation of the condition that equal points must have equal images.

DEFINITION 4.3.3. An open map f of orbispaces is an *embedding* if the functor F is *full*, i.e., if every element $g \in \mathcal{G}_2$ such that $\mathbf{s}(g), \mathbf{r}(g) \in F(\mathcal{X}_1)$ belongs to $F(\mathcal{G}_1)$.

If an open embedding of an orbispace $\mathcal{M}_1$ into an orbispace $\mathcal{M}_2$ is fixed, then we say that $\mathcal{M}_1$ is an *open sub-orbispace* of $\mathcal{M}_2$.

DEFINITION 4.3.4. Let $(\mathcal{G}'_i, \mathcal{X}'_i)$ and $(\mathcal{G}''_i, \mathcal{X}''_i)$ be atlases of the orbispace $\mathcal{M}_i$, where $i = 1, 2$. Two functors $F' : \mathcal{G}'_1 \longrightarrow \mathcal{G}'_2$ and $F'' : \mathcal{G}''_1 \longrightarrow \mathcal{G}''_2$ define the same open map $f : \mathcal{M}_1 \longrightarrow \mathcal{M}_2$ if and only if it is possible to extend the functors F' and F'' to a functor $F : \mathcal{G}'_1 \vee \mathcal{G}''_1 \longrightarrow \mathcal{G}'_2 \vee \mathcal{G}''_2$.

We have the following obvious corollary of Proposition 4.3.2.

PROPOSITION 4.3.5. *Two maps* $F' : \mathcal{X}'_1 \longrightarrow \mathcal{X}'_2$ *and* $F'' : \mathcal{X}''_1 \longrightarrow \mathcal{X}''_2$ *define the same open map* $f : \mathcal{M}_1 \longrightarrow \mathcal{M}_2$ *if and only if their union satisfies the condition of Proposition 4.3.2.* $\square$

Let $f : \mathcal{M}_1 \longrightarrow \mathcal{M}_2$ be an open map defined by a functor F. The map f induces the map $|f| : |\mathcal{M}_1| \longrightarrow |\mathcal{M}_2|$ of the underlying spaces by the rule

$$|f|(q_1(\tilde{x})) = q_2(F(\tilde{x})),$$

where q_1, q_2 are the uniformizing maps. Definition 4.3.1 implies that the map $|f|$ is well defined. It is also easy to see that the map $|f|$ does not depend on the choice of the atlases. If the open map is an embedding, then the induced map is injective.

We will usually use the same notation for the map $f : \mathcal{M}_1 \longrightarrow \mathcal{M}_2$ and for the induced map $|f| : |\mathcal{M}_1| \longrightarrow |\mathcal{M}_2|$.

Let $x_1 \in \mathcal{M}_1$ and $x_2 \in \mathcal{M}_2$ be such that $f(x_1) = x_2$. Choose their preimages $\tilde{x}_i \in q_i^{-1}(x_i)$ so that $F(\tilde{x}_1) = \tilde{x}_2$. Then the restriction f_{x_1} of the functor F onto the isotropy group of the point $\tilde{x}_1$ is a homomorphism from the isotropy group $\mathcal{G}_{x_1}$ of x_1 to the isotropy group $\mathcal{G}_{x_2}$ of x_1. If the map f is an embedding, then the homomorphism f_{x_1} is surjective.

Let $f_1 : \mathcal{M}_1 \longrightarrow \mathcal{M}_2$ and $f_2 : \mathcal{M}_2 \longrightarrow \mathcal{M}_3$ be two open maps of orbispaces. Passing to localizations, we can find such atlases $(\mathcal{G}_i, \mathcal{X}_i)$, for $i = 1, 2, 3$, that f_i is defined by a functor $F_i : \mathcal{G}_i \longrightarrow \mathcal{G}_{i+1}$ for $i = 1, 2$. Then the functor $F_2 \circ F_1 : \mathcal{G}_1 \longrightarrow \mathcal{G}_3$ defines the *composition* $f_2 \circ f_1 : \mathcal{M}_1 \longrightarrow \mathcal{M}_2$ of the maps. It is easy to prove, using Proposition 4.3.5, that the composition $f_2 \circ f_1$ depends only on the maps f_1 and f_2. It is also easy to see that composition of two open embeddings is an open embedding.

4.3.2. Equivalences as open maps. An embedding $f : \mathcal{M}_1 \longrightarrow \mathcal{M}_2$ is called *unbranched* if it is defined by a functor which is étale on the unit spaces of the atlases (or, equivalently, étale on the groupoids of changes of charts).

PROPOSITION 4.3.6. *An open embedding $f : \mathcal{M}_1 \longrightarrow \mathcal{M}_2$ is unbranched if and only if it induces an isomorphism of the isotropy groups in every point $x \in \mathcal{M}_1$.*

PROOF. It is obvious that if the map f is unbranched, then it induces isomorphisms of the isotropy groups.

Suppose that we have an open functor $F : \mathcal{G}_1 \longrightarrow \mathcal{G}_2$ inducing isomorphisms of the isotropy groups. We have to prove that the restriction of F onto the unit spaces $\mathcal{X}_1$ and $\mathcal{X}_2$ of the groupoids $\mathcal{G}_1$ and $\mathcal{G}_2$ is a local homeomorphism. Suppose that the contrary is true. Since the space $\mathcal{X}_1$ is locally compact, there is a point $x \in \mathcal{X}_1$ such that every neighborhood of x contains points y, z such that $y \neq z$ and $F(y) = F(z)$. Then we can find two sequences $y_n, z_n \in \mathcal{X}_1$ such that $y_n \neq z_n$, $F(y_n) = F(z_n)$ for every n and $\lim_{n\to\infty} y_n = \lim_{n\to\infty} z_n = x$.

By definition of an embedding, there exist $g_n \in \mathcal{G}_1$ such that $\mathbf{s}(g_n) = y_n, \mathbf{r}(g_n) = z_n$ and $F(g_n) = 1_{F(y_n)}$. The set $\{y_n, z_n\}_{n\geq 1} \cup \{x\}$ is compact in $\mathcal{X}_1$. The groupoid $\mathcal{G}_1$ is proper; hence the set $\{g_n\}$ is contained in some compact set. Therefore, we can find a convergent subsequence $\{g_{n_k}\}_{k\geq 1}$. Let g be its limit. Then we have $\lim_{k\to\infty} F(g_{n_k}) = F(g)$. Then the equality $F(g_{n_k}) = 1_{F(y_{n_k})}$ implies that $F(g) = F(1_x) = 1_{F(x)}$. We know that F induces an isomorphism of the isotropy groups; therefore, the last equality implies that $g = 1_x$. But the groupoid $\mathcal{G}_1$ is étale; i.e., the element 1_x has an open neighborhood containing units only, which contradicts the fact that all g_{n_k} are not units. □

DEFINITION 4.3.7. An open embedding $f : \mathcal{M}_1 \longrightarrow \mathcal{M}_2$ is called *equivalence* (or *isomorphism* of the orbispaces) if it is unbranched and surjective on the underlying spaces.

It is easy to see that if $F : (\mathcal{G}_1, \mathcal{X}_1) \longrightarrow (\mathcal{G}_2, \mathcal{X}_2)$ defines an open map which is an equivalence, then the set of the local homeomorphism of the form $F \circ H$, where H is a change of charts in $(\mathcal{G}_1, \mathcal{X}_1)$, is an equivalence of the groupoids $(\mathcal{G}_1, \mathcal{X}_1)$ and $(\mathcal{G}_2, \mathcal{X}_2)$ in the sense of Definition 4.2.3.

It is also not hard to prove that orbispaces $\mathcal{M}_1$ and $\mathcal{M}_2$ are defined by equivalent atlases if and only if there exists an equivalence $f : \mathcal{M}_1 \longrightarrow \mathcal{M}_2$.

4.4. Coverings and skew-products

4.4.1. Coverings.

DEFINITION 4.4.1. A *covering* $P : (\widehat{\mathcal{G}}, \widehat{\mathcal{X}}) \longrightarrow (\mathcal{G}, \mathcal{X})$ of étale groupoids is an étale surjective functor such that for every $g \in \mathcal{G}$ and for every $\widehat{x} \in F^{-1}(\mathbf{s}(g))$ there exists a unique element $\widehat{g} \in \widehat{\mathcal{G}}$ such that $\mathbf{s}(\widehat{g}) = \widehat{x}$ and $F(\widehat{g}) = g$. A covering is said to be *d-fold* if every point $x \in \mathcal{X}$ has exactly d preimages.

A *covering of orbispaces* is an open map defined by a covering of their atlases.

PROPOSITION 4.4.2. *Let $p : \widehat{\mathcal{M}} \longrightarrow \mathcal{M}$ be a d-fold covering of orbispaces defined by a functor $P : (\widehat{\mathcal{G}}, \widehat{\mathcal{X}}) \longrightarrow (\mathcal{G}, \mathcal{X})$. Then for every point $x \in \mathcal{M}$ and every $\widehat{x} \in p^{-1}(x)$ the induced homomorphism $p_{\widehat{x}} : \mathcal{G}_{\widehat{x}} \longrightarrow \mathcal{G}_x$ of the isotropy groups is* *injective. The following equality holds for every $x \in \mathcal{M}$:*

$$\sum_{\widehat{x} \in p^{-1}(x)} \frac{|\mathcal{G}_x|}{|\mathcal{G}_{\widehat{x}}|} = d.$$

PROOF. Let $z \in \mathcal{X}$ be a preimage of x. Injectivity of $p_{\widehat{x}}$ follows directly from Definition 4.4.1. It follows also that the isotropy group $\mathcal{G}_z \cong \mathcal{G}_x$ of the point z acts on the set $P^{-1}(z)$ by permutations. The orbits of this action are in bijective correspondence with the points of $p^{-1}(x)$, and the stabilizer of a point $y \in P^{-1}(z)$ is the isotropy of this point in $\widehat{\mathcal{G}}$. This implies the statement of the proposition. □

Let $p : \widehat{\mathcal{M}} \longrightarrow \mathcal{M}$ be a d-fold covering of orbispaces defined by a covering of the atlases $P : (\widehat{\mathcal{X}}, \widehat{\mathcal{G}}) \longrightarrow (\mathcal{X}, \mathcal{G})$. Let $\mathcal{U} = \{U_i\}_{i \in I}$ be an open cover of the space $\mathcal{X}$ such that there exists a homeomorphism $\nu_i : U_i \times \mathsf{D} \longrightarrow P^{-1}(U_i)$ such that $P(\nu_i(x, a)) = x$ for all $x \in U_i$ and $a \in \mathsf{D}$ (here $\mathsf{D} = \{1, 2, \ldots, d\}$ is a discrete set). Such a cover exists by definition.

Then the collection $\widehat{\mathcal{U}} = \left\{ \widehat{U}_{(i,a)} = \nu_i(U_i \times \{a\}) \right\}_{(i,a) \in I \times \mathsf{D}}$ is an open cover of the space $\widehat{\mathcal{X}}$, and we get a map between the localizations of the atlases

$$P_\mathcal{U} : \left(\widehat{\mathcal{G}}_{\widehat{\mathcal{U}}}, \widehat{\mathcal{X}}_{\widehat{\mathcal{U}}} \right) \longrightarrow (\mathcal{X}_\mathcal{U}, \mathcal{G}_\mathcal{U}) \; : \; ((i, a), g, (j, b)) \mapsto (i, P(g), j),$$

where $(i, a), (j, b) \in I \times \mathsf{D}$ and $g \in \widehat{\mathcal{G}}$. The map $P_\mathcal{U}$ is called the *localization of the covering* P.

It is easy to see that the localization $P_\mathcal{U}$ is a functor, defining the same covering as the functor P. Hence, every covering can be *graded* in the sense of the following definition.

DEFINITION 4.4.3. Let $(\mathcal{G}, \mathcal{X})$ be an atlas of an orbispace $\mathcal{M}$. A *covering, graded by a set* D *over the atlas* $(\mathcal{G}, \mathcal{X})$, is a $|\mathsf{D}|$-fold covering $p : \widehat{\mathcal{M}} \longrightarrow \mathcal{M}$, defined by the projection $P : \mathcal{X} \times \mathsf{D} \longrightarrow \mathcal{X} : (x, a) \mapsto x$, where $\mathcal{X} \times \mathsf{D}$ is the unit space of an atlas of the orbispace $\widehat{\mathcal{M}}$.

4.4.2. Cocycles and skew products. Suppose that we have a graded covering. Then for every $h \in \mathcal{G}$ and every preimage (x, a) of the point $x = \mathbf{s}(h)$ there exists a unique preimage $\widehat{h}$ of h such that $\mathbf{s}\left(\widehat{h}\right) = (x, a)$. Then $\mathbf{r}\left(\widehat{h}\right) = (\mathbf{r}(h), b)$, where $\sigma(h) : a \mapsto b$ is a permutation of the set D.

The map $\sigma : \mathcal{G} \longrightarrow \mathfrak{S}(\mathsf{D})$ is a continuous homomorphism (functor) of groupoids. A continuous homomorphism from a groupoid to a group is called a *cocycle*.

Let us consider the general situation. Let G be a topological group with a fixed continuous action on a topological space D. Suppose that we have a cocycle $\sigma : \mathcal{G} \longrightarrow G$, where $\mathcal{G}$ is an étale groupoid. We will denote the image of a point $a \in \mathsf{D}$ under the action of $\sigma(g)$ by $\sigma(g, a)$.

DEFINITION 4.4.4. The *skew product* groupoid $\mathcal{G} \rtimes \sigma$ is the direct product $\mathcal{G} \times \mathsf{D}$ of topological spaces with the multiplication

(4.2) $$(g_1, a_1) \cdot (g_2, a_2) = (g_1 g_2, a_2),$$

where the left-hand side product is defined if and only if the product $g_1 g_2$ is defined and $\sigma(g_2, a_2) = a_1$.

The space of units of the groupoid $\mathcal{G} \rtimes \sigma$ is $\mathcal{X} \times \mathsf{D}$, where $\mathcal{X}$ is the space of units of the groupoid $\mathcal{G}$. The source and the range maps are defined by the rules

(4.3) $\qquad \mathbf{s}(g, a) = (\mathbf{s}(g), a), \quad \mathbf{r}(g, a) = (\mathbf{r}(g), \sigma(g, a)).$

PROPOSITION 4.4.5. *Let $\mathcal{G}$ be an étale groupoid and let $\sigma : \mathcal{G} \longrightarrow G$ be a cocycle, where the topological group G acts continuously on D. Then the skew product $\mathcal{G} \rtimes \sigma$ is an étale groupoid.*

PROOF. It is easy to check that the skew product groupoid is well defined and that the multiplication and the inversion (given by $(g, a)^{-1} = \left(g^{-1}, \sigma\left(g^{-1}, a\right)\right)$) are continuous. We have to prove that the source and the range maps are local homeomorphisms. If (g, a) is an element of $\mathcal{G} \rtimes \sigma$, then it has a neighborhood of the form $U \times \mathsf{D}$, where U is an open $\mathcal{G}$-set containing g. Then, the restrictions of $\mathbf{s}, \mathbf{r} : \mathcal{G} \rtimes \mathsf{D} \longrightarrow \mathcal{X} \times \mathsf{D}$ onto $U \times \mathsf{D}$ are given by

$$\mathbf{s}(h, a) = (\mathbf{s}(h), a), \quad \mathbf{r}(h, a) = (\mathbf{r}(h), \sigma(h, a)),$$

and are homeomorphisms, since $\mathbf{s}|_U, \mathbf{r}|_U$ are homeomorphisms, $\sigma(h, a)$ is continuous on h and is a homeomorphism $\mathsf{D} \longrightarrow \mathsf{D}$ on a. $\square$

We have the following proposition, whose proof is straightforward.

PROPOSITION 4.4.6. *Let $\mathcal{G}$ be an étale groupoid and let $\sigma : \mathcal{G} \longrightarrow \mathfrak{S}(\mathsf{D})$ be a cocycle. Then the projection map $P : \mathcal{G} \rtimes \sigma \longrightarrow \mathcal{G} : (g, a) \mapsto g$ is a covering.*

Let $P : \left(\widehat{\mathcal{G}}, \mathcal{X} \times \mathsf{D}\right) \longrightarrow (\mathcal{G}, \mathcal{X})$ be a projection defining a graded covering, and let σ be the respective cocycle. Then $\widehat{\mathcal{G}} = \mathcal{G} \rtimes \sigma$ (i.e., the identical map on $\mathcal{X} \times \mathsf{D}$ induces an isomorphism of the groupoids $\left(\widehat{\mathcal{G}}, \mathcal{X} \times \mathsf{D}\right)$ and $(\mathcal{G} \rtimes \sigma, \mathcal{X} \times \mathsf{D})$). $\square$

The covering $P : \mathcal{G} \rtimes \sigma \longrightarrow \mathcal{G} : (g, a) \mapsto g$ is the *covering defined by the cocycle* σ.

In general, skew products $\mathcal{G} \rtimes \sigma$ together with the projection are *fiber bundles* over the orbispace defined by the atlas $\mathcal{G}$. For example, an *n-dimensional vector bundle* is the skew product $\mathcal{G} \rtimes \sigma$, where σ is a cocycle from $\mathcal{G}$ to the general linear group $\mathrm{GL}(n, \mathbb{R})$ (with the natural action on $\mathbb{R}^n$). If $\mathcal{G}$ is a groupoid of changes of charts in an atlas of an n-dimensional orbifold, then the derivative $D : \mathcal{G} \longrightarrow \mathrm{GL}(n, \mathbb{R})$ (which is defined in the obvious way on the germs of changes of charts) defines the atlas $\mathcal{G} \rtimes D$ of the *tangent bundle* $T\mathcal{M}$ of the orbifold.

If $(\mathcal{G}_1, \mathcal{X}_1)$ and $(\mathcal{G}_2, \mathcal{X}_2)$ are atlases of one orbispace $\mathcal{M}$, then cocycles $\sigma_1 : \mathcal{G}_1 \longrightarrow G$ and $\sigma_2 : \mathcal{G}_2 \longrightarrow G$ define the same fiber bundle if and only if σ_1 and σ_2 are restrictions of one cocycle $\sigma : \mathcal{G}_1 \vee \mathcal{G}_2 \longrightarrow G$.

4.4.3. Pull-back. Let $f : \mathcal{M}_1 \longrightarrow \mathcal{M}$ be an open map of orbispaces and let $p : \widehat{\mathcal{M}} \longrightarrow \mathcal{M}$ be a covering. Then there exists a unique orbispace $\widehat{\mathcal{M}}_1$, a covering $p_1 : \widehat{\mathcal{M}}_1 \longrightarrow \mathcal{M}_1$ and an open map $\widehat{f} : \widehat{\mathcal{M}}_1 \longrightarrow \widehat{\mathcal{M}}$ such that the diagram

(4.4)
$$\begin{array}{ccc} \widehat{\mathcal{M}}_1 & \xrightarrow{\widehat{f}} & \widehat{\mathcal{M}} \\ \downarrow{p_1} & & \downarrow{p} \\ \mathcal{M}_1 & \xrightarrow{f} & \mathcal{M} \end{array}$$

is commutative. The covering p_1 is the *pull-back* of p by f and is constructed in the following way.

Let $(\mathcal{G}_1, \mathcal{X}_1)$ and $(\mathcal{G}, \mathcal{X})$ be atlases of the orbispaces $\mathcal{M}_1$ and $\mathcal{M}$ such that the map f is defined by a functor $F : \mathcal{G}_1 \longrightarrow \mathcal{G}$ and the covering p is defined by a cocycle $\sigma : \mathcal{G} \longrightarrow \mathfrak{S}(\mathsf{D})$. We can find such atlases passing to localizations.

Then the composition $\sigma_1 = \sigma \circ F$ is a cocycle $\sigma_1 : \mathcal{G}_1 \longrightarrow \mathfrak{S}(\mathsf{D})$ defining the covering $p_1 : \widehat{\mathcal{M}}_1 \longrightarrow \mathcal{M}_1$, where $\widehat{\mathcal{M}}_1$ is the orbispace defined by the atlas $(\mathcal{G}_1 \rtimes \sigma_1, \mathcal{X}_1 \times \mathsf{D})$.

We also get a functor $\widehat{F} : \mathcal{G}_1 \rtimes \sigma \longrightarrow \mathcal{G} \rtimes \sigma$ acting by the rule

$$\widehat{F}(g, a) = (F(g), a).$$

It defines the open map $\widehat{f} : \widehat{\mathcal{M}}_1 \longrightarrow \widehat{\mathcal{M}}$.

If f is an embedding, then $\widehat{f}$ is obviously also an embedding and p_1 is called the *restriction* of p onto the sub-orbispace $\mathcal{M}_1$.

4.5. Partial self-coverings

A *partial self-covering* is a covering map $f : \mathcal{M}_1 \longrightarrow \mathcal{M}$ of an orbispace $\mathcal{M}$ by its open sub-orbispace $\mathcal{M}_1$. So, an embedding $\mathcal{M}_1 \hookrightarrow \mathcal{M}$ is fixed.

If we have two partial self-coverings $f_1 : \mathcal{M}_1 \longrightarrow \mathcal{M}$ and $f_2 : \mathcal{M}_2 \longrightarrow \mathcal{M}$, then their composition $f_2 \circ f_1 : \mathcal{M}_2 \longrightarrow \mathcal{M}$ is defined as the composition $f_2 \circ f_1^\circ$, where $f_1^\circ : \mathcal{M}_3 \longrightarrow \mathcal{M}_2$ is the restriction of f_1 onto the sub-orbispace $\mathcal{M}_2$ (i.e., the pull-back of f_1 by the embedding $\mathcal{M}_2 \hookrightarrow \mathcal{M}$). Then $\mathcal{M}_3$ is a sub-orbispace of $\mathcal{M}_1$ and thus is also a sub-orbispace of $\mathcal{M}$. See the diagram below.

$$\begin{array}{ccccc} \mathcal{M}_3 & \hookrightarrow & \mathcal{M}_1 & \hookrightarrow & \mathcal{M} \\ \downarrow f_1^\circ & & \downarrow f_1 & & \\ \mathcal{M}_2 & \hookrightarrow & \mathcal{M} & & \\ \downarrow f_2 & & & & \\ \mathcal{M} & & & & \end{array}$$

In particular, we can define for every partial self-covering $f : \mathcal{M}_1 \longrightarrow \mathcal{M}$ its iterates $f^n = f^{\circ n} : \mathcal{M}_n \longrightarrow \mathcal{M}$.

DEFINITION 4.5.1. Let $p : \mathcal{M}_1 \longrightarrow \mathcal{M}$ be a partial self-covering and let $e : \mathcal{M}_1 \hookrightarrow \mathcal{M}$ be the embedding. Suppose that $f : \mathcal{M}^\circ \longrightarrow \mathcal{M}$ is an open map and let $p^\circ : \mathcal{M}_1^\circ \longrightarrow \mathcal{M}^\circ$ be the pull-back of p by f. Then we have an open map $f_1 : \mathcal{M}_1^\circ \longrightarrow \mathcal{M}_1$ such that $p \circ f_1 = f \circ p^\circ$. Suppose that there exists an embedding $e_0 : \mathcal{M}_1^\circ \hookrightarrow \mathcal{M}^\circ$ making the diagram

$$\begin{array}{ccc} \mathcal{M}_1^\circ & \xrightarrow{f_1} & \mathcal{M}_1 \\ \downarrow e^\circ & & \downarrow e \\ \mathcal{M}^\circ & \xrightarrow{f} & \mathcal{M} \end{array}$$

commutative. We say then that $\mathcal{M}^\circ$ is *invariant under* p^{-1} and that the obtained partial self-covering p° of $\mathcal{M}^\circ$ by the sub-orbispace $\mathcal{M}_1^\circ$ is the *pull-back of p by f*.

If $f : \mathcal{M}^\circ \longrightarrow \mathcal{M}$ is an embedding, then the partial self-covering p° is called the *restriction* of p onto $\mathcal{M}^\circ$.

The case when f is an equivalence is considered in the next definition.

DEFINITION 4.5.2. Two partial self-coverings $p' : \mathcal{M}'_1 \longrightarrow \mathcal{M}'$ and $p'' : \mathcal{M}''_1 \longrightarrow \mathcal{M}''$ are said to be *conjugate* if there exist equivalences $f_1 : \mathcal{M}'_1 \longrightarrow \mathcal{M}''_1$ and $f : \mathcal{M}' \longrightarrow \mathcal{M}''$ such that the diagrams

$$\begin{array}{ccc} \mathcal{M}'_1 & \xrightarrow{f_1} & \mathcal{M}''_1 \\ \downarrow e' & & \downarrow e'' \\ \mathcal{M}' & \xrightarrow{f} & \mathcal{M}'' \end{array} \qquad \begin{array}{ccc} \mathcal{M}'_1 & \xrightarrow{f_1} & \mathcal{M}''_1 \\ \downarrow p' & & \downarrow p'' \\ \mathcal{M}' & \xrightarrow{f} & \mathcal{M}'' \end{array}$$

are commutative. Here $e' : \mathcal{M}'_1 \hookrightarrow \mathcal{M}'$ and $e'' : \mathcal{M}''_1 \hookrightarrow \mathcal{M}''$ are the embeddings.

4.6. The limit orbispace $\mathcal{J}_G$

4.6.1. Definition.

PROPOSITION 4.6.1. *Let (G, X) be a self-similar contracting action and let $\mathcal{X}_G$ be the respective limit G-space. Then the action of G on $\mathcal{X}_G$ is proper.*

PROOF. Let $\mathcal{T} \subset \mathcal{X}_G$ be the digit tile of the action. If $C \subset \mathcal{X}_G$ is compact, then it is a subset of $\bigcup_{h \in A} \mathcal{T} \cdot h$ for some finite set $A \subset G$ by Proposition 3.4.1 (we use only that $U_1(\zeta)$ is a neighborhood of ζ). If $C \cdot g \cap C \neq \emptyset$, then there exist $g_1, g_2 \in A$ such that $\mathcal{T} \cdot g_1 g \cap \mathcal{T} \cdot g_2 \neq \emptyset$. This implies that $g_1 g g_2^{-1} \in \mathcal{N}$, by Proposition 3.3.5, i.e., that $g \in A^{-1} \mathcal{N} A$. □

PROPOSITION 4.6.2. *Suppose that the action (G, X) is faithful. Let $g \in G$ and suppose that there exists an open set $U \subset \mathcal{X}_G$ such that $\zeta \cdot g = \zeta$ for every $\zeta \in U$. Then $g = 1$.*

PROOF. If $\zeta \in U$ is represented by a sequence $\ldots x_2 x_1 \cdot h$, then there exists $k \in \mathbb{N}$ such that the tile $\mathcal{T} \otimes x_k \ldots x_1 \cdot h$ is a subset of U.

Let $\{f_1, \ldots, f_m\}$ be the nucleus $\mathcal{N}$ of the action, where $f_1 \neq 1$. Find a word $v_1 \in \mathsf{X}^*$ such that $f_1(v_1) \neq v_1$ and then define v_i inductively to be a word of the form $v_{i-1} v$, where $v = \emptyset$ if $f_i|_{v_{i-1}} = 1$ and v is such that $f_i|_{v_{i-1}}(v) \neq v$ otherwise. Then we have $f_i(v_i) \neq v_i$ or $f_i|_{v_i} = 1$. At the end we get a word v_m such that for every $f_i \in \mathcal{N}$ we have $f_i(v_m) \neq v_m$ or $f_i|_{v_m} = 1$, since v_i is a beginning of v_m.

Consider now an arbitrary word $\ldots y_2 y_1 v_m x_k \ldots x_1 \cdot h$. It represents a point of $\mathcal{T} \otimes x_k \ldots x_1 \cdot h \subset U$; therefore $\ldots y_2 y_1 v_m x_k \ldots x_1 \cdot h$ is asymptotically equivalent to $\ldots y_2 y_1 v_m x_k \ldots x_1 \cdot hg$. This implies that there exists an element $f \in \mathcal{N}$ such that $f \cdot v_m x_k \ldots x_1 \cdot h = v_m x_k \ldots x_1 \cdot hg$. But by the choice of v_m, we have either $f \cdot v_m = u \cdot f'$ for $u \neq v_m$ and $f' \in \mathcal{N}$, or $f \cdot v_m = v_m \cdot 1$. The first case is impossible, and the second one implies that $hg = h$, i.e., that $g = 1$. □

DEFINITION 4.6.3. *Limit orbispace $\mathcal{J}_G$ is the orbispace defined by the atlas equal to the groupoid of germs of the right action of G on $\mathcal{X}_G$.*

Let us denote the groupoid of germs of the action of G on $\mathcal{X}_G$ by $(\mathcal{G}_G, \mathcal{X}_G)$. The action of G on $\mathcal{X}_G$ is right, while all groupoids in our notations act on their spaces of units from the left. Therefore, in order to keep uniform notation for all groupoids, the element (g, ξ) of the groupoid of (germs of) the action corresponds to the transformation $\zeta \mapsto \zeta \cdot g^{-1}$.

Proposition 4.6.1 implies that $(\mathcal{G}_G, \mathcal{X}_G)$ is proper. Proposition 4.6.2 implies that the groupoid of germs in our case coincides with the groupoid of the action, which in turn shows that it is Hausdorff. Thus the limit orbispace is rigid.

The action of G on X defines the *action cocycle* on the groupoid $\mathcal{G}_G$, i.e., the functor $\sigma : \mathcal{G}_G \longrightarrow \mathfrak{S}(\mathsf{X})$ by the natural rule

$$\sigma(g,\xi)(x) = g(x).$$

Recall that the skew product $\mathcal{G}_G \rtimes \sigma$ acts on the space $\mathcal{X}_G \times \mathsf{X}$, and its elements are triples (ξ, g, x), where $\xi \in \mathcal{X}_G$ and $g \in G$ (here (g, ξ) is an element of the groupoid $\mathcal{G}_G$, i.e., the germ of the transformation $\zeta \mapsto \zeta \cdot g^{-1}$ in a neighborhood of $\xi \in \mathcal{X}_G$).

We have

$$\mathbf{s}(\xi, g, x) = (\xi, x) \in \mathcal{X}_G \times \mathsf{X}$$
$$\mathbf{r}(\xi, g, x) = (\xi \cdot g^{-1}, g(x)) \in \mathcal{X}_G \times \mathsf{X}.$$

Multiplication is defined by the formula

$$(\xi_1, g_1, x_1) \cdot (\xi_2, g_2, x_2) = (\xi_2, g_1 g_2, x_2),$$

where the product is defined if and only if $\xi_1 = \xi_2 \cdot g_2^{-1}$ and $g_2(x_2) = x_1$. Let us denote by $\mathcal{J}_G^\circ$ the orbispace defined by the atlas $(\mathcal{X}_G \times \mathsf{X}, \mathcal{G}_G \rtimes \sigma)$.

The projection $P_\mathsf{s} : (\xi, g, x) \mapsto (g, \xi)$ of the skew product $(\mathcal{G}_G \rtimes \sigma, \mathcal{X}_G \times \mathsf{X})$ onto $(\mathcal{G}_G, \mathcal{X}_G)$ defines a covering of the orbispace $\mathcal{J}_G$ by the orbispace $\mathcal{J}_G^\circ$. We denote the covering by $\mathsf{s} : \mathcal{J}_G^\circ \longrightarrow \mathcal{J}_G$.

Let us define a functor $E_\mathcal{J} : (\mathcal{G}_G \rtimes \sigma, \mathcal{X}_G \times \mathsf{X}) \longrightarrow (\mathcal{G}_G, \mathcal{X}_G)$ by the formula

$$E_\mathcal{J}(\xi, g, x) = (g|_x, \xi \otimes x).$$

Direct computations show that $E_\mathcal{J}$ is a functor. It is open by Lemma 3.4.2.

It is easy to see that if the action of G on X^* is recurrent, then the functor $E_\mathcal{J}$ defines an open embedding $\mathcal{J}_G^\circ \hookrightarrow \mathcal{J}_G$, which is identical on the underlying space $\mathcal{J}_G$. Therefore, we assume in the recurrent case that the orbispace $\mathcal{J}_G^\circ$ is an open sub-orbispace of $\mathcal{J}_G$. Then the covering $\mathsf{s} : \mathcal{J}_G^\circ \longrightarrow \mathcal{J}_G$ acts on the underlying spaces as the shift $\mathsf{s} : \mathcal{J}_G \longrightarrow \mathcal{J}_G$, defined in 3.6.2.

THEOREM 4.6.4. *The partial self-covering* $\mathsf{s} : \mathcal{J}_G^\circ \longrightarrow \mathcal{J}_G$ *depends only on the self-similarity bimodule* $\mathfrak{M}$.

PROOF. Let $\mathsf{X} = \{x_1, x_2, \ldots, x_d\}$ and $\mathsf{Y} = \{y_1, y_2, \ldots, y_d\}$ be two bases of the bimodule $\mathfrak{M}$. Then (possibly after changing the indexing) there exists a collection $\{r_1, r_2, \ldots, r_d\} \subset G$ such that $y_i = x_i \cdot r_i$.

Let us define a map $F : \mathcal{X}_G \times \mathsf{Y} \longrightarrow \mathcal{X}_G \times \mathsf{X}$ of the unit spaces of the skew product groupoids by the formula $F(\xi, y_i) = (\xi, x_i)$. Then F is a homeomorphism and can be extended to a functor $F(\xi, g, y_i) = (\xi, g, x_i)$ (it is a functor, since $g(x_i) = x_j$ implies $g(y_i) = y_j$). It follows directly from the definitions that this functor is an equivalence (it is even an isomorphism of the groupoids).

The functor F obviously agrees with the projections P_s, i.e., $P_\mathsf{s}(F(\xi, y_i)) = P_\mathsf{s}(\xi, y_i) = \xi$. On the other hand, $E_\mathcal{J}(F(\xi, y_i)) = E_\mathcal{J}(\xi, x_i) = \xi \otimes x_i$, and $E_\mathcal{J}(\xi, y_i) = \xi \otimes y_i = \xi \otimes x_i \cdot r_i$. Thus $E_\mathcal{J}(F(\xi, y_i))$ differs from $E_\mathcal{J}(\xi, y_i)$ by the action of an element of the group G, so that the functors $E_\mathcal{J} \circ F$ and $E_\mathcal{J}$ are equivalent, i.e., define the same map of the orbispaces. $\square$

REMARK. It follows that the partial self-covering $\mathsf{s} : \mathcal{J}_G^\circ \longrightarrow \mathcal{J}_G$ depends only on the right G-space $\mathcal{X}_G$ and the self-similarity $\mathcal{X}_G \otimes \mathfrak{M} = \mathcal{X}_G$, i.e., that it is uniquely defined in the conditions of Theorem 3.4.13.

4.6.2. Non-faithful actions. We will have to consider also the case when the action (G, X) is not faithful. Then the action of G on $\mathcal{X}_G$ may not be rigid; i.e., the groupoid of germs may be different from the groupoid of the action. The limit orbispaces $\mathcal{J}_G$ and $\mathcal{J}_G^\circ$ are defined then using the groupoid of germs.

PROPOSITION 4.6.5. *Suppose that we have a recurrent contracting action of a group G_1. Let G be the quotient of G_1 by the kernel of the action. Then the partial self-covering $\mathsf{s}_1 : \mathcal{J}_{G_1}^\circ \longrightarrow \mathcal{J}_{G_1}$ is a restriction of the partial self-covering $\mathsf{s} : \mathcal{J}_G^\circ \longrightarrow \mathcal{J}_G$. Moreover, the respective embeddings $\mathcal{J}_{G_1} \hookrightarrow \mathcal{J}_G$ and $\mathcal{J}_{G_1}^\circ \hookrightarrow \mathcal{J}_G^\circ$ are homeomorphisms of the underlying spaces.*

Suppose in addition that the canonical epimorphism $\pi : G_1 \longrightarrow G$ is such that $\pi(g_1 g_2 g_3) = 1$ implies $g_1 g_2 g_3 = 1$ whenever g_i are elements of the nucleus of G_1. Then the partial self-coverings are conjugate.

Compare the conditions of the second part of the proposition with the definition of the *covering group* (see Definition 2.13.3).

PROOF. We have to prove that there exist embeddings of orbispaces $e : \mathcal{J}_{G_1} \longrightarrow \mathcal{J}_G$ and $e^\circ : \mathcal{J}_{G_1}^\circ \longrightarrow \mathcal{J}_G^\circ$ such that the diagrams of the orbispace mappings

(4.5)
$$\begin{array}{ccc} \mathcal{J}_{G_1}^\circ & \xrightarrow{e^\circ} & \mathcal{J}_G^\circ \\ \downarrow p_1 & & \downarrow p \\ \mathcal{J}_{G_1} & \xrightarrow{e} & \mathcal{J}_G \end{array} \quad \text{, and} \quad \begin{array}{ccc} \mathcal{J}_{G_1}^\circ & \xrightarrow{e^\circ} & \mathcal{J}_G^\circ \\ \downarrow \epsilon_1 & & \downarrow \epsilon \\ \mathcal{J}_{G_1} & \xrightarrow{e} & \mathcal{J}_G \end{array}$$

are commutative. Here $\epsilon_1 : \mathcal{J}_{G_1}^\circ \longrightarrow \mathcal{J}_{G_1}$ and $\epsilon : \mathcal{J}_G^\circ \longrightarrow \mathcal{J}_G$ are the embeddings.

The spaces $\mathcal{X}_{G_1}$ and $\mathcal{X}_G$ are quotients of the spaces $\mathsf{X}^{-\omega} \cdot G_1$ and $\mathsf{X}^{-\omega} \cdot G$ by the asymptotic equivalence relation. Let $\pi : G_1 \longrightarrow G$ be the canonical epimorphism. Consider the map
$$\mathcal{E}(\ldots x_2 x_1 \cdot g) = \ldots x_2 x_1 \cdot \pi(g).$$
It is easy to see that $\mathcal{E}$ agrees with the asymptotic equivalence relation and the group actions, so that it defines a continuous map $\mathcal{E} : \mathcal{X}_{G_1} \longrightarrow \mathcal{X}_G$ such that $\mathcal{E}(\xi \cdot g) = \mathcal{E}(\xi) \cdot \pi(g)$ for all $\xi \in \mathcal{X}_{G_1}$ and $g \in G_1$. It follows that $\mathcal{E}$ can be extended to a functor $\mathcal{E} : \mathcal{G}_1 \longrightarrow \mathcal{G}$ between the groupoids of germs. It acts by the rule
$$\mathcal{E}(g, \xi) = (\pi(g), \mathcal{E}(\xi))$$
for all $g \in G_1$ and $\xi \in \mathcal{X}_{G_1}$. It is also straightforward that the map
$$\mathcal{E}^\circ(\xi, g, x) = (\mathcal{E}(\xi), \pi(g), x)$$
is a functor $\mathcal{E}^\circ : \mathcal{G}_1 \rtimes \sigma \longrightarrow \mathcal{G} \rtimes \sigma$ between the atlases of the orbispaces $\mathcal{J}_{G_1}^\circ$ and $\mathcal{J}_G^\circ$. Direct computations show that $\mathcal{E}$ and $\mathcal{E}^\circ$ define open maps $e : \mathcal{J}_{G_1} \longrightarrow \mathcal{J}_G$ and $e^\circ : \mathcal{J}_{G_1}^\circ \longrightarrow \mathcal{J}_G^\circ$ such that the diagrams (4.5) are commutative.

The functors $\mathcal{E} : \mathcal{G}_1 \longrightarrow \mathcal{G}$ and $\mathcal{E}^\circ : \mathcal{G}_1^\circ \longrightarrow \mathcal{G}^\circ$ define embeddings, since they are surjective.

Suppose now that the epimorphism π is such that $\pi(g_1 g_2 g_3) = 1$ implies $g_1 g_2 g_3 = 1$ whenever g_i are elements of the nucleus. We have to prove that the functors are equivalences, i.e., that they are étale. Let us prove that $\mathcal{E}$ is étale; the case of $\mathcal{E}^\circ$ will easily follow.

Let $\mathcal{N}_1$ and $\mathcal{N}$ be the nuclei of the actions (G_1, X) and (G, X), respectively. Take any $\xi = \ldots x_2 x_1 \cdot g \in \mathcal{X}_{G_1}$ and consider its neighborhood $U_1(\xi)$ (see Proposition 3.4.1). It is sufficient to prove that the restriction of $\mathcal{E}$ onto $U_1(\xi)$ is injective, since the neighborhood $U_1(\xi)$ is compact.

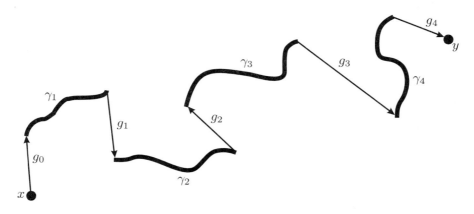

FIGURE 4.2. A $\mathcal{G}$-path

Suppose that we have $\zeta_1, \zeta_2 \in U_1(\xi)$ such that $\mathcal{E}(\zeta_1) = \mathcal{E}(\zeta_2)$. Then $\zeta_1 = \ldots y_2 y_1 \cdot h_1 g$ and $\zeta_2 = \ldots z_2 z_1 \cdot h_2 g$, for some $h_1, h_2 \in \mathcal{N}_1$ and $\ldots y_2 y_1, \ldots z_2 z_1 \in \mathsf{X}^{-\omega}$.

We have $\ldots y_2 y_1 \cdot \pi(h_1 g) = \ldots z_2 z_1 \cdot \pi(h_2 g)$ in $\mathcal{X}_G$, i.e., $\ldots y_2 y_1 \cdot \pi(h_1) = \ldots z_2 z_1 \cdot \pi(h_2)$. It follows that there exists a bounded sequence $g_k \in G$ such that $\pi(g_k) \cdot y_k \ldots y_1 \cdot \pi(h_1) = z_k \ldots z_1 \cdot \pi(h_2)$. Then $g_k \cdot y_k \ldots y_1 = z_k \ldots z_1 \cdot g'_k$, where $\pi(g'_k h_1) = \pi(h_2)$. But $g'_k \in \mathcal{N}_1$ for all sufficiently big k; therefore $g'_k h_1 = h_2$ and $\ldots y_2 y_1 \cdot h_1 g = \ldots z_2 z_1 \cdot h_2 g$ in $\mathcal{X}_{G_1}$. $\square$

4.7. Paths in an orbispace

4.7.1. $\mathcal{G}$-paths and their homotopy.
We again follow [**BH99**].

Let $(\mathcal{G}, \mathcal{X})$ be an étale groupoid. A $\mathcal{G}$-*path* $\gamma = g_k \cdot \gamma_k \cdots g_1 \cdot \gamma_1 \cdot g_0$ from a point $x \in \mathcal{X}$ to a point $y \in \mathcal{X}$ consists of:
(1) a subdivision $a = t_0 \le t_1 \le \ldots \le t_k = b$ of a real interval $[a, b]$;
(2) continuous maps $\gamma_i : [t_{i-1}, t_i] \longrightarrow \mathcal{X}$, $i = 1, 2, \ldots, k$;
(3) elements $g_i \in \mathcal{G}$, $i = 0, 1, \ldots, k$ such that $\mathbf{s}(g_i) = \gamma_i(t_i)$ for $i = 1, 2, \ldots, k$, $\mathbf{r}(g_i) = \gamma_{i+1}(t_i)$ for $i = 0, 1, \ldots, k-1$, $\mathbf{s}(g_0) = x$ and $\mathbf{r}(g_k) = y$.

See Figure 4.2 for a picture of a $\mathcal{G}$-path.

Two paths are *equivalent* if one can pass from one to the other using the following operations and their inverses.
(1) *Subdivision*: add a new division point $t'_i \in [t_{i-1}, t_i]$ together with the unit element $g'_i = \gamma_i(t'_i) = 1_{\gamma_i(t'_i)}$ and replace γ_i by $\gamma|_{[t'_i, t_i]} \cdot g'_i \cdot \gamma|_{[t_{i-1}, t'_i]}$.
(2) For each $i = 1, \ldots, k$ choose a continuous function $h_i : [t_{i-1}, t_i] \longrightarrow \mathcal{G}$ such that $\mathbf{s}(h_i(t)) = \gamma_i(t)$ for all $t \in [t_{i-1}, t_i]$ and replace γ_i by $\gamma'_i(t) = \mathbf{r}(h_i(t))$ for $i = 1, \ldots, k$, g_i by $g'_i = h_{i+1}(t_i) g_i h_i(t_i)^{-1}$ for $i = 1, \ldots, k-1$, g_0 by $g'_0 = h_1(t_0) g_0$ and g_k by $g'_k = g_k h_k(t_k)^{-1}$. (See Figure 4.3.)

DEFINITION 4.7.1. Two paths γ and γ' (parametrized by the interval $[0, 1]$) are *homotopic* if one can pass from the first to the second by a finite sequence of the following operations:
(1) equivalence of paths;
(2) *elementary homotopies*: an elementary homotopy between two paths γ and γ' is a family, parametrized by $s \in [s_0, s_1]$, of paths $\gamma^s = g_k^s \cdots g_0^s$ over

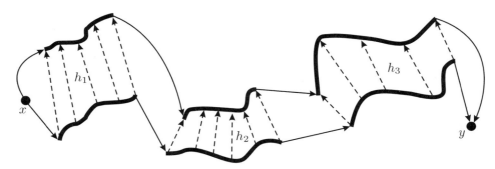

FIGURE 4.3. Equivalence of paths

the subdivision $0 = t_0^s \leq t_1^s \leq \cdots \leq t_k^s = 1$, where t_i^s, γ_i^s and g_i^s depend continuously on the parameter s, the elements g_0^s and g_k^s are independent of s and $\gamma^{s_0} = \gamma$, $\gamma^{s_1} = \gamma'$.

The set of homotopy classes of the paths is a groupoid under the natural operation of multiplication of paths. It is called the *fundamental groupoid* and is denoted $\pi_1(\mathcal{G})$.

REMARK. We multiply the $\mathcal{G}$-paths in the same order as we compose functions: if γ_1, γ_2 are two paths parametrized by the interval $[0, 1]$, then their product $\gamma_1 \gamma_2$ is the path equal to γ_2 on $[0, 1/2]$ and to γ_1 on $[1/2, 1]$ after reparametrization. This is made to agree with the multiplication of the elements of groupoids.

The space of units of the fundamental groupoid is identified in the natural way with $\mathcal{X}$. We have $\mathbf{s}(\gamma)$ equal to the beginning of the path γ and $\mathbf{r}(\gamma)$ equal to the end of the path γ.

The isotropy group of a point $x \in \mathcal{X}$ in the fundamental groupoid is called the *fundamental group* of the étale groupoid and is denoted $\pi_1(\mathcal{G}, x)$.

An étale groupoid $(\mathcal{G}, \mathcal{X})$ is said to be *path-connected* if for any two points $x, y \in \mathcal{X}$ there exists a $\mathcal{G}$-path starting in x and ending in y. An étale groupoid is path-connected if and only if its fundamental groupoid is transitive. If the groupoid is path-connected, then the fundamental group $\pi_1(\mathcal{G}, x)$ does not depend, up to an isomorphism, on the point x.

An orbispace is path-connected if some (and thus every) groupoid of changes of charts is path-connected. An orbispace is path-connected if and only if its underlying space is such. The fundamental group $\pi_1(\mathcal{M})$ of a path-connected orbispace $\mathcal{M}$ is the fundamental group $\pi_1(\mathcal{G}, x)$, where $(\mathcal{G}, \mathcal{X})$ is its atlas and $x \in \mathcal{X}$. Since the group $\pi_1(\mathcal{G}, x)$ does not depend on the choice of $(\mathcal{G}, \mathcal{X})$ and $x \in \mathcal{X}$, the fundamental group of a path-connected orbispace is well defined.

The limit orbispaces $\mathcal{J}_G$ and $\mathcal{J}_G^\circ$ are path-connected and locally path-connected if the action is recurrent (see Corollary 3.5.3).

See [**BH99**] for more on fundamental groups of orbispaces and étale groupoids.

4.7.2. Induced homomorphisms. Let $F : \mathcal{G}_1 \longrightarrow \mathcal{G}_2$ be a continuous functor and let $\gamma = g_k \cdot \gamma_k \cdots \gamma_1 \cdot g_0$ be a path in $\mathcal{G}_1$. Then

$$F(\gamma) = F(g_k) \cdot F \circ \gamma_k \cdots F(g_1) \cdot F \circ \gamma_1 \cdot F(g_0)$$

is a $\mathcal{G}_2$-path called the *image of γ under F*.

It is easy to see that the images of equivalent (homotopic) paths are equivalent (resp. homotopic).

Consequently, every continuous functor $F : (\mathcal{G}_1, \mathcal{X}_1) \longrightarrow (\mathcal{G}_2, \mathcal{X}_2)$ induces a functor $F_* : \pi_1(\mathcal{G}_1) \longrightarrow \pi_1(\mathcal{G}_2)$ of the fundamental groupoids by the rule $F_*(\gamma) = F(\gamma)$.

If $f : \mathcal{M}_1 \longrightarrow \mathcal{M}_2$ is an open map of path-connected orbispaces given by a functor $F : \mathcal{G}_1 \longrightarrow \mathcal{G}_2$, then we get the *induced* homomorphism of the fundamental groups $F_* : \pi_1(\mathcal{G}_1, t) \longrightarrow \pi_1(\mathcal{G}_1, F(t))$. The obtained homomorphism $f_* : \pi_1(\mathcal{M}_1) \longrightarrow \pi_1(\mathcal{M}_2)$ is defined up to a conjugacy in $\pi_1(\mathcal{M}_2)$.

4.7.3. The universal covering. An atlas $(\mathcal{G}, \mathcal{X})$ of an orbispace $\mathcal{M}$ is said to be *locally simply connected* if the space $\mathcal{X}$ is locally simply connected. It is easy to see that if some atlas of an orbispace is locally simply connected, then everyone of its atlases is such. We say that an orbispace is locally simply connected if some (and thus every) atlas of the orbispace is locally simply connected.

Suppose that $\mathcal{M}$ is locally simply connected and let $(\mathcal{G}, \mathcal{X})$ be its atlas. Let us choose a point $t \in \mathcal{X}$ and let $\mathcal{X}_t$ be the set of all homotopy classes of $\mathcal{G}$-paths starting in t.

Let γ be any element of $\mathcal{X}_t$ and let $z \in \mathcal{X}$ be its end. For every simply connected neighborhood U of z, let $U(\gamma)$ be the set of the $\mathcal{G}$-paths of the form $\gamma' \cdot \gamma$, where γ' is a usual path in U starting in z. Since U is simply connected, the map

$$\gamma' \cdot \gamma \mapsto \text{end of } \gamma'$$

is a bijection $U(\gamma) \longrightarrow U$. We introduce a topology on $\mathcal{X}_t$ declaring the collection of the sets of the form $U(\gamma)$ to be a fundamental system of neighborhoods of the point γ.

Let $\mathcal{G}_t$ be the set of pairs (g, γ), where $\gamma \in \mathcal{X}_t$ and $g \in \mathcal{G}$ are such that $\mathsf{s}(g)$ is the end of γ. We introduce a groupoid structure on $\mathcal{G}_t$ putting

$$\mathsf{s}(g, \gamma) = \gamma, \quad \mathsf{r}(g, \gamma) = g\gamma$$

and

$$(g_1, \gamma_1) \cdot (g_2, \gamma_2) = (g_1 g_2, \gamma_2),$$

where the product is defined if and only if $\gamma_1 = g_2 \gamma_2$.

Let U be a simply connected neighborhood of the end z of $\gamma \in \mathcal{X}_t$ and let $H : U \longrightarrow V$ be a change of charts in $(\mathcal{G}, \mathcal{X})$ (i.e., an open $\mathcal{G}$-set). The set of the elements $(g', \gamma'\gamma)$, where g' is the germ (H, z') and z' is the end of the path $\gamma' \subset U$, is a neighborhood of (H, z) in a natural topology on $\mathcal{G}_t$.

The groupoid $(\mathcal{G}_t, \mathcal{X}_t)$ is an atlas of an orbispace $\widehat{\mathcal{M}}$. The map $P : (g, \gamma) \longrightarrow g$ is a covering map of groupoids defining a covering $p : \widehat{\mathcal{M}} \longrightarrow \mathcal{M}$. The orbispace $\widehat{\mathcal{M}}$ and the covering p do not depend on the choice of the atlas $(\mathcal{G}, \mathcal{X})$ and are called the *universal covering* of $\mathcal{M}$.

The universal covering can also be defined in the classical way as the universal object in the category of coverings (see [**BH99**]).

If the universal covering $\widehat{\mathcal{M}}$ has no singular points, then the orbispace $\mathcal{M}$ is called *developable*. It is easy to deduce from the construction of the universal covering that $\mathcal{M}$ is developable if and only if for any unit $x \in \mathcal{X}$ and every non-unit element $g \in \mathcal{G}_x$ of the isotropy group the $\mathcal{G}$-loop (g) at x is not contractible, i.e., iff the isotropy groups are faithfully represented in the fundamental group.

The fundamental group $\pi_1(\mathcal{G}, t)$ acts naturally on $\mathcal{X}_t$ by the right multiplication. This action commutes with the left multiplication by $\mathcal{G}$. It follows that the natural right action of $\pi_1(\mathcal{G}, t)$ induces an action of $\pi_1(\mathcal{M}) = \pi_1(\mathcal{G}, t)$ on $\widehat{\mathcal{M}}$. It is easy to see that if $\mathcal{M}$ is developable, then $\mathcal{M}$ coincides with the orbispace $\widehat{\mathcal{M}}\big/\pi_1(\mathcal{M})$. See more on developability of the orbispaces in [**BH99**].

If $\mathcal{M}$ is not developable, then $\widehat{\mathcal{M}}$ has singular points, and we arrive at a special atlas $\left(\mathcal{X}_t, \tilde{\mathcal{G}}_t\right)$ of $\mathcal{M}$, where $\tilde{\mathcal{G}}_t$ consists of triples (g, ζ, γ), where $g \in \mathcal{G}$, $\zeta \in \mathcal{X}_t$ and $\gamma \in \pi_1(\mathcal{G}, t)$ are such that $\mathbf{s}(g)$ is the end of ζ.

We introduce on $\tilde{\mathcal{G}}_t$ the topology of a subset of the direct product $\mathcal{G} \times \mathcal{X}_t \times \pi_1(\mathcal{G}, t)$, where the fundamental group is taken with the discrete topology.

Then $\tilde{\mathcal{G}}_t$ is a groupoid with respect to the multiplication

(4.6) $\qquad (g_1, \zeta_1, \gamma_1)(g_2, \zeta_2, \gamma_2) = (g_1 g_2, \zeta_2, \gamma_1 \gamma_2),$

where the product is defined if and only if $\mathbf{s}(g_1, \zeta_1, \gamma_1) = \mathbf{r}(g_2, \zeta_2, \gamma_2)$ (see below).

We identify a point $\gamma \in \mathcal{X}_t$ with the unit $(1_z, \gamma, 1)$, where $1_z \in \mathcal{G}$ is equal to the beginning of γ and 1 is the unit of the fundamental group.

Then the source and the range maps are given by the formulae

$$\mathbf{s}(g, \zeta, \gamma) = \zeta, \qquad \mathbf{r}(g, \zeta, \gamma) = g\zeta\gamma^{-1}.$$

PROPOSITION 4.7.2. *The map $E : \mathcal{X}_t \longrightarrow \mathcal{X}$ mapping a path $\gamma \in \mathcal{X}_t$ to its end is an equivalence of groupoids.*

PROOF. It follows directly from the definition that E is a local homeomorphism and that it induces a surjective map of the spaces of orbits. Therefore, it is sufficient to prove that E can be extended to a functor and that the functor is full.

The functor is obviously the map $E : (g, \zeta, \gamma) \mapsto g$. If $x_1 = E(\zeta_1)$, $x_2 = E(\zeta_2)$ are points of $\mathcal{X}$ belonging to one $\mathcal{G}$-orbit, then for every $g \in \mathcal{G}$ such that $\mathbf{s}(g) = x_1, \mathbf{r}(g) = x_2$, we have the element $(g, \zeta_1, \zeta_2^{-1} g \zeta_1)$ such that $\mathbf{s}(g, \zeta_1, \zeta_2^{-1} g \zeta_1) = \zeta_1$, $\mathbf{r}(g, \zeta_1, \zeta_2^{-1} g \zeta_1) = g\zeta_1 \cdot (\zeta^{-1} g^{-1} \zeta_2) = \zeta_2$ and $E(g, \zeta_1, \zeta_2^{-1} g \zeta_1) = g$. Hence, E is a full functor. □

DEFINITION 4.7.3. *The groupoid $\left(\mathcal{X}_t, \tilde{\mathcal{G}}_t\right)$, defined above, is called the* derived atlas *of $(\mathcal{X}, \mathcal{G})$.*

For example, if $\mathcal{M}$ is a manifold, seen as an orbifold with the trivial atlas (i.e., the groupoid containing only units), then the derived atlas will be the groupoid of the action of the fundamental group on the universal covering of $\mathcal{M}$.

4.7.4. Preimages of paths under coverings. In the same way as for the topological spaces, paths in orbispaces can be lifted to the covering orbispace (see [**BH99**], p. 611).

Notation. Let $P : (\mathcal{G}_1, \mathcal{X}_1) \longrightarrow (\mathcal{G}, \mathcal{X})$ be a covering map of étale groupoids and let γ be a $\mathcal{G}$-path. Then we denote by $P^{-1}(\gamma)[x]$ (the equivalence class of) the preimage of γ under P, which starts in the point x. The point x must be a preimage of the beginning of the path γ.

LEMMA 4.7.4. *Let $p : \mathcal{M}_1 \longrightarrow \mathcal{M}$ be a covering of orbispaces given by a covering $P : (\mathcal{G}_1, \mathcal{X}_1) \longrightarrow (\mathcal{G}, \mathcal{X})$ of their atlases. Let the map $p_*^{-1} : \pi_1(\mathcal{M}) \dashrightarrow \pi_1(\mathcal{M}_1)$ be given by*

$$p_*^{-1}(\gamma) = P^{-1}(\gamma)[x_1]$$

and defined on the subgroup of the loops $\gamma \in \pi_1(\mathcal{G}, x)$ for which the path $P^{-1}(\gamma)[x_1]$ is also a loop. Here $x \in \mathcal{X}$ and $x_1 \in P^{-1}(x)$ are arbitrary and $\pi_1(\mathcal{M})$, $\pi_1(\mathcal{M}_1)$ are identified with $\pi_1(\mathcal{G}, x)$ and $\pi_1(\mathcal{G}_1, x_1)$, respectively.

Then p_*^{-1} is a virtual homomorphism which is uniquely determined, up to a conjugacy of virtual homomorphisms, by the map $p : \mathcal{M}_1 \longrightarrow \mathcal{M}$ only.

For the notion of conjugate virtual homomorphisms see Definition 2.5.4.

PROOF. The fact that p_*^{-1} is a virtual homomorphism, even that it is an isomorphism of a subgroup of index d in $\pi_1(\mathcal{G}, x)$ with $\pi_1(\mathcal{G}_1, x_1)$, is classical and follows directly from the definition of a covering (or from the explicit formula for the preimage of a path, which will be given below). Let us show that its conjugacy class does not depend on the choice of the atlases, functors and basepoints.

It is sufficient to prove that it does not depend on the choice of the basepoint, since we can always pass to the union of the atlases.

Take some $x, y \in \mathcal{X}$ and $x_1 \in P^{-1}(x)$, $y_1 \in P^{-1}(y)$. Choose a $\mathcal{G}_1$-path ℓ_1 from y_1 to x_1 and let $\ell = P(\ell_1)$ be its image. It is a $\mathcal{G}$-path from y to x.

We can identify $\pi_1(\mathcal{G}, x)$ with $\pi_1(\mathcal{G}, y)$ using the isomorphism $\gamma \mapsto \ell^{-1}\gamma\ell$. This isomorphism is defined uniquely, up to a conjugation. Both groups $\pi_1(\mathcal{G}, x)$ and $\pi_1(\mathcal{G}, y)$ are identified with $\pi_1(\mathcal{M})$ in a unique, up to a conjugation, way. Similarly, we identify $\pi_1(\mathcal{G}_1, x_1)$ with $\pi_1(\mathcal{G}_1, y_1)$ using the path ℓ_1, which is also canonical, up to a conjugation.

Then, for $\gamma \in \pi_1(\mathcal{G}, x)$:
$$P^{-1}(\ell^{-1}\gamma\ell)[y_1] = \ell_1^{-1} P^{-1}(\gamma)[x_1]\ell_1;$$
therefore p_* remains the same whatever basepoints we take if we assume the described identifications of the fundamental groups. Consequently, its conjugacy class remains the same for any choice of the basepoints and identifications. □

Let us show an explicit formula for the lift of a path in the case of a graded covering.

PROPOSITION 4.7.5. *Let $(\mathcal{G}, \mathcal{X})$ be an étale groupoid and let $\sigma : \mathcal{G} \longrightarrow \mathfrak{S}$ (D) be a cocycle. Denote by $P : (\mathcal{G} \rtimes \sigma, \mathcal{X} \times \mathsf{D}) \longrightarrow (\mathcal{G}, \mathcal{X})$ the respective covering map.*

Let $\gamma = g_k \cdot \gamma_k \cdots g_1 \cdot \gamma_1 \cdot g_0$ be a $\mathcal{G}$-path starting at a point $x = \mathsf{s}(g_0)$. Then for every preimage $x' = (x, a)$ of x under P we have
$$P^{-1}(\gamma)[x'] = (g_k, a_k) \cdot (\gamma_k, a_{k-1}) \cdots (\gamma_2, a_1) \cdot (g_1, a_1) \cdot (\gamma_1, a_0) \cdot (g_0, a_0)$$
where $a_0 = a$ and $a_{m+1} = \sigma(g_m, a_m)$ for $m = 0, \ldots, k-1$, (g_m, a_m) are elements of $\mathcal{G} \rtimes \sigma$ and $(\gamma_m(t), a_{m-1})$ denotes a function from $[t_{m-1}, t_m]$ to $\mathcal{X} \times \mathsf{D}$.

PROOF. It is sufficient to check that $(g_k, a_k) \cdot (\gamma_k, a_{k-1}) \cdots (\gamma_1, a_0) \cdot (g_0, a_0)$ is a $\mathcal{G} \rtimes \sigma$-path and that its image under P is equal to γ. Both facts are checked directly using the definitions. □

4.7.5. Monodromy action.
If x is the beginning and y is the end of the path γ, then the end of the path $P^{-1}(\gamma)[(x, a)]$ is a preimage (y, b) of the point y. The map $\sigma(\gamma) : a \mapsto b$ is a permutation, which can be computed explicitly in our case (see Proposition 4.7.5) as the *integral of the cocycle* σ *along* γ:

(4.7) $$\sigma(\gamma) = \sigma(g_k) \cdots \sigma(g_1) \sigma(g_0).$$

It is not hard to prove that the permutation $\sigma(\gamma)$ depends only on the homotopy class of the path γ and that the map $\gamma \mapsto \sigma(\gamma)$ is a cocycle on the fundamental groupoid of the atlas. In particular, we get an action of the fundamental group $\pi_1(\mathcal{G}, x)$ on the set of preimages of the point x, which is called the *monodromy action on the covering* $p: \mathcal{M}_1 \longrightarrow \mathcal{M}$.

One can prove, in the same way as for the usual topological spaces, that the monodromy action does not depend, up to a conjugacy of group actions, on the choice of the basepoint x (and on the choice of the atlases). In particular, the kernel of the monodromy action depends only on the covering p.

CHAPTER 5

Iterated Monodromy Groups

It is most natural to define iterated monodromy groups in the general context of orbispaces and their coverings. However, one can avoid using orbispaces in practical computations of iterated monodromy groups and in many applications. Readers not interested in the orbispace theory may read this chapter, omitting the subsections with titles written in *italic*. However, most proofs are presented only for the general case of orbispaces.

5.1. Definition of iterated monodromy groups

5.1.1. Definition. Let $f : \mathcal{M}_1 \longrightarrow \mathcal{M}$ be a covering of a path-connected and locally path-connected (orbi)space $\mathcal{M}$ by a path-connected open sub-(orbi)space. Then the *iterated monodromy group* is the quotient

$$\mathrm{IMG}\,(f) = \pi_1(\mathcal{M}) \bigg/ \bigcap_{n \geq 1} K_n ,$$

where K_n is the kernel of the *monodromy action* of $\pi_1(\mathcal{M})$ on the nth iterate $f^n : \mathcal{M}_n \longrightarrow \mathcal{M}$ (see 5.1.2, 4.7.5 and the next subsection).

The *profinite (or closed) iterated monodromy group* $\overline{\mathrm{IMG}}(f)$ is the completion of the group $\pi_1(\mathcal{M})$ with respect to the sequence of subgroups K_n.

The kernels K_n are normal subgroups of finite index. The iterated monodromy group is a dense subgroup of the profinite iterated monodromy group. In particular, iterated monodromy groups are residually finite.

We always assume, for the sake of simplicity, that the domains $\mathcal{M}_n$ of the iterates $f^n : \mathcal{M}_n \longrightarrow \mathcal{M}$ are path-connected.

5.1.2. Tree of preimages. Here we repeat the definition of the iterated monodromy group using a natural faithful action on the rooted *tree of preimages*, which is constructed as follows.

We consider here the case of usual topological spaces. The general case of orbispaces will be considered later.

Choose a basepoint $t \in \mathcal{M}$. The nth level of the tree of preimages T is the set $f^{-n}(t)$ of the preimages of t under the nth iteration of f. Every vertex $z \in f^{-n}(t)$ is connected by an edge with $f(z) \in f^{-(n-1)}(t)$.

If the covering $f : \mathcal{M}_1 \longrightarrow \mathcal{M}$ is d-fold, then every vertex $z \in f^{-(n-1)}(t)$ is connected exactly to d vertices of the level $f^{-n}(t)$. These vertices are the f-preimages of z. Thus T is a d-regular rooted tree.

If $\gamma \in \pi_1(\mathcal{M}, t)$ is a loop starting and ending in t, then for every n and $z \in f^{-n}(t)$ there exists precisely one f^n-preimage $\gamma_z = f^{-n}(\gamma)[z]$ of γ starting in z.

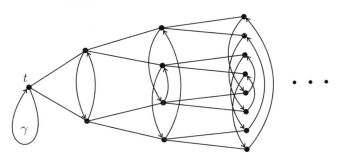

FIGURE 5.1. Iterated monodromy action

Let $\gamma(z)$ be the end of γ_z. Then the map

$$z \mapsto \gamma(z)$$

is a permutation of the level $f^{-n}(t)$ of the tree T (see Figure 5.1). It is, by definition, the *monodromy action* of γ on $f^{-n}(t)$.

If $z \in f^{-n}(t)$ and γ_z is the f^n-preimage of γ starting in z, then $f(\gamma_z)$ is an f^{n-1}-preimage of γ starting in $f(z)$. This proves that the permutation $z \mapsto \gamma(z)$ is an automorphism of the tree T.

We get in this way an action of the fundamental group $\pi_1(\mathcal{M}, t)$ on the tree T. This action is called the *iterated monodromy action* of $\pi_1(\mathcal{M})$. The quotient of the fundamental group by the kernel of the iterated monodromy action is, by definition, the *iterated monodromy group* IMG (f).

The profinite iterated monodromy group $\overline{\mathrm{IMG}}(f)$ is the closure of the iterated monodromy group IMG $(f) \le \operatorname{Aut} T$ in the automorphism group of the rooted tree T.

It is not hard to prove that the iterated monodromy action is defined uniquely, up to conjugacy of the actions (i.e., does not depend on the choice of the basepoint t). We will prove this later in a more general setting of orbispaces.

5.1.3. Tree of preimages for orbispaces. Let $p : \mathcal{M}_1 \longrightarrow \mathcal{M}$ be a partial d-fold self-covering of an orbispace $\mathcal{M}$ and let $p^n : \mathcal{M}_n \longrightarrow \mathcal{M}$ be its nth iteration. Choose some point $t \in \mathcal{M}$. Its *tree of preimages* T is a rooted tree of groups (in the sense of J.-P. Serre [**Ser80**]), which is constructed in the following way.

The nth level of the tree T is the set of preimages $p^{-n}(t)$. Two vertices $z_1 \in p^{-n}(t)$ and $z_2 \in p^{-(n-1)}(t)$ are connected by an edge if and only if $z_2 = p(z_1)$. The group attached to a vertex $z \in p^{-n}(t)$ is its isotropy group G_z in $\mathcal{M}_n$. The covering $p : \mathcal{M}_n \longrightarrow \mathcal{M}_{n-1}$ induces an injective homomorphism $p_* : G_z \longrightarrow G_{p(z)}$ of the isotropy groups for every $z \in \mathcal{M}_n$. Therefore, the group G_e attached to an edge $e = (z, p(z))$, where $z \in p^{-n}(t)$, is the isotropy group G_z of z in $\mathcal{M}_n$ with the identical isomorphism $G_e \longrightarrow G_z$ and the embedding $p_* : G_e \longrightarrow G_{p(z)}$.

Lemma 4.4.2 implies that the universal covering $\widetilde{T}$ of the tree T is a d-regular rooted tree.

This construction is easy in the case when the basepoint t is non-singular. Then the tree T is a usual d-regular tree (i.e., all the vertex and edge groups are trivial) and the universal covering $\widetilde{T}$ coincides with T.

5.1. DEFINITION OF ITERATED MONODROMY GROUPS

But if t is singular, then this definition of the tree T and its universal covering is not precise enough, since the isotropy groups and the homomorphisms $p_* : G_z \longrightarrow G_{p(z)}$ are defined only when some atlases of the orbispaces $\mathcal{M}_n$ are given. Let us make our definition more explicit.

Fix some $n \in \mathbb{N}$ and find some atlases $(\mathcal{G}_k, \mathcal{X}_k)$ of the orbispaces $\mathcal{M}_k$ for $0 \le k \le n$ so that the coverings $p : \mathcal{M}_k \longrightarrow \mathcal{M}_{k-1}$ are defined by functors $P_k : \mathcal{M}_k \longrightarrow \mathcal{M}_{k-1}$ for every $k = 1, \ldots, n$. We can do it for every n, but it is in general not clear how to do this for $n = \infty$, i.e., for all $\mathcal{M}_k$ simultaneously.

Let x be a preimage of the basepoint t in $\mathcal{X}_0$. Denote by P^k the functor $P_1 \circ \cdots \circ P_k$ defining the covering $p^k : \mathcal{M}_k \longrightarrow \mathcal{M}$. Let $L_k = (P^k)^{-1}(x)$ and $L_0 = \{x\}$. We get a rooted tree $\widetilde{T}_n$ consisting of $n+1$ levels L_k, $k = 0, 1, \ldots, n$ in which a vertex $z \in L_k$ is connected to the vertex $P_k(z) \in L_{k-1}$.

The fundamental group $\pi_1(\mathcal{G}_0, x)$ acts on the levels of the tree $\widetilde{T}_n$ by the monodromy action: if $\gamma \in \pi_1(\mathcal{G}_0, x)$ and $z \in L_k$, then $\gamma(z)$ is the end of the path $(P^k)^{-1}(\gamma)[z]$.

It follows from the equality
$$P_k\left((P_1 \circ \cdots \circ P_k)^{-1}(\gamma)[z]\right) = (P_1 \circ \cdots \circ P_{k-1})^{-1}(\gamma)[P_k(z)]$$
that the monodromy action of $\pi_1(\mathcal{G}_0, x)$ is an action by automorphisms of the tree $\widetilde{T}_n$.

The tree $\widetilde{T}_n$ and the monodromy action of $\pi_1(\mathcal{M})$ are constructed using specific atlases. We have to show that these objects are nevertheless canonically defined.

Suppose that $(\mathcal{G}'_k, \mathcal{X}'_k)$, $k = 0, 1, \ldots, n$ and $P'_k : \mathcal{G}'_k \longrightarrow \mathcal{G}'_{k-1}$ is another choice of the atlases and the functors defining $p : \mathcal{M}_k \longrightarrow \mathcal{M}_{k-1}$. Let $x' \in \mathcal{X}'_0$ be a preimage of t. Let $\widetilde{T}'_n$ be the tree defined by these data. Let $(\mathcal{G}_k \vee \mathcal{G}'_k, \mathcal{X}_k \sqcup \mathcal{X}'_k)$ be the respective unions of the atlases. We will denote the union of the functors P_k and P'_k by P_k (this will not lead to confusion).

Choose a $\mathcal{G}_0 \vee \mathcal{G}'_0$-path ℓ starting in x and ending in x'. It defines an isomorphism $\lambda : \pi_1(\mathcal{G}_0, x) \longrightarrow \pi_1(\mathcal{G}'_0, x')$ by $\lambda(\gamma) = \ell\gamma\ell^{-1}$. It is the standard identification of $\pi_1(\mathcal{G}_0, x)$ with $\pi_1(\mathcal{G}'_0, x')$. Given such an identification, we define also an identification τ of the trees $\widetilde{T}_n$ and $\widetilde{T}'_n$.

If $z \in L_k$ is a vertex of $\widetilde{T}_n$, then the corresponding vertex $\tau(z)$ is defined as the end of the $\mathcal{G}_0 \vee \mathcal{G}'_0$-path
$$(P_1 \circ \cdots \circ P_k)^{-1}(\ell)[z].$$

Take any $\gamma \in \pi_1(\mathcal{G}_0, x)$ and let z' be an arbitrary point of the ith level of the tree $\widetilde{T}'_n$. Then
$$(P^k)^{-1}\left(\ell\gamma\ell^{-1}\right)[z']$$
$$= (P^k)^{-1}(\ell)\left[\gamma\left(\tau^{-1}(z')\right)\right] \cdot (P^k)^{-1}(\gamma)\left[\tau^{-1}(z')\right] \cdot (P^k)^{-1}(\ell^{-1})[z'].$$

Hence the end $\lambda(\gamma)(z')$ of the path $(P^k)^{-1}\left(\ell\gamma\ell^{-1}\right)[z']$ is equal to the image of the end $\gamma\left(\tau^{-1}(z')\right)$ of the path $(P^k)^{-1}(\gamma)\left[\tau^{-1}(z')\right]$ under τ, i.e., $\tau\gamma\tau^{-1} = \lambda(\gamma)$. Consequently the identifications λ and τ agree with each other; therefore the above construction of the tree $\widetilde{T}_n$ and the action of $\pi_1(\mathcal{M})$ on it is canonical.

The uniformizing maps $q_k : \mathcal{X}_k \longrightarrow \mathcal{M}_k$ induce natural projection maps from $\widetilde{T}_n$ onto the subtree consisting of the first n levels of T. These projections also agree with the identification τ, since z and $\tau(z)$ belong to the same $\mathcal{G}_k \vee \mathcal{G}'_k$-orbit.

As a direct limit of the $\pi_1(\mathcal{M})$-sets $\widetilde{T}_n$ we get an infinite rooted d-regular tree $\widetilde{T}$ together with an action of $\pi_1(\mathcal{M})$ and a projection onto T. The isotropy group $\mathcal{G}_t = \mathcal{G}_x \leq \pi_1(\mathcal{G}_0, x)$ acts on $\widetilde{T}$, and the orbits of this action are exactly the fibers of the projection $q : \widetilde{T} \longrightarrow T$. Hence $\widetilde{T}$ is the universal covering of the graph of groups $T = \mathcal{G}_t \backslash \widetilde{T}$.

The action of $\pi_1(\mathcal{M})$ on the nth level of the tree $\widetilde{T}$ is conjugate to the monodromy action on the covering $p^n : \mathcal{M}_n \longrightarrow \mathcal{M}$ and the action of $\pi_1(\mathcal{M})$ on the tree $\widetilde{T}$ is the *iterated monodromy action*.

5.1.4. The bimodule of a partial self-covering (non-singular case).

Let $p : \mathcal{M}_1 \longrightarrow \mathcal{M}$ be a partial self-covering. Choose a basepoint $t \in \mathcal{M}$. Let $\mathfrak{M}(p)$ be the set of the homotopy classes of the paths in $\mathcal{M}$ starting in t and ending in a point of $p^{-1}(t)$. Then the set $\mathfrak{M}(p)$ has a natural structure of a $\pi_1(\mathcal{M}, t)$-bimodule. The right action is the natural one:

$$\ell \cdot \gamma = \ell\gamma,$$

where $\ell \in \mathfrak{M}(p)$ and $\gamma \in \pi_1(\mathcal{M}, t)$.

The path $\ell\gamma$ is a well defined element of $\mathfrak{M}(p)$, since the end of γ is the beginning of ℓ. See the remark on page 132 after Definition 4.7.1 about the order of multiplication of paths.

The left action is obtained by taking preimages of the loops under the covering:

$$\gamma \cdot \ell = p^{-1}(\gamma)[z]\ell,$$

where z is the end of ℓ and $p^{-1}(\gamma)[z]$ denotes the unique p-preimage of γ starting in z.

The bimodule $\mathfrak{M}(p)$ will be the main tool of computation of IMG(p). It is essentially the main object encoding the "action" of the self-covering on the fundamental group.

Two paths $\ell_1, \ell_2 \in \mathfrak{M}(p)$ belong to the same orbit of the right action if and only if their ends coincide. (If the ends of ℓ_1 and ℓ_2 coincide, then $\ell_1^{-1}\ell_2$ is a loop based at t and $\ell_1 \cdot \ell_1^{-1}\ell_2 = \ell_2$.) Hence we have d orbits and a collection $\mathsf{X} = \{\ell_1, \ldots, \ell_d\}$ is a basis of $\mathfrak{M}(p)$ if and only if the ends of ℓ_i are pairwise different and are all the p-preimages of t. It is also easy to see that the right action of $\pi_1(\mathcal{M})$ on $\mathfrak{M}(p)$ is free. Hence, $\mathfrak{M}(p)$ is a d-fold covering bimodule (see Definition 2.1.2).

The general definition. Let us define the bimodule $\mathfrak{M}(p)$ for self-coverings of orbispaces. It will be, as usual, a bit more technical than the definition for non-singular spaces.

Let $p : \mathcal{M}_1 \longrightarrow \mathcal{M}$ be a partial self-covering of a path-connected and locally path-connected orbispace. Let us choose atlases $(\mathcal{G}, \mathcal{X})$ and $(\mathcal{G}', \mathcal{X}')$ of $\mathcal{M}$ and an atlas $(\mathcal{G}_1, \mathcal{X}_1)$ of $\mathcal{M}_1$ such that the covering p is defined by a covering functor $P : \mathcal{G}_1 \longrightarrow \mathcal{G}$ and the embedding $\mathcal{M}_1 \hookrightarrow \mathcal{M}$ is defined by a functor $E : \mathcal{G}_1 \longrightarrow \mathcal{G}'$. Let $(\mathcal{G} \vee \mathcal{G}', \mathcal{X} \sqcup \mathcal{X}')$ be the union of the atlases of $\mathcal{M}$.

Let us choose a basepoint $t \in \mathcal{X} \subset \mathcal{X} \sqcup \mathcal{X}'$ and identify $\pi_1(\mathcal{M})$ with $\pi_1(\mathcal{G}, t) = \pi_1(\mathcal{G} \vee \mathcal{G}', t)$.

The elements of the $\pi_1(\mathcal{M})$-bimodule $\mathfrak{M}(p)$ are the pairs (ℓ, z), where $z \in P^{-1}(t)$ and ℓ is a homotopy class of a $\mathcal{G} \vee \mathcal{G}'$-path starting in t and ending in $E(z)$. The second coordinate z is just a label used for the case when E is not injective on $P^{-1}(t)$.

If $\gamma \in \pi_1(\mathcal{G}, t)$ is a $\mathcal{G}$-loop based in t, then
$$(\ell, z) \cdot \gamma = (\ell\gamma, z)$$
and
$$\gamma \cdot (\ell, z) = \left(E\left(P^{-1}(\gamma)[z]\right)\ell, z'\right),$$
where z' is the end of the path $P^{-1}(\gamma)[z]$. Recall that $P^{-1}(\gamma)[z]$ is the P-preimage of γ, starting in z.

The following is straightforward.

PROPOSITION 5.1.1. *Let $p : \mathcal{M}_1 \longrightarrow \mathcal{M}$ be a d-fold partial self-covering. Then the $\pi_1(\mathcal{M})$-bimodule $\mathfrak{M}(p)$ is a d-fold covering bimodule. It is irreducible if $\mathcal{M}_1$ is path-connected. A collection $\{(\ell_1, z_1), \ldots, (\ell_d, z_d)\}$ is a basis of $\mathfrak{M}(p)$ if and only if $\{z_1, \ldots, z_d\} = P^{-1}(t)$. If X is a basis of $\mathfrak{M}(p)$, then the associated action of the fundamental group $\pi_1(\mathcal{M})$ on $\mathsf{X} \subset \mathsf{X}^*$ is conjugate to the monodromy action on p.* □

We will prove later that the bimodule $\mathfrak{M}(p)$ does not depend on the choices we have made when constructing it (Proposition 5.1.2).

5.1.5. The associated virtual endomorphism. Fix an element $\ell \in \mathfrak{M}(p)$, i.e., a path ℓ starting in the basepoint t and ending in its preimage z. Let ϕ be the virtual endomorphism of $\pi_1(\mathcal{M})$ $(= \pi_1(\mathcal{G}, t) = \pi_1(\mathcal{G} \vee \mathcal{G}', t))$ associated to $\mathfrak{M}(p)$ and ℓ. Its domain is, by definition, the set of loops $\gamma \in \pi_1(\mathcal{M})$ such that $p^{-1}(\gamma)[z]$ is also a loop (we have to take $P^{-1}(\gamma)[z]$ in the case of an orbispace). Thus $\mathrm{Dom}\,\phi$ is an index d subgroup isomorphic to the fundamental group of $\mathcal{M}_1$. Action of ϕ on its domain is given by
$$\phi(\gamma) = \ell^{-1} p^{-1}(\gamma)[z]\ell$$
or
$$\phi(\gamma) = \ell^{-1} E\left(P^{-1}(\gamma)[z]\right)\ell$$
for orbispaces.

We say that ϕ is the *virtual endomorphism associated to the partial self-covering* $p : \mathcal{M}_1 \longrightarrow \mathcal{M}$.

PROPOSITION 5.1.2. *The virtual endomorphism ϕ of $\pi_1(\mathcal{M})$ is uniquely determined, up to a conjugacy, by the partial self-covering $p : \mathcal{M}_1 \longrightarrow \mathcal{M}$ and is conjugate to the composition $e_* \circ p_*^{-1}$, where $e : \mathcal{M}_1 \hookrightarrow \mathcal{M}$ is the embedding.*

The $\pi_1(\mathcal{M})$-bimodule $\mathfrak{M}(p)$ is isomorphic to $\phi(\pi_1(\mathcal{M}))\pi_1(\mathcal{M})$ and is determined uniquely (up to an isomorphism of bimodules) by the self-covering p.

PROOF. The virtual endomorphism ϕ is the composition of the homomorphisms

(5.1) $\pi_1(\mathcal{M}) = \pi_1(\mathcal{G}, t) \xdashrightarrow{P_*^{-1}} \pi_1(\mathcal{G}_1, z) \xrightarrow{E_*} \pi_1(\mathcal{G} \vee \mathcal{G}', E(z)) \xrightarrow{L} \pi_1(\mathcal{G}, t) = \pi_1(\mathcal{M}),$

where P_*^{-1} is the isomorphism $\gamma \mapsto P^{-1}(\gamma)[z]$ of a subgroup of finite index in $\pi_1(\mathcal{M})$ with $\pi_1(\mathcal{M}_1)$, E_* is the homomorphism induced by the functor E and L is the isomorphism of $\pi_1(\mathcal{G} \vee \mathcal{G}', E(z))$ with $\pi_1(\mathcal{G} \vee \mathcal{G}', t) = \pi_1(\mathcal{G}, t)$ given by the path ℓ.

The first statement follows from Lemma 4.7.4. The second one follows from the decomposition (5.1) and Propositions 2.5.6 and 2.5.8. □

5.2. Standard self-similar actions of $\mathrm{IMG}(p)$ on X^*

In this section we give an effective method of computation of a faithful self-similar action of an iterated monodromy group.

5.2.1. Construction of a standard action.
The tree of preimages T (its universal cover $\widetilde{T}$, for orbispaces) defined by a partial self-covering $p : \mathcal{M}_1 \longrightarrow \mathcal{M}$ is a d-regular rooted tree. Therefore T is isomorphic to the tree of words X^* over an alphabet X of d letters.

We will prove later that the nth tensor power $\mathfrak{M}(p)^{\otimes n}$ of the bimodule $\mathfrak{M}(p)$ is isomorphic to the bimodule $\mathfrak{M}(p^n)$ of the nth iteration of p.

The isomorphism $\ell : \mathfrak{M}(p)^{\otimes n} \longrightarrow \mathfrak{M}(p^n)$ is defined inductively by

$$(5.2) \qquad \ell(v_1 \otimes v_2) = p^{-n_2}\left(\ell(v_1)\right)[z]\ell(v_2),$$

where $v_1 \in \mathfrak{M}(p)^{\otimes n_1}$, $v_2 \in \mathfrak{M}(p)^{\otimes n_2}$, and z is the end of $\ell(v_2)$.

Recall that a set of paths $\{\ell_1, \ldots, \ell_d\}$ is a basis of $\mathfrak{M}(p)$ if and only if the paths ℓ_i start in t and end in t_i, where $\{t_1, \ldots, t_d\} = p^{-1}(t)$. So, if $\mathsf{X} = \{x_1 = \ell_1, \ldots, x_d = \ell_d\}$ is a basis of the bimodule $\mathfrak{M}(p)$, then $\ell(\mathsf{X}^n)$ is a basis of $\mathfrak{M}(p^n)$. In particular, the map

$$v \mapsto \text{ end of } \ell(v)$$

is a bijection $\Lambda : \mathsf{X}^n \longrightarrow p^{-n}(t)$.

Let us prove the following direct description of the obtained bijection $\Lambda : \mathsf{X}^* \longrightarrow T$ for the case of non-singular spaces. The general case will be treated later using tensor products of permutational bimodules.

PROPOSITION 5.2.1. *Let $p : \mathcal{M}_1 \longrightarrow \mathcal{M}$ be a partial self-covering of topological spaces and let $T = \bigsqcup_{n \geq 0} f^{-n}(t)$ be the tree of preimages. Choose an alphabet X, a bijection $\Lambda : \mathsf{X} \longrightarrow p^{-1}(t)$ and paths $\ell(x)$ starting in t and ending in $\Lambda(x)$ for every $x \in \mathsf{X}$. Define the map $\Lambda : \mathsf{X}^* \longrightarrow T$ inductively by the rule that $\Lambda(xv)$ for $x \in \mathsf{X}$ and $v \in \mathsf{X}^n$ is the end of the path*

$$p^{-n}\left(\ell(x)\right)[\Lambda(v)].$$

Then $\Lambda : \mathsf{X}^* \longrightarrow T$ is an isomorphism of rooted trees.

PROOF. If we know that $\Lambda(v)$ is adjacent to $\Lambda(vy)$ in T, i.e., that $p\left(\Lambda(vy)\right) = \Lambda(v)$, then

$$p\left(p^{-n-1}\left(\ell(x)\right)[\Lambda(vy)]\right) = p^{-n}\left(\ell(x)\right)[\Lambda(v)].$$

Hence the end $p\left(\Lambda(xvy)\right)$ of the path on the left hand side is equal to the end $\Lambda(xv)$ of the path on the right hand side of the equality. This proves that $\Lambda(xv)$ and $\Lambda(xvy)$ are adjacent in the tree T and, by induction, Λ is a morphism of the rooted trees. See Figure 5.2 for the obtained picture of the paths and the action of p.

We leave it to the reader to prove that Λ is injective (or that it is surjective), which will imply that it is an isomorphism. $\square$

The action of $\pi_1(\mathcal{M})$ on a basis X of the bimodule $\mathfrak{M}(p)$ is conjugate to the monodromy action of $\pi_1(\mathcal{M})$ on $p : \mathcal{M}_1 \longrightarrow \mathcal{M}$, by definition of the bimodule $\mathfrak{M}(p)$. Since $\mathfrak{M}(p)^{\otimes n}$ is isomorphic to $\mathfrak{M}(p^n)$, the iterated monodromy action of $\pi_1(\mathcal{M})$ on T is conjugated by Λ with the self-similar action $(\pi_1(\mathcal{M}), \mathfrak{M}(p), \mathsf{X})$. This self-similar action is called the *standard action* of $\pi_1(\mathcal{M})$ on X^*.

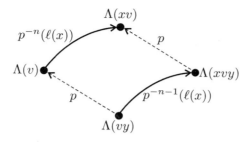

FIGURE 5.2. Isomorphism $\Lambda : \mathsf{X}^* \longrightarrow T$

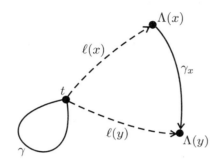

FIGURE 5.3. Recurrent formula of the standard action

Let us give a direct proof of the following proposition for the case of usual topological spaces directly, using the definition of the isomorphism $\Lambda : \mathsf{X}^* \longrightarrow T$ given in Proposition 5.2.1. It will be proved once more for the general case using permutational bimodules (Proposition 5.2.4).

PROPOSITION 5.2.2. *Let $\Lambda : \mathsf{X}^* \longrightarrow T$ be the isomorphism of the rooted trees defined by a bijection $\Lambda : \mathsf{X} \longrightarrow p^{-1}(t)$ and a collection of paths $\ell(x)$ as in Proposition 5.2.1. The standard action is, by definition, the action of $\pi_1(\mathcal{M}, t)$ (or of $\mathrm{IMG}(p)$) on X^* obtained by conjugation by Λ of the iterated monodromy action on T. Then the standard action is self-similar and is given by the recurrent formula*

$$(5.3) \qquad \gamma(xv) = y\left(\ell(y)^{-1}\gamma_x \ell(x)\right)(v),$$

where $\gamma_x = p^{-1}(\gamma)[\Lambda(x)]$, $v \in \mathsf{X}^*$ and y is such that $\Lambda(y)$ is the end of γ_x.

See Figure 5.3, where the loop $\ell(y)^{-1}\gamma_x \ell(x)$ is drawn.

PROOF. Suppose that $\gamma(xv) = yu$, i.e., that $\gamma(\Lambda(xv)) = \Lambda(yu)$ in the iterated monodromy action, for $v, u \in \mathsf{X}^n$ and $x, y \in \mathsf{X}$. This implies first of all that $\gamma(\Lambda(x)) = \Lambda(y)$, i.e., that the path $\gamma_x = p^{-1}(\gamma)[\Lambda(x)]$ ends in the point $\Lambda(y)$.

The point $\Lambda(xv)$ is, by the definition of Λ, the end of the path $p^{-n}(\ell(x))[\Lambda(v)]$, and $\Lambda(yu)$ is the end of $p^{-n}(\ell(y))[\Lambda(u)]$. Denote

$$L = \left(p^{-n}(\ell(y))[\Lambda(u)]\right)^{-1} \cdot p^{-(n+1)}(\gamma)[\Lambda(xv)] \cdot p^{-n}(\ell(x))[\Lambda(v)].$$

Then L is a well defined path starting in $\Lambda(v)$ and ending in $\Lambda(u)$. The path $p^n(L)$ is a loop starting and ending in t and is equal to

$$\ell(y)^{-1} \cdot p^{-1}(\gamma)[\Lambda(x)] \cdot \ell(x) = \ell(y)^{-1}\gamma_x \ell(x).$$

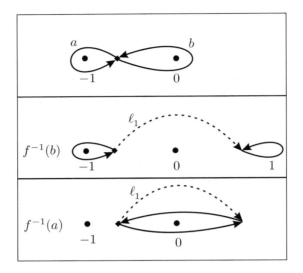

FIGURE 5.4. Computation of the group IMG (z^2-1)

We get that
$$\ell(y)^{-1}\gamma_x\ell(x)\left(\Lambda(v)\right)=\Lambda(u)$$
in the iterated monodromy action, i.e., that $\ell(y)^{-1}\gamma_x\ell(x)(v)=u$ in the standard action on X^*. □

5.2.2. An example of computation of IMG (p). Consider the polynomial z^2-1 as a covering of the space $\mathcal{M}=\mathbb{C}\setminus\{0,-1\}$ by its open subset $\mathcal{M}_1=\mathbb{C}\setminus\{0,-1,1\}$. Here $\{0,1\}$ is the *post-critical set* of p, i.e., the orbit of the *critical value* -1. We have to delete the post-critical set when we want to get a partial self-covering.

Choose $t=\frac{1-\sqrt{5}}{2}$ as a basepoint of $\mathcal{M}$. It has two preimages: itself and $-t$. Choose the path $\ell_0=\ell(x_0)$ to be the trivial path at t and $\ell_1=\ell(x_1)$ to be the path connecting t with $-t$ above the real axis, as the dotted path shown in Figure 5.4. Let a and b be the generators of $\pi_1(\mathcal{M},t)$ equal to the loops going in the positive direction around the points -1 and 0 respectively. The loops a and b are shown in the upper part of Figure 5.4.

The preimages of the loops a and b are shown on the two lower parts of Figure 5.4. It follows that
$$a\cdot x_0=x_1\cdot b,\quad a\cdot x_1=x_0\cdot 1,$$
$$b\cdot x_0=x_0\cdot a,\quad b\cdot x_1=x_1\cdot 1,$$
so that the group IMG (z^2-1) is generated by the automaton with the Moore diagram shown in Figure 3.2 on page 111.

5.2.3. Tensor products correspond to compositions.

PROPOSITION 5.2.3. *Let* $p_1:\mathcal{M}_1\longrightarrow\mathcal{M}$ *and* $p_2:\mathcal{M}_2\longrightarrow\mathcal{M}$ *be partial self-coverings. Then*
$$\mathfrak{M}(p_1\circ p_2)\cong\mathfrak{M}(p_1)\otimes\mathfrak{M}(p_1).$$

PROOF. One can define the isomorphism $\ell : \mathfrak{M}(p_1) \otimes \mathfrak{M}(p_2) \longrightarrow \mathfrak{M}(p_1 \circ p_2)$ by rewriting (5.2) (page 142) in appropriate atlases of the orbispaces. The proof that Λ is an isomorphism is straightforward in the case of a non-singular topological space, though rather technical for orbispaces. Therefore, we prefer a less technical proof which uses associated virtual endomorphisms.

The partial self-covering is defined on a sub-orbispace $\mathcal{M}_2^\circ$ of $\mathcal{M}_2$ and is equal to the composition $p_1 \circ p_2^\circ$, where $p_2^\circ : \mathcal{M}_2^\circ \longrightarrow \mathcal{M}_1$ is the restriction of p_2 onto $\mathcal{M}_1$. We have the following diagram from the definition of a pullback (where the square is commutative).

(5.4)
$$\begin{array}{ccccc} \mathcal{M}_2^\circ & \hookrightarrow & \mathcal{M}_2 & \hookrightarrow & \mathcal{M} \\ \downarrow p_2^\circ & & \downarrow p_2 & & \\ \mathcal{M}_1 & \hookrightarrow & \mathcal{M} & & \\ \downarrow p_1 & & & & \\ \mathcal{M} & & & & \end{array}$$

Composition of the embeddings $\mathcal{M}_2^\circ \hookrightarrow \mathcal{M}_2 \hookrightarrow \mathcal{M}$ is the embedding of $\mathcal{M}_2^\circ$ into $\mathcal{M}$.

It follows from the construction of the pullback of a covering (see 4.4.3) that we can find atlases $(\mathcal{G}, \mathcal{X})$; $(\mathcal{G}', \mathcal{X}')$, $(\mathcal{G}'', \mathcal{X}'')$ of $\mathcal{M}$, atlases $(\mathcal{G}_1, \mathcal{X}_1)$ and $(\mathcal{G}_2, \mathcal{X}_2)$ of $\mathcal{M}_1$ and $\mathcal{M}_2$, respectively; and an atlas $(\mathcal{G}_2^\circ, \mathcal{X}_2^\circ)$ of $\mathcal{M}_2^\circ$ such that the embeddings and the coverings from the diagram (5.4) are defined by functors

$$\begin{array}{ccccc} \mathcal{G}_2^\circ & \xrightarrow{E_1^\circ} & \mathcal{G}_2 & \xrightarrow{E_2} & \mathcal{G}'' \\ \downarrow P_2^\circ & & \downarrow P_2 & & \\ \mathcal{G}_1 & \xrightarrow{E_1} & \mathcal{G}' & & \\ \downarrow P_1 & & & & \\ \mathcal{G} & & & & \end{array}$$

Choose a basepoint $t \in \mathcal{X}$ and its preimage $t_1 \in P_1^{-1}(t) \subset \mathcal{X}_1$, and denote $t' = E_1(t_1) \in \mathcal{X}'$. Choose some $t_2^\circ \in (P_2^\circ)^{-1}(t_1) \subset \mathcal{X}_2^\circ$ and let $t_2 = E_1^\circ(t_2^\circ)$. Denote $t'' = E_2(t_2)$. Then t_2 belongs to $P_2^{-1}(t')$ (see Figure 5.5). Let ℓ_1 be a $\mathcal{G} \vee \mathcal{G}' \vee \mathcal{G}''$-path from t to t' and let ℓ_2 be a $\mathcal{G} \vee \mathcal{G}' \vee \mathcal{G}''$-path from t' to t''.

We identify $\pi_1(\mathcal{M})$ with $\pi_1(\mathcal{G}, t) = \ell_1^{-1} \cdot \pi_1(\mathcal{G}', t') \cdot \ell_1 = \ell_1^{-1} \ell_2^{-1} \cdot \pi_1(\mathcal{G}'', t'') \cdot \ell_2 \ell_1$. Let ϕ_i be the virtual endomorphism of $\pi_1(\mathcal{M})$ associated to p_i, $i = 1, 2$. Then for every $\gamma \in \text{Dom}\, \phi_2 \circ \phi_1$ we have

$$\phi_2(\phi_1(\gamma)) = \phi_2\left(\ell_1^{-1} \cdot E_1 \circ P_1^{-1}(\gamma)[t_1] \cdot \ell_1\right)$$
$$= \ell_1^{-1} \ell_2^{-1} \cdot E_2 \circ P_2^{-1}\left(E_1 \circ P_1^{-1}(\gamma)[t_1]\right)[t_2] \cdot \ell_2 \ell_1$$
$$= \ell_1^{-1} \ell_2^{-1} \cdot (E_2 \circ E_1^\circ) \circ (P_1 \circ P_2^\circ)^{-1}(\gamma)[t_2^\circ] \cdot \ell_2 \ell_1,$$

which implies that $\phi_2 \circ \phi_1$ is associated to $p_1 \circ p_2$. Now Proposition 2.8.4 and uniqueness, up to conjugacy, of the associated virtual endomorphism finishes the proof. □

5.2.4. The standard action. Let us summarize the obtained results on computation of the standard action on X^* in the next proposition.

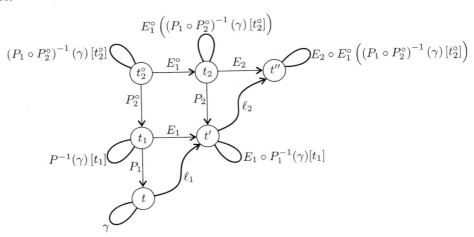

Figure 5.5

Proposition 5.2.4. *Let $p : \mathcal{M}_1 \longrightarrow \mathcal{M}$ be a partial self-covering defined by a functor $P : \mathcal{G}_1 \longrightarrow \mathcal{G}$, and let $E : \mathcal{G}_1 \longrightarrow \mathcal{G}'$ be a functor defining the embedding $\mathcal{M}_1 \hookrightarrow \mathcal{M}$. Let $\mathfrak{M}(p)$ be the corresponding $\pi_1(\mathcal{G}, t)$-bimodule. Choose a basis $\mathsf{X} = \{x_1 = (\ell_1, z_1), \ldots, x_d = (\ell_d, z_d)\}$ of $\mathfrak{M}(p)$, where z_i are the P-preimages of the basepoint t and ℓ_i are $\mathcal{G} \vee \mathcal{G}'$-paths from t to $E(z_i)$. Then the monodromy action of the fundamental group $\pi_1(\mathcal{M})$ on the covering $p^n : \mathcal{M}_n \longrightarrow \mathcal{M}$ is conjugate to the restriction of the associated standard self-similar action $(\pi_1(\mathcal{M}), \mathfrak{M}(p), \mathsf{X})$ onto $\mathsf{X}^n \subset \mathsf{X}^*$.*

For every $\gamma \in \pi_1(\mathcal{G}, t)$, $x_i \in \mathsf{X}$ and $v \in \mathsf{X}^*$, the following equality holds for the standard action of $\pi_1(\mathcal{M})$ (and of $\mathrm{IMG}(p)$) on X^*:

$$(5.5) \qquad \gamma(x_i v) = x_j \left(\ell_j^{-1} \gamma_i \ell_i \right)(v),$$

where $\gamma_i = E \circ P^{-1}(\gamma)[z_i]$ and the path $x_j = \ell_j \in \mathsf{X}$ ends in the same point z_j as γ_i does.

Here and further $E \circ P^{-1}(\gamma)[z_i]$ denotes $E\left(P^{-1}(\gamma)[z_i]\right)$.

Proof. It follows directly from Propositions 5.1.1 and 5.2.3. The recurrent formula for the standard action follows directly from the definition of the bimodule $\mathfrak{M}(p)$ (see 5.1.4) and from the definition of the associated self-similar action (see 2.3.2). □

Corollary 5.2.5. *The iterated monodromy group $\mathrm{IMG}(p)$ is isomorphic to the quotient of $\pi_1(\mathcal{M})$ by the kernel of the self-similar action defined by the bimodule $\mathfrak{M}(p)$.* □

5.3. Iterated monodromy groups of limit dynamical systems

Let us show that contracting recurrent groups can be reconstructed from the action of the shift of their limit orbispace $\mathcal{J}_G$.

5.3. ITERATED MONODROMY GROUPS OF LIMIT DYNAMICAL SYSTEMS

THEOREM 5.3.1. *Let (G, X) be a faithful contracting recurrent action of a finitely generated group G. Then (G, X) is a standard action of the iterated monodromy group $\mathrm{IMG}\,(\mathsf{s})$ of the partial self-covering $\mathsf{s} : \mathcal{J}_G^\circ \longrightarrow \mathcal{J}_G$. In particular, $\mathrm{IMG}\,(\mathsf{s})$ is isomorphic to G.*

PROOF. The orbispaces $\mathcal{J}_G^\circ$ and $\mathcal{J}_G$ are path-connected and locally path-connected by Theorem 3.5.1. The orbispace $\mathcal{J}_G$ is the orbispace of the action of G on $\mathcal{X}_G$. Let $(\mathcal{G}_G, \mathcal{X}_G)$ be the corresponding atlas. Then $(\mathcal{G}_G \rtimes \sigma, \mathcal{X}_G \times \mathsf{X})$ is an atlas of $\mathcal{J}_G^\circ$, where σ is the cocycle $\sigma\left((g, \xi), x\right) = g(x)$ (see Section 4.6).

If $\gamma = (g_k, \xi_k) \cdot \gamma_k \cdots (g_1, \xi_1) \cdot \gamma_1 \cdot (g_0, \xi_0)$ is a $\mathcal{G}_G$-path, then the element

$$\varphi(\gamma) = g_k \cdots g_1 g_0 \in G$$

depends only on the homotopy class of γ (see Section 4.7 for the definition of $\mathcal{G}$-paths and their homotopy). Restriction of the map φ onto the fundamental group $\pi_1(\mathcal{G}_G, \xi_0)$ is therefore a surjective homomorphism of groups $\varphi : \pi_1(\mathcal{G}_G, \xi_0) \longrightarrow G$. Note that the group $\pi_1(\mathcal{G}_G, \xi_0)$ is usually "wild" (for example, it is often uncountable).

Let $\mathfrak{M}(\mathsf{s})$ be the $\pi_1(\mathcal{G}_G, \xi_0)$-bimodule of the covering s. It consists of the pairs (ℓ, x), where $x \in \mathsf{X}$ and ℓ is (a homotopy class of) a $\mathcal{G}_G$-path starting in ξ_0 and ending in the point $E_\mathcal{J}(\xi_0, x) = \xi_0 \otimes x$. Let us define the map $F : \mathfrak{M}(\mathsf{s}) \longrightarrow \mathfrak{M}$, where $\mathfrak{M} = \mathsf{X} \cdot G$ is the self-similarity bimodule by $F(\ell, x) = x \cdot \varphi(\ell)$.

It is easy to see that for all $m = (\ell, x) \in \mathfrak{M}(\mathsf{s})$ and $\gamma \in \pi_1(\mathcal{G}_G, \xi_0)$ we have $F(m \cdot \gamma) = F(m) \cdot \varphi(\gamma)$.

Let us prove that $F(\gamma \cdot m) = \varphi(\gamma) \cdot F(m)$. We have

$$\gamma \cdot (\ell, x) = \left(E_\mathcal{J} \circ P_\mathsf{s}^{-1}(\gamma)\left[(\xi_0, x)\right]\ell, \quad \sigma(\gamma, x)\right)$$

(here $\sigma(\gamma, x)$ is the monodromy action of γ on X). If

$$\gamma = (g_k, \xi_k) \cdot \gamma_k \cdots (g_1, \xi_1) \cdot \gamma_1 \cdot (g_0, \xi_0),$$

then

$$P_\mathsf{s}^{-1}(\gamma)\left[(\xi_0, x)\right] = (\xi_k, \ g_k, \ g_{k-1} \cdots g_1 g_0(x)) \cdots (\xi_2, g_2, g_1 g_0(x))$$
$$\cdot (\gamma_2, g_0(x)) \cdot (\xi_1, g_1, g_0(x)) \cdot (\gamma_1, x) \cdot (\xi_0, g_0, x)$$

and $\sigma(\gamma, x) = g_k \cdots g_1 g_0(x) = \varphi(\gamma)(x)$, by Proposition 4.7.5. Hence

$$E_\mathcal{J} \circ P_\mathsf{s}^{-1}(\gamma)\left[(\xi_0, x)\right] =$$
$$\left(\left(g_k|_{g_{k-1} \cdots g_1 g_0(x)}, \xi_k \otimes g_{k-1} \cdots g_1 g_0(x)\right) \cdots \left(g_2|_{g_1 g_0(x)}, \xi_2 \otimes g_1 g_0(x)\right)\right.$$
$$\left. \cdot (\gamma_2 \otimes g_0(x)) \cdot \left(g_1|_{g_0(x)}, \xi_1 \otimes g_0(x)\right) \cdot (\gamma_1 \otimes x) \cdot \left(g_0|_x, \xi_0 \otimes x\right)\right).$$

The product $\varphi\left(E_\mathcal{J} \circ P_\mathsf{s}^{-1}(\gamma)\left[(\xi_0, x)\right]\right)$ of the elements of the group G appearing in this path is

$$g_k|_{g_{k-1} \cdots g_1 g_0(x)} \cdots g_2|_{g_1 g_0(x)} g_1|_{g_0(x)} g_0|_x = \left(g_k \cdots g_2 g_1 g_0\right)|_x = \varphi(\gamma)|_x.$$

Consequently

$$F(\gamma \cdot (\ell, x)) = \varphi(\gamma)(x) \cdot (\varphi(\gamma)|_x \varphi(\ell)) = \varphi(\gamma) \cdot (x \cdot \varphi(\ell));$$

i.e., $F(\gamma \cdot m) = \varphi(\gamma) \cdot F(m)$ for all $m \in \mathfrak{M}(\mathsf{s})$ and $\gamma \in \pi_1(\mathcal{G}_G, \xi_0)$.

It follows now from the construction of the self-similar action associated to a permutation bimodule that $\varphi(\gamma)(w) = \gamma(w)$ for all $\gamma \in \pi_1(\mathcal{G}_G, \xi_0)$ and $w \in \mathsf{X}^*$, i.e., that the standard action $(\mathrm{IMG}(\mathsf{s}), \mathsf{X})$ coincides with the action (G, X), where the isomorphism $\mathrm{IMG}(\mathsf{s}) \longrightarrow G$ is induced by the homomorphism φ. □

5.4. Length structures and expanding maps

5.4.1. Length structure on topological spaces. Let $|x - y|$ be a metric on a space $\mathfrak{X}$. We say that a curve $\gamma : [a, b] \longrightarrow \mathfrak{X}$ is *rectifiable* if its *length* $l(\gamma)$ is finite, where

$$l(\gamma) = \sup \left(|\gamma(t_0) - \gamma(t_1)| + |\gamma(t_1) - \gamma(t_2)| + \cdots + |\gamma(t_{n-1}) - \gamma(t_n)| \right),$$

where supremum is taken over all partitions $a = t_0 < t_1 < \ldots < t_n = b$ of the interval $[a, b]$.

We say that the metric is a *length structure* on $\mathfrak{X}$ if the distance $|x - y|$ is equal to the infimum of the length of the curves connecting x and y.

A classical example of a space with a length structure is a Riemannian manifold with the usual notions of length of a curve and distance between points.

5.4.2. Length structures on orbispaces. A *quasi-metric* on a set M is a function $|x - y|$ from $M \times M$ to $[0, +\infty]$ such that $|x - x| = 0$ for every $x \in M$, $|x - y| = |y - x|$ and $|x - y| + |y - z| \geq |x - z|$ for all $x, y, z \in M$. We assume that $+\infty + t = t + \infty = +\infty$ for all $t \in [0, +\infty]$. A quasi-metric is called *finite* if $|x - y| \neq +\infty$ for all $x, y \in M$. It is called *positive* if $|x - y| \neq 0$ for all $x \neq y$. Thus a positive finite quasi-metric is a metric.

If we have a quasi-metric on a set M, then the corresponding topological space $(M, |x - y|)$ is defined by the fundamental system of open sets

$$B(x, r) = \{y \in M \,:\, |x - y| < r\}.$$

The *Hausdorff space* $(M, |x - y|)_H$ defined by the quasi-metric is the quotient of the topological space $(M, |x - y|)$ by the equivalence relation $x \sim y \iff |x - y| = 0$.

If $(M, |x - y|)$ is a space with a positive quasi-metric, then the *length* $l(\gamma)$ of a path $\gamma : [a, b] \to M$ is defined as $\sup \sum_{i=0}^{k-1} |\gamma(t_i) - \gamma(t_{i+1})|$, where supremum is taken over all partitions $t_0 = a < t_1 < t_2 < \ldots < t_k = b$ of the segment $[a, b]$. A positive quasi-metric is a *length quasi-metric* (and the space is called a *length space*) if distance between two points is equal to the infimum of the lengths of all paths connecting them.

DEFINITION 5.4.1. A *length structure* on a path-connected orbispace $\mathcal{M}$ is a positive quasi-metric $|x - y|$ on the unit space $\mathcal{X}$ of its atlas $(\mathcal{G}, \mathcal{X})$ such that

  (i) the quasi-metric defines the original topology on $\mathcal{X}$;
  (ii) it is a length quasi-metric (here the usual paths in $\mathcal{X}$ and not the $\mathcal{G}$-paths are considered);
  (iii) every change of charts $h : U \to V$ is a local isometry, i.e., for every $x \in U$ there exists a neighborhood U' of x such that $h|_{U'}$ is an isometry;
  (iv) the quasi-metric $|x - y|_l$ on the underlying space $|\mathcal{M}|$ (defined below) is a metric compatible with the topology on $|\mathcal{M}|$.

Two length structures on an orbispace are equivalent if their union satisfies condition (iii) of the definition (the distance between two points in different atlases is equal to infinity).

Suppose that we have a length quasi-metric on $\mathcal{X}$ satisfying condition (iii). The length $l(\gamma)$ of a $\mathcal{G}$-path $\gamma = g_k \cdot \gamma_k \cdots g_1 \cdot \gamma_1 \cdot g_0$ is equal by definition to the sum $\sum_{i=1}^{k} l(\gamma_k)$ of the lengths of the paths γ_i. It is easy to see that the lengths of equivalent paths are equal.

Then the quasi-metric $|x - y|_l$ on $\mathcal{X}$ is defined as the infimum of the lengths of $\mathcal{G}$-paths connecting x to y. It follows from the definitions that $|x - y|_l = 0$ if x and y belong to one $\mathcal{G}$-orbit. Therefore, the quasi-metric $|x - y|_l$ induces a quasi-metric on $|\mathcal{M}|$. It is the quasi-metric mentioned in (iv).

Let $\mathcal{M}_1$ be an open sub-orbispace of an orbispace $\mathcal{M}$, and let $E : (\mathcal{G}_1, \mathcal{X}_1) \longrightarrow (\mathcal{G}, \mathcal{X})$ be the embedding functor. Suppose that we have a length structure on $(\mathcal{G}, \mathcal{X})$. If the orbispace $\mathcal{M}_1$ is path-connected, then we also get a length structure on $\mathcal{M}_1$. Namely, if γ is a path in $\mathcal{X}_1$, then its length is equal by definition to the length of the path $E(\gamma)$ in the space $\mathcal{X}$. Then the distance between two points $x, y \in \mathcal{X}_1$ is equal, by definition, to the infimum of lengths of paths in $\mathcal{X}_1$ connecting x and y.

An orbispace $\mathcal{M}$ with a length structure is said to be *complete* if its underlying space $|\mathcal{M}|$ is complete with respect to the induced metric $|x - y|_l$.

5.4.3. Expanding self-coverings.
Let $p : \mathcal{M}_1 \longrightarrow \mathcal{M}$ be a partial self-covering of a path-connected (orbi)space $\mathcal{M}$ by an open path-connected subspace (sub-orbispace) $\mathcal{M}_1$. Suppose that we have a length structure on $\mathcal{M}$. Then we have the induced length structures on the domains $\mathcal{M}_n$ of the iterations p^n.

DEFINITION 5.4.2. A partial self-covering $p : \mathcal{M}_1 \longrightarrow \mathcal{M}$ is *expanding* if there exists a constant λ, $0 < \lambda < 1$ and for every path γ there exists $c > 0$ such that every preimage of γ under p^n has length not greater than $c \cdot \lambda^n$.

Let $p : \mathcal{M}_1 \longrightarrow \mathcal{M}$ be an expanding self-covering. Suppose that the metric $|x - y|_l$ is complete on $|\mathcal{M}|$. Then the *Julia set* of p is the set of the accumulation points of $\bigcup_{n \geq 1} p^{-n}(t_0) \subset |\mathcal{M}|$, where $t_0 \in \mathcal{M}$ is arbitrary. We consider here p as a usual partial map of the underlying space. Hence the Julia set is just a closed subset of the underlying space $|\mathcal{M}|$. We will introduce an orbispace structure on it later.

This is not the classical definition of the Julia set, but it will coincide with the classical one in the case of a *sub-hyperbolic* rational function (see Subsection 6.4.4).

PROPOSITION 5.4.3. *The Julia set $\mathcal{J}_p$ of an expanding map $p : \mathcal{M}_1 \longrightarrow \mathcal{M}$ does not depend on the choice of t_0. The Julia set is completely invariant with respect to p, i.e., $p(\mathcal{J}_p) = p^{-1}(\mathcal{J}_p) = \mathcal{J}_p$.*

PROOF. Suppose that $t_1 \in \mathcal{M}$ is another point and let γ be a path of finite length connecting t_0 with t_1. Then for every $x \in p^{-n}(t_0)$ there exists a p^n-preimage of the path γ, starting in x. Let y be the end of this preimage. Then y belongs to $p^{-n}(t_1)$. The length of the path γ' is not greater than $c \cdot \lambda^n$ for some constants $c > 0$, $0 < \lambda < 1$. Therefore $|x - y|_l \leq c\lambda^n$, which implies that the set of the accumulation points of $\bigcup_{n=0}^{\infty} p^{-n}(t_1)$ is equal to the set of the accumulation points of $\bigcup_{n=0}^{\infty} p^{-n}(t_0)$.

We have

$$p\left(\bigcup_{n=0}^{\infty} p^{-n}(t_0)\right) = \bigcup_{n=0}^{\infty} p^{-n}(p(t_0)) = \bigcup_{n=0}^{\infty} p^{-n}(t_0) \cup \{p(t_0)\};$$

hence $p(\mathcal{J}_p) = \mathcal{J}_p$.

We also have
$$p^{-1}\left(\bigcup_{n=0}^{\infty} p^{-n}(t_0)\right) \cup \{t_0\} = \bigcup_{n=0}^{\infty} p^{-n}(t_0);$$
hence $p^{-1}(\mathcal{J}_p) = \mathcal{J}_p$. □

DEFINITION 5.4.4. A partial self-covering $p : \mathcal{M}_1 \longrightarrow \mathcal{M}$ is *uniformly expanding on its Julia set* $\mathcal{J}_p$ if it is expanding and for all $R > 0$ and $\epsilon > 0$ there exists n_0 such that for every path γ of length $< R$ starting and ending in $\mathcal{J}_p$ every preimage of γ under p^n for $n \geq n_0$ has length less than ϵ.

5.4.4. The Julia set as an orbispace. The *Julia orbispace* is the orbispace defined by the restriction of an atlas $(\mathcal{G}, \mathcal{X})$ of $\mathcal{M}$ onto the full preimage of the Julia set $\mathcal{J}_p$ in $\mathcal{X}$ (and passing, if necessary, to the faithful quotient of the respective groupoid, i.e., to the groupoid of germs of the associate pseudogroup). We will denote the Julia orbispace also by $\mathcal{J}_p$.

By Proposition 5.4.3, the Julia set is a subset of $\mathcal{M}_1$. We also can consider the restriction of the atlas of $\mathcal{M}_1$ onto the full preimage of $\mathcal{J}_p$. We will get in this way an atlas of an orbispace $\mathcal{J}_p^\circ$. The orbispaces $\mathcal{J}_p^\circ$ and $\mathcal{J}_p$ have the same underlying spaces, but in general $\mathcal{J}_p^\circ$ is only an open sub-orbispace of $\mathcal{J}_p$.

We get a partial self-covering $p : \mathcal{J}_p^\circ \longrightarrow \mathcal{J}_p$, called the *restriction of p onto its Julia orbispace*.

5.5. Limit spaces of iterated monodromy groups

5.5.1. Faithfully represented isotropy groups. Let $p : \mathcal{M}_1 \longrightarrow \mathcal{M}$ be a partial self-covering. Let $(\mathcal{G}, \mathcal{X})$ be an atlas of $\mathcal{M}$ and choose a basepoint $t \in \mathcal{X}$. Let $x \in \mathcal{M}$ and let $\mathcal{G}_x$ be the isotropy group of x. If we choose a path γ from the basepoint t to a preimage of x in $\mathcal{X}$, then we define a homomorphism $I_x = I_{x,\gamma} : \mathcal{G}_x \longrightarrow \pi_1(\mathcal{M}, t) : g \mapsto \gamma^{-1} \cdot g \cdot \gamma$. The homomorphism I_x is unique up to a conjugation in $\pi_1(\mathcal{M})$. We know that $\mathcal{M}$ is developable if and only if I_x is injective for every $x \in \mathcal{M}$ (see Subsection 4.7.3).

DEFINITION 5.5.1. We say that the *isotropy group* $\mathcal{G}_x$ of $\mathcal{M}$ is *faithfully represented in* IMG (p) if the composition of I_x with the canonical epimorphism

$$\pi_1(\mathcal{M}) \longrightarrow \mathrm{IMG}\,(p)$$

is injective.

In particular, if all the isotropy groups of $\mathcal{M}$ are faithfully represented in IMG (p), then the orbispace $\mathcal{M}$ is developable.

It is relatively simple to check whether the isotropy groups are faithfully represented in IMG (p).

LEMMA 5.5.2. *Suppose that* $p : \mathcal{M}_1 \longrightarrow \mathcal{M}$ *is a partial self-covering of an orbispace $\mathcal{M}$ and let $x \in \mathcal{M}$ be any point. Let T be the preimage tree of the point x and let $\widetilde{T}$ be its universal covering. Then the following conditions are equivalent:*

(1) *The isotropy group $\mathcal{G}_x$ of x is faithfully represented in* IMG (p).
(2) *The iterated monodromy action of $\mathcal{G}_x$ on $\widetilde{T}$ is faithful.*
(3) *The intersection of the images of the vertex groups of T in $\mathcal{G}_x$ is trivial.*

5.5.2. Julia sets as limit spaces of iterated monodromy groups.

THEOREM 5.5.3. *Let $p : \mathcal{M}_1 \longrightarrow \mathcal{M}$ be a partial self-covering of a path-connected and locally simply connected orbispace $\mathcal{M}$ with a complete length structure. Suppose that the fundamental group of $\mathcal{M}$ is finitely generated, p is uniformly expanding on its Julia set and the isotropy groups of $\mathcal{M}$ are faithfully represented in* IMG (p).

Then the associated bimodule $\mathfrak{M}(p)$ of the self-covering is hyperbolic, and the next partial self-coverings are conjugate:

- *the restriction $p : \mathcal{J}_p^\circ \longrightarrow \mathcal{J}_p$ of p onto the Julia orbispace,*
- *the shift $\mathsf{s} : \mathcal{J}_{\pi_1(\mathcal{M})}^\circ \longrightarrow \mathcal{J}_{\pi_1(\mathcal{M})}$,*
- *the shift $\mathsf{s} : \mathcal{J}_{\mathrm{IMG}(p)}^\circ \longrightarrow \mathcal{J}_{\mathrm{IMG}(p)}$,*

where $\mathcal{J}_{\pi_1(\mathcal{M})}$ and $\mathcal{J}_{\mathrm{IMG}(p)}$ are constructed using the hyperbolic bimodule $\mathfrak{M}(p)$ and the standard contracting action of IMG (p), *respectively.*

In particular, the limit dynamical system $(\mathcal{J}_{\mathrm{IMG}(p)}, \mathsf{s})$ and the dynamical system $(\mathcal{J}_p, p)$ are topologically conjugate.

PROOF. The isotropy groups of $\mathcal{M}$ are faithfully represented in IMG (p); hence $\mathcal{M}$ is developable. Let $\widehat{\mathcal{M}}$ be the universal covering. Then the groupoid of germs of the natural action of $\pi_1(\mathcal{M})$ on $\widehat{\mathcal{M}}$ is an atlas of $\mathcal{M}$. If $\widehat{\mathcal{J}}$ is the preimage of the Julia set $\mathcal{J}_p$ in $\widehat{\mathcal{M}}$, then $\widehat{\mathcal{J}}$ is $\pi_1(\mathcal{M})$-invariant and the groupoid of germs of the action of $\pi_1(\mathcal{M})$ on $\widehat{\mathcal{J}}$ is an atlas of the Julia orbispace $\mathcal{J}_p$.

Let us prove at first that the standard action of $\pi_1(\mathcal{M})$ is contracting. Suppose that the standard action is defined by a collection of paths $\ell(\mathsf{X}) = \{\ell(x) : x \in \mathsf{X}\}$. Let λ and c be the constants as in Definition 5.4.2, where c is chosen common for all the elements of $\ell(\mathsf{X})$.

It follows from the definition of the standard action that for $\gamma \in \pi_1(\mathcal{M})$ and $v \in \mathsf{X}^n$ the restriction $\gamma|_v$ is a loop whose image in $|\mathcal{M}|$ is of the form

$$\rho = \alpha_0^{-1} \alpha_1^{-1} \cdots \alpha_n^{-1} \cdot \gamma_n \cdot \beta_n \cdots \beta_1 \beta_0,$$

where $\gamma_n \in p^{-n}(\gamma)$ and $\alpha_k, \beta_k \in p^{-k}(\ell(\mathsf{X}))$. There exists, by Definition 5.4.2, a constant $c_1 > 0$ not depending on n such that the length of the loop ρ is not greater than

$$c_1 \lambda^n + 2c \left(\lambda^{n-1} + \lambda^{n-2} + \cdots + \lambda + 1 \right) < c_1 \lambda^n + \frac{2c}{1 - \lambda}.$$

Consequently, for all n big enough the restrictions $\gamma|_v$ in the words v of length n are defined by loops of length less than $R = 1 + 2c/(1 - \lambda)$. The universal cover $\widehat{\mathcal{M}}$ of $\mathcal{M}$ is a complete locally compact Hausdorff length space. Therefore, by Hopf-Rinow theorem (see [**HR32**] and [**BH99**], p. 35), the set of the ends of the lifts to $\widehat{\mathcal{M}}$ of the loops which have length less than R is compact and hence finite. Consequently, the set of the elements of $\pi_1(\mathcal{M})$ defined by the loops of length less than R is finite and the standard action is contracting.

We have to prove the conjugacy of the partial self-coverings. Let us show that the right $\pi_1(\mathcal{M})$-space $\widehat{\mathcal{J}}$ satisfies the conditions of Theorem 3.4.13.

The Julia set $\mathcal{J}_p$ is a bounded closed subset of a complete length space $\mathcal{M}$; hence it is compact by Hopf-Rinow theorem. This implies that the action of $\pi_1(\mathcal{G}, t)$ on $\widehat{\mathcal{J}}$ is co-compact.

It remains to prove that the right $\pi_1(\mathcal{M})$-space $\widehat{\mathcal{J}}$ is self-similar with a contracting self-similarity.

Let us describe the self-similarity. Choose some atlases $(\mathcal{G}, \mathcal{X})$, $(\mathcal{G}', \mathcal{X}')$ of $\mathcal{M}$ and an atlas $(\mathcal{G}_1, \mathcal{X}_1)$ of $\mathcal{M}_1$ such that the covering p and the embedding $\mathcal{M}_1 \hookrightarrow \mathcal{M}$ are defined by functors $P : \mathcal{G}_1 \longrightarrow \mathcal{G}$ and $E : \mathcal{G}_1 \longrightarrow \mathcal{G}'$. Choose a basepoint $t \in \mathcal{X}$ and construct the bimodule $\mathfrak{M}(p)$ using these data.

Let $(\mathcal{G}_t, \mathcal{X}_t)$ and $((\mathcal{G} \vee \mathcal{G}')_t, (\mathcal{X} \sqcup \mathcal{X}')_t)$ be the respective atlases of $\widehat{\mathcal{M}}$, constructed in 4.7.3 on page 133. Let $(\mathcal{G}_t^J, \mathcal{X}_t^J)$ and $\left((\mathcal{G} \vee \mathcal{G}')_t^J, (\mathcal{X} \sqcup \mathcal{X}')_t^J\right)$ be restrictions of these atlases onto the preimage of $\widehat{\mathcal{J}}$ in them.

Then the self-similarity of the right $\pi_1(\mathcal{M})$-space $\widehat{\mathcal{J}}$ is defined in the following way.

LEMMA 5.5.4. *For every $\gamma \in \mathcal{X}_t^J$ and $(\ell, z) \in \mathfrak{M}(p)$ put*

$$\Phi(\gamma \otimes (\ell, z)) = E \circ P^{-1}(\gamma)[z]\ell \in (\mathcal{X} \sqcup \mathcal{X}')_t^J.$$

Then Φ defines an $\mathfrak{M}(p)$-self-similarity structure on $\widehat{\mathcal{J}}$.

PROOF. It is easy to see that $\Phi(\gamma \otimes (\ell, z))$ is an element of $(\mathcal{X} \sqcup \mathcal{X}')_t^J$ and that if γ_1 and γ_2 belong to one $\mathcal{G}_t^J$-orbit, then $\Phi(\gamma_1 \otimes (\ell, z))$ and $\Phi(\gamma_2 \otimes (\ell, z))$ belong to one $(\mathcal{G} \vee \mathcal{G}')_t^J$-orbit. Hence, for every $y \in \widehat{\mathcal{J}}$ and $(\ell, z) \in \mathfrak{M}(p)$ the image of $\Phi(y \otimes m)$ is a well defined point of $\widehat{\mathcal{J}}$.

Let us show that $\Phi : \widehat{\mathcal{J}} \otimes_{\pi_1(\mathcal{G}, t)} \mathfrak{M}(p) \longrightarrow \widehat{\mathcal{J}}$ is a self-similarity structure on the right $\pi_1(\mathcal{M}, t)$-space $\widehat{\mathcal{J}}$. We have to prove that Φ is well defined, agrees with the action of $\pi_1(\mathcal{M})$ and is a homeomorphism.

We have for every $\gamma \in \mathcal{X}_t^J$, $g \in \pi_1(\mathcal{G}, t)$ and $(\ell, z) \in \mathfrak{M}(p)$

$$\Phi(\gamma \cdot g \otimes (\ell, z)) = E \circ P^{-1}(\gamma \cdot g)[z]\ell$$
$$= E \circ P^{-1}(\gamma)[z'] \left(E \circ P^{-1}(g)[z] \cdot \ell\right)$$
$$= E \circ P^{-1}(\gamma)[z'] \left(g \cdot (\ell, z)\right) = \Phi(\gamma \otimes g \cdot (\ell, z)),$$

where z' is the end of $E \circ P^{-1}(g)[z]$. Hence $\Phi : \widehat{\mathcal{J}} \otimes_{\pi_1(\mathcal{G}, t)} \mathfrak{M}(p) \longrightarrow \widehat{\mathcal{J}}$ is well defined.

The right action of $\pi_1(\mathcal{M}) = \pi_1(\mathcal{G}, t)$, both on $\mathfrak{M}$ and on $\widehat{\mathcal{J}}$, is multiplication of paths; hence Φ agrees with the right action of $\pi_1(\mathcal{M})$.

Let us prove that Φ is injective. Suppose that

$$\Phi(\gamma_1 \otimes (\ell_1, z_1)) = \Phi(\gamma_2 \otimes (\ell_2, z_2)).$$

We have then

(5.6) $$E \circ P^{-1}(\gamma_1)[z_1] \cdot \ell_1 = E \circ P^{-1}(\gamma_2)[z_2] \cdot \ell_2.$$

Thus, the endpoints of $E \circ P^{-1}(\gamma_1)[z_1]$ and $E \circ P^{-1}(\gamma_2)[z_2]$ coincide; hence

$$\left(E \circ P^{-1}(\gamma_2)[z_2]\right)^{-1} \left(E \circ P^{-1}(\gamma_1)[z_1]\right)$$

is a well defined path from $E(z_1)$ to $E(z_2)$. The functor E is full; therefore there exists a change of charts $h \in \mathcal{G}_1$ such that $\mathbf{s}(h)$ is the end of $P^{-1}(\gamma_1)[z_1]$, $\mathbf{r}(h)$ is the end of $P^{-1}(\gamma_2)[z_2]$ and $E(h)$ is the unit at the end of $E \circ P^{-1}(\gamma_i)[z_i]$ for $i = 1, 2$. Then

$$\left(E \circ P^{-1}(\gamma_2)[z_2]\right)^{-1}\left(E \circ P^{-1}(\gamma_1)[z_1]\right) = E \circ P^{-1}\left(\gamma_2^{-1} P(h) \gamma_1\right)[z_1]$$

and $P^{-1}\left(\gamma_2^{-1} P(h) \gamma_1\right)[z_1]$ is a $\mathcal{G}_1$-path from z_1 to z_2. Consequently $\alpha = \gamma_2^{-1} P(h) \gamma_1$ is a $\mathcal{G}$-loop starting and ending in t.

The paths $\gamma_2 \cdot \alpha = P(h)\gamma_1 \in \mathcal{X}_t^J$ and $\gamma_1 \in \mathcal{X}_t^J$ represent the same point of $\widehat{\mathcal{J}}$. We also have

$$\alpha \cdot (\ell_1, z_1) = \left(E \circ P^{-1}\left(\gamma_2^{-1} P(h) \gamma_1\right)[z_1] \ell_1, z_2\right)$$
$$= \left(\left(\left(E \circ P^{-1}(\gamma_2)[z_2]\right)^{-1}\left(E \circ P^{-1}(\gamma_1)[z_1]\right)\right) \ell_1, z_2\right) = (\ell_2, z_2)$$

by (5.6).

Consequently, $\gamma_1 \otimes (\ell_1, z_1)$ and $\gamma_2 \otimes (\ell_2, z_2)$ represent the same point of the tensor product $\widehat{\mathcal{J}} \otimes \mathfrak{M}(p)$ and Φ is injective.

Let us prove that Φ is surjective. Take an arbitrary $\gamma \in \mathcal{X}_t^J$. The Julia set is completely invariant under p. Therefore, there exists a point $\zeta \in \mathcal{X}_1$ such that $E(\zeta)$ and the end of γ belong to one $\mathcal{G} \vee \mathcal{G}'$-orbit. Then the point ζ belongs to the preimage of $\mathcal{J}$ in $\mathcal{X}_1$. Let $g \in \mathcal{G} \vee \mathcal{G}'$ be such that $\mathbf{r}(g) = E(\zeta)$ and $\mathbf{s}(g)$ is the end of γ.

Choose a $\mathcal{G}_1$-path α_1 starting in a P-preimage z of t and ending in ζ. Denote $\alpha = P(\alpha_1)$. Then α starts in t and ends in $P(\zeta)$. Take the element (ℓ, z) of $\mathfrak{M}(p)$, where

$$\ell = E(\alpha_1)^{-1} g\gamma.$$

It is a well defined element of $\mathfrak{M}(p)$, since $E(\alpha_1)$ starts in $E(z)$ and ends in $E(\zeta)$, $\mathbf{r}(g) = E(\zeta)$, $\mathbf{s}(g)$ is the end of γ and γ starts in t. But then

$$\Phi(\alpha \otimes (\ell, z)) = E\left(P^{-1}(\alpha)[z]\right) \ell = E(\alpha_1) E(\alpha_1)^{-1} g\gamma = g\gamma,$$

which proves that Φ is surjective. $\square$

Let us prove that the constructed self-similarity is contracting with respect to the natural uniformity on the metric space $\widehat{\mathcal{J}}$. Since $\pi_1(\mathcal{M})$ acts on $\widehat{\mathcal{J}}$ by isometries, the action is uniformly equicontinuous. It also follows from Hopf-Rinow theorem that a relation V on $\widehat{\mathcal{J}}$ is bounded if and only if $R(V) = \sup_{(\xi_1, \xi_2) \in V} d(\xi_1, \xi_2) < \infty$. Let U be an arbitrary entourage. Then there exists $\epsilon > 0$ such that $(\zeta_1, \zeta_2) \in U$ whenever the distance between ζ_1 and ζ_2 is less than ϵ. By Definition 5.4.4 there exists n_0 such that if a path γ starting and ending in the Julia set has length less than R, then every one of its p^n-preimages has length less than ϵ if $n > n_0$.

Let $x_i = (\ell_i, z_i)$ be elements of $\mathfrak{M}(p)$ and let γ_1, γ_2 be paths starting in the basepoint and ending in the Julia set, such that $\gamma_2 = \gamma \gamma_1$, where γ has length less than R. Then $\Phi(\gamma_i \otimes x_{n-1} \ldots x_1 x_0)$ is the end of a path of the form

$$\gamma_{(n,i)} \widetilde{\ell}_{n-1} \ldots \widetilde{\ell}_1 \widetilde{\ell}_0,$$

where $\gamma_{(n,i)} \in p^{-n}(\gamma_i)$ and the paths $\widetilde{\ell}_k \in p^{-k}(\ell_k)$ depend only on the word $x_{n-1} \ldots x_1 x_0$. It follows that the distance between the points $\Phi(\gamma_1 \otimes x_{n-1} \ldots x_1 x_0)$ and $\Phi(\gamma_2 \otimes x_{n-1} \ldots x_1 x_0)$ of $\widehat{\mathcal{J}}$ is not greater than ϵ for all $n \geq n_0$. Consequently, $V \otimes v \subset U$ for all $v \in \mathfrak{M}^{\otimes n}$, $n \geq n_0$.

Theorem 3.4.13 together with the remark on page 129 imply now that the partial self-coverings $p : \mathcal{J}_p^\circ \longrightarrow \mathcal{J}_p$ and $\mathbf{s} : \mathcal{J}_{\pi_1(\mathcal{M})}^\circ \longrightarrow \mathcal{J}_{\pi_1(\mathcal{M})}$ are conjugate.

We know that the shift $\mathsf{s} : \mathcal{J}^\circ_{\pi_1(\mathcal{M})} \longrightarrow \mathcal{J}_{\pi_1(\mathcal{M})}$ is a restriction of the shift $\mathsf{s} : \mathcal{J}^\circ_{\mathrm{IMG}(p)} \longrightarrow \mathcal{J}_{\mathrm{IMG}(p)}$ and that the embeddings $\mathcal{J}_{\pi_1(\mathcal{M})} \hookrightarrow \mathcal{J}_{\mathrm{IMG}(p)}$ and $\mathcal{J}^\circ_{\pi_1(\mathcal{M})} \hookrightarrow \mathcal{J}^\circ_{\mathrm{IMG}(p)}$ induce homeomorphisms of the underlying spaces (see Proposition 4.6.5). Hence, conjugacy of the shifts follows from Proposition 4.3.6. □

Conjugacy of the partial self-coverings $p : \mathcal{J}^\circ_p \longrightarrow \mathcal{J}_p$ and $\mathsf{s} : \mathcal{J}^\circ_{\pi_1(\mathcal{M})} \longrightarrow \mathcal{J}_{\pi_1(\mathcal{M})}$ holds also without the condition that the isotropy groups are faithfully represented in IMG (p). Note that we have actually proved it for developable orbispaces $\mathcal{M}$ (this was the only implication of the faithfulness of the isotropy groups that we have used).

5.5.3. Local connectivity of the Julia sets of "geometrically finite" expanding maps.
We get "for free" the following result.

COROLLARY 5.5.5. *Suppose that $p : \mathcal{M}_1 \longrightarrow \mathcal{M}$ is an expanding partial self-covering of a complete orbispace such that $\pi_1(\mathcal{M})$ is finitely-generated and the inclusion $e : \mathcal{M}_1 \hookrightarrow \mathcal{M}$ induces a surjective homomorphism $e_* : \pi_1(\mathcal{M}_1) \longrightarrow \pi_1(\mathcal{M})$. Then the Julia set of p is connected and locally connected.*

PROOF. A direct corollary of Theorem 5.5.3 and Theorem 3.5.1. □

5.6. Iterated monodromy group of a pull-back

For the notion of a pull-back of a partial self-covering see Definition 4.5.1.

PROPOSITION 5.6.1. *Let $p : \mathcal{M}_1 \longrightarrow \mathcal{M}$ be a partial self-covering and let $p^\circ : \mathcal{M}_1^\circ \longrightarrow \mathcal{M}^\circ$ be its pull-back by an open map $f : \mathcal{M}^\circ \longrightarrow \mathcal{M}$. Then there exists an embedding $f_{img} : \mathrm{IMG}(p^\circ) \longrightarrow \mathrm{IMG}(p)$ such that the diagram*

$$\begin{array}{ccc} \pi_1(\mathcal{M}^\circ) & \stackrel{f_*}{\longrightarrow} & \pi(\mathcal{M}) \\ \downarrow & & \downarrow \\ \mathrm{IMG}(p^\circ) & \stackrel{f_{img}}{\longrightarrow} & \mathrm{IMG}(p) \end{array}$$

is commutative, where f_* is the homomorphism of the fundamental groups induced by f, and $\pi_1(\mathcal{M}^\circ) \longrightarrow \mathrm{IMG}(p^\circ)$ and $\pi_1(\mathcal{M}) \longrightarrow \mathrm{IMG}(p)$ are the canonical epimorphisms.

Moreover, f_{img} agrees with the respective standard actions of $\mathrm{IMG}(p^\circ)$ and $\mathrm{IMG}(p)$.

In particular, if f_* is surjective, then f_{img} is an isomorphism.

PROOF. By definition of a pull-back of a self-covering, we have embeddings $e : \mathcal{M}_1 \hookrightarrow \mathcal{M}$, $e^\circ : \mathcal{M}_1^\circ \hookrightarrow \mathcal{M}^\circ$ and an open map $f_1 : \mathcal{M}_1^\circ \longrightarrow \mathcal{M}_1$ such that the diagrams

$$\begin{array}{ccc} \mathcal{M}_1^\circ & \stackrel{f_1}{\longrightarrow} & \mathcal{M}_1 \\ \downarrow p^\circ & & \downarrow p \\ \mathcal{M}^\circ & \stackrel{f}{\longrightarrow} & \mathcal{M} \end{array} \qquad \begin{array}{ccc} \mathcal{M}_1^\circ & \stackrel{f_1}{\longrightarrow} & \mathcal{M}_1 \\ \downarrow e^\circ & & \downarrow e \\ \mathcal{M}^\circ & \stackrel{f}{\longrightarrow} & \mathcal{M} \end{array}$$

are commutative. Moreover, we can choose such atlases of orbispaces that f_1, f, p and p° are given by a commutative diagram of functors and atlases:

(5.7)
$$\begin{array}{ccc} \mathcal{G}_1^\circ & \xrightarrow{F_1} & \mathcal{G}_1 \\ \downarrow{\scriptstyle P^\circ} & & \downarrow{\scriptstyle P} \\ \mathcal{G}^\circ & \xrightarrow{F} & \mathcal{G} \end{array}$$

Let us choose the basepoints t, t_1, t° and t_1° in the atlases so that they agree with the commutative diagram, i.e., that $t = F(t^\circ)$, $t^\circ = P^\circ(t_1^\circ)$, $t_1 = F_1(t_1^\circ)$ and $t = P(t_1)$. Then the virtual endomorphism associated to the self-covering p is given by

$$\phi(\gamma) = e_* \left(P^{-1}(\gamma)[t_1] \right),$$

where the right-hand side is defined up to a conjugation in $\pi_1(\mathcal{G}, t)$ (see the definition of the induced homomorphism in 4.7.2).

Similarly, the virtual endomorphism ϕ° associated to the self-covering p° is given by

$$\phi^\circ(\gamma) = e_*^\circ \left(P^{\circ-1}(\gamma)[t_1^\circ] \right).$$

The domains of ϕ and ϕ° are the sets of the loops γ such that $P^{-1}(\gamma)[t_1]$ (resp. $P^{\circ-1}(\gamma)[t_1^\circ]$) are again loops.

Let $F_* : \pi_1(\mathcal{G}^\circ, t^\circ) \longrightarrow \pi_1(\mathcal{G}, t)$ be the homomorphism of the fundamental groups induced by the functor F. It follows from (5.7) that

$$F_*(\operatorname{Dom} \phi^\circ) \leq \operatorname{Dom} \phi,$$

since the image of a loop under F_1 is a loop. We also have

$$F_*^{-1}(\operatorname{Dom} \phi) \leq \operatorname{Dom} \phi^\circ,$$

since if this inclusion is not true, then F_1 maps two different preimages of t to one point, which contradicts the construction of the pull-back.

We get then, using commutativity of the diagrams above:

$$F_*\left(\phi^\circ(\gamma)\right) = F_*\left(e_*^\circ \left(P^{\circ-1}(\gamma)[t_1^\circ]\right)\right) = g^{-1} \cdot e_* \left(F_1\left(P^{\circ-1}(\gamma)[t_1^\circ]\right)\right) \cdot g$$
$$= g^{-1} \cdot e_* \left(P^{-1}\left(F(\gamma)\right)[t_1]\right) \cdot g = g^{-1} \phi(\gamma) g$$

for some fixed $g \in \pi_1(\mathcal{M})$ and any $\gamma \in \pi_1(\mathcal{M}^\circ)$. Thus, if we take the virtual endomorphism $g^{-1} \cdot \phi \cdot g$, which is also associated to p and has the same domain as ϕ, then we can apply Proposition 2.7.6 to the virtual endomorphisms $g^{-1} \cdot \phi \cdot g$ and ϕ° and the homomorphism $f_* : \pi_1(\mathcal{M}^\circ) \longrightarrow \pi_1(\mathcal{M})$. □

A useful application of Proposition 5.6.1 is that it gives a way to avoid singular points in computation of the iterated monodromy group. If $\mathcal{M}^\circ$ and $\mathcal{M}_1^\circ$ have no singular points, i.e., are usual topological spaces, and the homomorphism $f_* : \pi_1(\mathcal{M}^\circ) \longrightarrow \pi_1(\mathcal{M})$ is surjective, then $\operatorname{IMG}(p^\circ) \cong \operatorname{IMG}(p)$ and we can compute the iterated monodromy group (and the standard action) inside $\mathcal{M}^\circ$.

We will see some other applications of Proposition 5.6.1 in the next chapter.

5.7. The limit solenoid and inverse limits of self-coverings

5.7.1. The limit solenoid.
Let (G, X) be a contracting self-similar action and let $\mathfrak{M} = \mathsf{X} \cdot G$ be the associated hyperbolic bimodule. A natural way to define a two-sided infinite tensor power $\mathfrak{M}^{\otimes \mathbb{Z}}$ is to define it as the tensor product $\mathfrak{M}^{\otimes -\omega} \otimes \mathfrak{M}^{\otimes \omega}$ of the right G-module $\mathfrak{M}^{\otimes -\omega} = \mathcal{X}_G$ with the left G-module $\mathfrak{M}^{\otimes \omega} = \mathsf{X}^\omega$.

Recall that the tensor product of the G-modules is defined as the quotient of the direct product $\mathfrak{M}^{\otimes -\omega} \times \mathfrak{M}^{\otimes \omega}$ by the equivalence relation

$$\xi \cdot g \otimes w = \xi \otimes g \cdot w,$$

where $\xi \in \mathcal{X}_G = \mathfrak{M}^{\otimes -\omega}$ and $w \in \mathsf{X}^\omega = \mathfrak{M}^{\otimes \omega}$. If we transform the right action of G on $\mathfrak{M}^{\otimes -\omega}$ into a left action by the usual agreement

$$g \cdot \xi = \xi \cdot g^{-1},$$

then the tensor product $\mathfrak{M}^{\otimes -\omega} \otimes \mathfrak{M}^{\otimes \omega}$ becomes the quotient of the direct product by the diagonal left action of G.

DEFINITION 5.7.1. The *limit solenoid* $\mathcal{S}_G = \mathfrak{M}^{\otimes \mathbb{Z}}$ of a contracting action (G, X) is the topological space equal to the tensor product $\mathfrak{M}^{\otimes -\omega} \otimes \mathfrak{M}^{\otimes \omega}$, i.e., the space of the orbits of the left action of G on the direct product $\mathcal{X}_G \times \mathsf{X}^{-\omega}$ given by

$$g(\xi \times w) = \xi \cdot g^{-1} \times g(w).$$

If we have a preferred basis (alphabet) X, then we may define the limit space $\mathcal{X}_G$ as the quotient of the direct product $\mathsf{X}^{-\omega} \cdot G$ by the asymptotic equivalence relation (see Proposition 3.2.6). This gives us a more handy definition of the solenoid.

DEFINITION 5.7.2. Let $\mathsf{X}^{\mathbb{Z}}$ be the set of two-sided infinite sequences of the form $\ldots x_{-2} x_{-1} . x_0 x_1 \ldots$ over the alphabet X (here the dot marks the place between the coordinate number -1 and the coordinate number 0). We introduce on $\mathsf{X}^{\mathbb{Z}}$ the direct product topology of the discrete sets X. Two sequences

$$\ldots x_{-2} x_{-1} . x_0 x_1 \ldots, \qquad \ldots y_{-2} y_{-1} . y_0 y_1 \ldots \in \mathsf{X}^{\mathbb{Z}}$$

are *asymptotically equivalent* if there exists a bounded sequence $\{g_k\}_{k \geq 0}$ such that

$$g_k(x_k x_{k+1} \ldots) = y_k y_{k+1} \ldots$$

with respect to the action of G on X^ω for all $k \in \mathbb{Z}$.

One can prove, in the same way as Lemma 3.2.3, that two sequences $\xi_1, \xi_2 \in \mathsf{X}^{\mathbb{Z}}$ are asymptotically equivalent if and only if there exists a two-sided infinite path in the Moore diagram of the nucleus such that ξ_1 is read on the left halves and ξ_2 is read on the right halves of the labels of the arrows.

PROPOSITION 5.7.3. *Every point of $\mathcal{S}_G$ can be written in the form*

$$\ldots x_{-2} x_{-1} \otimes x_0 x_1 \ldots,$$

where $x_0 x_1 \ldots \in \mathsf{X}^\omega$ and $\ldots x_{-2} x_{-1} \in \mathsf{X}^{-\omega}$ (more pedantically, we should have written $\xi \in \mathcal{T} \subset \mathcal{X}_G$ instead of $\ldots x_{-2} x_{-1}$, but we as usual identify the points of $\mathcal{X}_G$ with the sequences representing them).

The sequences

$$\ldots x_{-2} x_{-1} \otimes x_0 x_1 \ldots \quad \text{and} \quad \ldots y_{-2} y_{-1} \otimes y_0 y_1 \ldots$$

represent the same point of $\mathcal{S}_G$ if and only if the sequences

$$\ldots x_{-2} x_{-1} . x_0 x_1 \ldots \quad \text{and} \quad \ldots y_{-2} y_{-1} . y_0 y_1 \ldots$$

are asymptotically equivalent in $\mathsf{X}^{\mathbb{Z}}$. *The topological space* $\mathcal{S}_G$ *is homeomorphic to the quotient of the topological space* $\mathsf{X}^{\mathbb{Z}}$ *by the asymptotic equivalence relation.*

PROOF. Every point of $\mathcal{X}_G$ can be written as a sequence $\ldots x_{-2}x_{-1} \cdot g \in \mathsf{X}^{-\omega} \cdot G$ by Proposition 3.2.5. Hence, every element of $\mathcal{S}_G$ can be represented by

$$\ldots x_{-2}x_{-1} \cdot g \otimes y_0 y_1 \ldots \sim \ldots x_{-2}x_{-1} \otimes g \cdot y_0 y_1 \ldots = \ldots x_{-2}x_{-1} \otimes x_0 x_1 \ldots,$$

where $x_0 x_1 \ldots = g(y_0 y_1 \ldots)$ with respect to the action of G on X^{ω}.

Two sequences of $\mathsf{X}^{\mathbb{Z}}$ represent the same point of $\mathcal{S}_G$ if and only if there exists $g \in G$ such that

$$\ldots x_{-2}x_{-1} = \ldots y_{-2}y_{-1} \cdot g$$

in $\mathcal{X}_G$ and

$$g(x_0 x_1 \ldots) = y_0 y_1 \ldots$$

in X^{ω}.

The first equality is equivalent to existence of a left-infinite path γ_1 in the Moore diagram of the nucleus, which ends in g and is such that $\ldots x_{-2}x_{-1}$ and $\ldots y_{-2}y_{-1}$ are read along γ_1 on the left and the right halves of the labels, respectively.

The second equality means that the Moore diagram has a right-infinite path γ_1 starting in g such that $x_0 x_1 \ldots$ and $y_0 y_1 \ldots$ are read along γ_2 on the left and the right halves of the labels, respectively. □

The definition of $\mathcal{S}_G$ in terms of the sequences over the alphabet X shows that the two-sided shift

$$\ldots x_{-2}x_{-1} \cdot x_0 x_1 x_2 \ldots \mapsto \ldots x_{-2}x_{-1}x_0 \cdot x_1 x_2 \ldots$$

preserves the asymptotic equivalence relation on $\mathsf{X}^{\mathbb{Z}}$ and induces a homeomorphism $\mathsf{e} : \mathcal{S}_G \longrightarrow \mathcal{S}_G$. Its inverse will be denoted $\widehat{\mathsf{s}}$ and called the *natural extension of* $\mathsf{s} : \mathcal{J}_G \longrightarrow \mathcal{J}_G$, for reasons which will be clear a bit later.

5.7.2. Orbispace structure on $\mathcal{S}_G$. Let $(\mathcal{G}_G, \mathcal{X}_G)$ be the groupoid of the action of G on $\mathcal{X}_G$. Recall that $(g, \xi) \in \mathcal{G}_G$ denotes the germ of the map $\zeta \mapsto \zeta \cdot g^{-1}$ at the point ξ. For $(g, \xi) \in \mathcal{G}_G$ and $w \in \mathsf{X}^{\omega}$ define

(5.8) $$\sigma_\omega(g, \xi, w) = \sigma_\omega((g, \xi), w) = g(w).$$

LEMMA 5.7.4. *The map* $\sigma_\omega : \mathcal{G}_G \longrightarrow \operatorname{Aut} \mathsf{X}^*$ *is a well defined continuous cocycle.*

PROOF. The element g is uniquely determined by its germ (g, ξ) by Proposition 4.6.2. Its functoriality and continuity is easy to check. □

PROPOSITION 5.7.5. *The skew-product* $(\mathcal{G}_G \rtimes \sigma_\omega, \mathcal{X}_G \times \mathsf{X}^{\omega})$ *is a locally compact étale Hausdorff groupoid whose space of orbits is canonically isomorphic to* $\mathcal{S}_G$; *i.e., the map*

$$(\xi, w) \mapsto \xi \otimes w$$

induces a homeomorphism between the space of orbits and $\mathcal{S}_G$.

For the definition of skew-product see Subsection 4.4.2, page 125.

PROOF. The elements of the groupoid $\mathcal{G}_G \rtimes \sigma_\omega$ are the triples (g, ξ, w), where $g \in G, \xi \in \mathcal{X}_G$ and $w \in \mathsf{X}^\omega$. We have
$$\mathbf{s}(g, \xi, w) = (\xi, w), \qquad \mathbf{r}(g, \xi, w) = \left(\xi \cdot g^{-1}, g(w)\right)$$
and
$$(g_1, \xi_1, w_1)(g_2, \xi_2, w_2) = (g_1 g_2, \xi_2, w_2).$$

The groupoid $\mathcal{G}_G \rtimes \sigma_\omega$ is Hausdorff and locally compact as a direct product of a locally compact Hausdorff groupoid $\mathcal{G}_G$ (see Proposition 4.6.2) and a compact Hausdorff space X^ω. It is étale by Proposition 4.4.5. (Note that the groupoid of germs of the action of G on X^ω is usually not Hausdorff.)

The formulae for the source and the range maps $\mathbf{s}, \mathbf{r}$ on $\mathcal{G}_G \rtimes \sigma_\omega$ show that two points of $\mathcal{X}_G \times \mathsf{X}^\omega$ belong to one $\mathcal{G}_G \rtimes \sigma_\omega$-orbit if and only if they belong to one orbit with respect to the diagonal left action of G, i.e., if they represent one point of $\mathcal{S}_G$. $\square$

DEFINITION 5.7.6. The groupoid $(\mathcal{G}_G \rtimes \sigma_\omega, \mathcal{J}_G \times \mathsf{X}^\omega)$, where σ_ω is the cocycle given by (5.8), defines the orbispace structure on $\mathcal{S}_G$.

Thus the limit solenoid $\mathcal{S}_G$ is a fiber bundle over the limit orbispace $\mathcal{J}_G$ with fibers homeomorphic to the boundary X^ω of the tree X^* on which G acts by the self-similar action.

Note that the groupoid $(\mathcal{G}_G \rtimes \sigma_\omega, \mathcal{J}_G \times \mathsf{X}^\omega)$ coincides with the groupoid of the (germs of the) diagonal action of G on $\mathcal{J}_G \times \mathsf{X}^\omega$. In other words, the map $((g, \xi), w) \mapsto (g, (\xi, w))$ is an isomorphism of the groupoids.

Let us now show that the shift $\mathsf{e} : \mathcal{S}_G \longrightarrow \mathcal{S}_G$ is induced by an embedding.

PROPOSITION 5.7.7. Let $E_\mathcal{S} : \mathcal{G}_G \rtimes \sigma_\omega \longrightarrow \mathcal{G}_G \rtimes \sigma_\omega$ be given by
$$(5.9) \qquad E_\mathcal{S}(g, \xi, x_0 x_1 x_2 \ldots) = (g|_{x_0}, (\xi \otimes dx_0), x_1 x_2 \ldots).$$
Then $E_\mathcal{S}$ is a well defined open functor. If the action is recurrent, then $E_\mathcal{S}$ defines an embedding of orbispaces.

PROOF. Recall that every point of $\mathfrak{M}^{\otimes \omega}$ is written in the form of a sequence $x_0 x_1 \ldots \in \mathsf{X}^\omega$ in a unique way (Proposition 2.4.1), so that the letter x_0 is well defined.

Functoriality of $E_\mathcal{S}$ is checked by direct computation. It is obviously continuous and it is open by Lemma 3.4.2.

The shift $x_0 x_1 \ldots \mapsto x_1 x_2 \ldots$ is surjective on X^ω. If the action is recurrent, then the map $g \mapsto g|_{x_0}$ is surjective on G and the map $\xi \mapsto \xi \otimes x_0$ is surjective on $\mathcal{X}_G$. This implies that the functor $E_\mathcal{S}$ is surjective and a fortiori is full. $\square$

Similar arguments as in Theorem 4.6.4 show that the embedding $\mathsf{e} : \mathcal{S}_G \longrightarrow \mathcal{S}_G$ defined by the functor $E_\mathcal{S}$ depends only on the self-similarity bimodule $\mathfrak{M}$, i.e., does not depend on the choice of the basis X.

5.7.3. Solenoid as an inverse limit.

PROPOSITION 5.7.8. *The space $\mathcal{S}_G$ is homeomorphic to the inverse limit of the topological spaces*
$$\mathcal{J}_G \xleftarrow{\mathsf{s}} \mathcal{J}_G \xleftarrow{\mathsf{s}} \cdots.$$
The map $\mathsf{e} : \mathcal{S}_G \longrightarrow \mathcal{S}_G$ *acts on the inverse limit by*
$$\mathsf{e}(\xi_1, \xi_2, \ldots) = (\xi_2, \xi_3, \ldots).$$

Its inverse $\widehat{\mathsf{s}}$ *acts by*

$$\widehat{\mathsf{s}} : (\xi_1, \xi_2, \ldots) = (\mathsf{s}(\xi_1), \mathsf{s}(\xi_2), \ldots) = (\mathsf{s}(\xi_1), \xi_1, \xi_2, \ldots),$$

i.e., is the natural extension of s *on* $\mathcal{S}_G$.

PROOF. We have the following infinite commutative diagram:

$$\begin{array}{ccccccc} \mathsf{X}^{-\omega} & \stackrel{\sigma}{\longleftarrow} & \mathsf{X}^{-\omega} & \stackrel{\sigma}{\longleftarrow} & \mathsf{X}^{-\omega} & \cdots \\ \downarrow \pi & & \downarrow \pi & & \downarrow \pi & \\ \mathcal{J}_G & \stackrel{\mathsf{s}}{\longleftarrow} & \mathcal{J}_G & \stackrel{\mathsf{s}}{\longleftarrow} & \mathcal{J}_G & \cdots \end{array}$$

where σ is the shift on the space $\mathsf{X}^{-\omega}$ and π is the canonical quotient map. Obviously the limit of the first row of the diagram is homeomorphic to $\mathsf{X}^{\mathbb{Z}}$, where the homeomorphism maps the sequence

$$\ldots x_{-2}x_{-1}, \quad \ldots x_{-2}x_{-1}x_0, \quad \ldots x_{-2}x_{-1}x_0 x_1, \quad \ldots$$

representing a point of the limit, to the point $\ldots x_{-2}x_{-1} \cdot x_0 x_1 \ldots \in \mathsf{X}^{\mathbb{Z}}$.

Consequently, the limit of the lower row is the image of $\mathsf{X}^{\mathbb{Z}}$ under the quotient map, which is the limit of the quotient maps π in the inverse spectrum. This quotient map carries two elements,

$$\ldots x_{-2}x_{-1}, \quad \ldots x_{-2}x_{-1}x_0, \quad \ldots x_{-2}x_{-1}x_0 x_1, \quad \ldots,$$
$$\ldots y_{-2}y_{-1}, \quad \ldots y_{-2}y_{-1}y_0, \quad \ldots y_{-2}y_{-1}y_0 y_1, \quad \ldots,$$

to equal points if and only if for every $n \geq -1$ we have

$$\pi(\ldots x_{n-1}x_n) = \pi(\ldots y_{n-1}y_n).$$

But it follows from the description of the asymptotic equivalence relations on $\mathsf{X}^{\mathbb{Z}}$ and $\mathsf{X}^{-\omega}$ that this is equivalent to the condition that the sequences

$$\ldots x_{-2}x_{-1} \cdot x_0 x_1 \ldots, \quad \ldots y_{-2}y_{-1} \cdot y_0 y_1 \ldots$$

are asymptotically equivalent. Thus the inverse limit of the lower row of the commutative diagram is homeomorphic to $\mathcal{S}_G$.

The statements about the maps e and $\widehat{\mathsf{s}}$ are straightforward. $\square$

5.7.4. Leaves and tiles. The proof of the following proposition is similar to the proof of Proposition 3.3.10 (after obvious changes).

PROPOSITION 5.7.9. *If a contracting action* (G, X) *is level-transitive, then the limit solenoid* $\mathcal{S}_G$ *is connected.* $\square$

DEFINITION 5.7.10. For every $w \in \mathsf{X}^{\omega}$ the *tile* $\mathcal{T} \otimes w$ is the image of the set $\mathcal{T} \otimes w \subset \mathcal{X}_G \times \mathsf{X}^{\omega}$ in $\mathcal{S}_G$ and the *leaf* $\mathcal{X}_G \otimes w$ is the image of the set $\mathcal{X}_G \otimes w \subset \mathcal{X}_G \times \mathsf{X}^{\omega}$ in $\mathcal{S}_G$.

Since $\mathcal{X}_G = \bigcup_{g \in G} \mathcal{T} \cdot g$, the leaf $\mathcal{X}_G \otimes w$ is equal to the union $\bigcup_{g \in G} \mathcal{T} \otimes g(w)$. Consequently, the following four conditions are equivalent.
(1) The tiles $\mathcal{T} \otimes w_1$ and $\mathcal{T} \otimes w_2$ belong to one leaf.
(2) The leaves $\mathcal{X}_G \otimes w_1$ and $\mathcal{X}_G \otimes w_2$ coincide.
(3) The leaves $\mathcal{X}_G \otimes w_1$ and $\mathcal{X}_G \otimes w_2$ intersect.
(4) w_1 and w_2 belong to one orbit of the action of G on X^{ω}.

Note that the leaves do not depend on the choice of the basis X (though the tiles do).

Similar arguments as in the proof of Proposition 3.3.5 show that two tiles, $\mathcal{T} \otimes w_1$ and $\mathcal{T} \otimes w_2$, intersect if and only if there exists an element g of the nucleus such that $g(w_1) = w_2$.

This implies that if the self-similar action (G, X) is recurrent, then the adjacency graph of a leaf $\mathcal{X}_G \otimes w$ is isomorphic to the Schreier graph of the action of G on the orbit $G(w)$.

One of the properties of the Schreier graphs is their self-similarity, which can be interpreted in our terms as the action of the map e on the leaves of $\mathcal{S}_G$. We have:
$$\mathcal{T} \otimes w = \bigcup_{x \in \mathsf{X}} \mathsf{e}\left(\mathcal{T} \otimes (xw)\right);$$
i.e., every tile $\mathcal{T} \otimes w$ is a union of $|\mathsf{X}|$ "similar" tiles. If the action (G, X) is recurrent, then the orbit of xw does not depend on x and the map e maps the leaf $\mathcal{X}_G \otimes xw$ onto the leaf $\mathcal{X}_G \otimes w$.

5.7.5. Relation with iterated monodromy groups. Let $p : \mathcal{M}_1 \longrightarrow \mathcal{M}$ be an expanding partial self-covering of a topological space $\mathcal{M}$. We may consider then the inverse sequence
$$\mathcal{M} \xleftarrow{p} \mathcal{M}_1 \xleftarrow{p} \mathcal{M}_2 \xleftarrow{p} \cdots,$$
where $\mathcal{M}_n$ denotes the domain of the nth iteration of p. Let $\mathcal{M}_\omega$ be the projective limit of this sequence. We have a natural projection map $p_\omega : \mathcal{M}_\omega \longrightarrow \mathcal{M}$ and the maps
$$\mathsf{e}\left(\xi_0, \xi_1, \ldots\right) = (\xi_1, \xi_2, \ldots)$$
and
$$\widehat{p}\left(\xi_0, \xi_1, \ldots\right) = (p(\xi_0), p(\xi_1), \ldots) = (p(\xi_0), \xi_0, \xi_1, \ldots).$$

The first map is a map from $\mathcal{M}_\omega$ to itself, while the map $\widehat{p}$ is defined on the subset equal to the preimage of $\mathcal{M}_1$ in $\mathcal{M}_\omega$.

The projection $p_\omega : \mathcal{M}_\omega \longrightarrow \mathcal{M}$ is a fiber bundle such that the fundamental group of $\mathcal{M}$ acts on the fibers by an action which is topologically conjugate with the iterated monodromy action on X^ω. Consequently, the iterated monodromy group $\mathrm{IMG}(p)$ may be naturally interpreted as the *holonomy group* of the bundle $p_\omega : \mathcal{M}_\omega \longrightarrow \mathcal{M}$.

The preimage of the Julia set in $\mathcal{M}_\omega$ is homeomorphic, by Theorem 5.5.3 and Proposition 5.7.8, to the limit solenoid $\mathcal{S}_{\mathrm{IMG}(p)}$.

If all the domains $\mathcal{M}_n$ are path-connected, then the leaves of $\mathcal{S}_{\mathrm{IMG}(p)}$ are in a natural bijective correspondence with the path-connected components of the projective limit $\mathcal{M}_\omega$, since two preimages of a basepoint t in $\mathcal{M}_\omega$ belong to one path-connected component if and only if they belong to one orbit of the action of $\mathrm{IMG}(p)$ on the fiber $p_\omega^{-1}(t)$.

Projective limits of this sort and their leaves in the case of rational iterations were defined and studied by M. Lyubich and Y. Minsky in [**LM97**]. They proved, for example, that in the case of a post-critically finite rational map the natural conformal structure of the leaves is Euclidean. See also the work of V. Kaimanovich and M. Lyubich [**KL05**].

CHAPTER 6

Examples and Applications

6.1. Expanding self-coverings of orbifolds

We consider in this section the case when the self-covering $p : \mathcal{M} \longrightarrow \mathcal{M}$ is everywhere defined.

6.1.1. Theorems of M. Shub and M. Gromov.
If $p : \mathcal{M} \longrightarrow \mathcal{M}$ is a self-covering, then the virtual endomorphism $\phi_p : \pi_1(\mathcal{M}) \dashrightarrow \pi_1(\mathcal{M})$ associated to it is an isomorphism of a finite-index subgroup $\operatorname{Dom} \phi_p < \pi_1(\mathcal{M})$ with $\pi_1(\mathcal{M})$ (see Subsection 5.1.5 on page 141). Its inverse is an injective endomorphism $p_* : \pi_1(\mathcal{M}) \longrightarrow \pi_1(\mathcal{M})$, which is the homomorphism induced by p. Both ϕ_p and p_* are uniquely defined up to a conjugation in $\pi_1(\mathcal{M})$.

The kernel of the iterated monodromy action of the fundamental group $\pi_1(\mathcal{M})$ is equal, by Proposition 2.7.5, to the subgroup

$$N_p = \bigcap_{k \geq 1} \bigcap_{g \in \pi_1(\mathcal{M})} g^{-1} \cdot p_*^k(\pi_1(\mathcal{M})) \cdot g.$$

The iterated monodromy group $\operatorname{IMG}(p)$ is isomorphic to the quotient $\pi_1(\mathcal{M})/N_p$.

DEFINITION 6.1.1. An endomorphism $p : \mathcal{M} \longrightarrow \mathcal{M}$ of a compact Riemannian orbifold is *expanding* if there exist constants $c > 0$ and $\lambda > 1$ such that $\|Dp^n(\vec{v})\| \geq c\lambda^n \|\vec{v}\|$ for every tangent vector $\vec{v} \in T\mathcal{M}$ and every $n \geq 1$.

It is easy to see that if an endomorphism of a compact Riemannian orbifold is expanding, then it is also expanding in the sense of Definition 5.4.2.

The following properties of expanding endomorphisms of Riemannian *manifolds* were proved by M. Shub and J. Franks [**Shu69, Shu70**].

THEOREM 6.1.2 (M. Shub, J. Franks). *Suppose that the endomorphism $p : \mathcal{M} \longrightarrow \mathcal{M}$ of a compact Riemannian manifold $\mathcal{M}$ is expanding. Then the following is true.*

(1) *The universal covering of the space $\mathcal{M}$ is diffeomorphic to $\mathbb{R}^n$.*
(2) *The fundamental group $\pi_1(\mathcal{M})$ is torsion-free and has polynomial growth.*
(3)
$$\bigcap_{k \geq 1} p_*^k(\pi_1(\mathcal{M})) = \{1\}.$$

The statements of this theorem (except for (3) and absence of torsion in the fundamental group) also follow from the results which we are going to prove in a more general setting in Subsection 6.1.2. Absence of torsion follows from (1) and the well known fact that a finite cyclic group cannot act freely on $\mathbb{R}^n$ (a theorem of P.A. Smith). Statement (3) is an easy consequence of expansion: there exists $R > 0$ such that every non-trivial element of $\pi_1(\mathcal{M})$ is defined by a loop of length greater than R, but then every non-trivial element of $p_*^k(\pi_1(\mathcal{M}))$ is defined by a loop of length $> c\lambda^k R$.

Theorems 5.5.3 and 6.1.2 imply the following result.

THEOREM 6.1.3. *Suppose that $p : \mathcal{M} \longrightarrow \mathcal{M}$ is an expanding endomorphism of a compact Riemannian manifold $\mathcal{M}$. Then the iterated monodromy group* IMG (p) *is isomorphic to the fundamental group $\pi_1(\mathcal{M})$. Standard actions of the iterated monodromy group are contracting, and the limit dynamical system $(\mathcal{J}_{\mathrm{IMG}(p)}, \mathsf{s})$ is topologically conjugate to the system $(\mathcal{M}, p)$.*

PROOF. The only thing that remains to be proved is that the Julia set of p is the whole manifold $\mathcal{M}$, i.e., that the set $\bigcup_{n\ge 1} p^{-n}(x)$ is dense in $\mathcal{M}$ for some (and thus all) x. But this follows easily from expansion.

The space $\mathcal{M}$ is compact; therefore there exists l such that the distance from every point $x \in \mathcal{M}$ to the Julia set of p is less than l. The Julia set $\mathcal{J}_p$ is completely invariant; hence the distance from every point of $p^{-n}(x)$ to $\mathcal{J}_p$ is less than $c^{-1}\lambda^{-n}l$, by definition of an expanding self-covering. But $p^{-n}(\mathcal{M}) = \mathcal{M}$, so we conclude that $\mathcal{J}_p = \mathcal{M}$. □

We will also prove this theorem later in a more general setting without using the results of M. Shub.

Theorems 6.1.3, 5.5.3 and 4.6.4 imply the following result (see [**Shu69**], Theorems 4 and 5), which was proved by M. Shub using other methods.

THEOREM 6.1.4 (M. Shub). *An expanding endomorphism $p : \mathcal{M} \longrightarrow \mathcal{M}$ is determined uniquely, up to a topological conjugacy, by the action of the homomorphism p_* on the fundamental group $\pi_1(\mathcal{M})$.*

M. Gromov in [**Gro81**] proved a conjecture of M. Shub (see [**Shu70, Hir70**]) using his theorem on groups of polynomial growth. M. Shub's conjecture describes all possible expanding endomorphisms of Riemannian manifolds.

Let L be a connected and simply connected nilpotent Lie group and let $\mathrm{Aff}(L)$ be the group of diffeomorphisms of L generated by the translations $x \mapsto g \cdot x$ and automorphisms of L. Take a subgroup $G < \mathrm{Aff}(L)$ acting freely and properly on L. Suppose that the quotient $\mathcal{M} = L/G$ is compact. Then $\mathcal{M}$ is a manifold. If an expanding endomorphism P of L conjugates G with a subgroup of G, then P induces an expanding map $p : \mathcal{M} \longrightarrow \mathcal{M}$. Such maps p are called *expanding endomorphisms* of the *infra-nil-manifold* $\mathcal{M}$.

Note that an endomorphism of a Lie group is expanding if and only if its derivative at the unit is an expanding linear map.

THEOREM 6.1.5 (M. Gromov). *Every expanding map of a compact manifold is topologically conjugate to an expanding endomorphism of an infra-nil-manifold.*

6.1.2. Singular case. Now let $p : \mathcal{M} \longrightarrow \mathcal{M}$ be an expanding self-covering of a compact orbispace. The associated virtual endomorphism ϕ will be an isomorphism between $\mathrm{Dom}\,\phi$ and $\pi_1(\mathcal{M})$.

We are going to prove the following theorem, using the theory of self-similar groups.

THEOREM 6.1.6. *Let $\phi : G \dashrightarrow G$ be a surjective and injective contracting virtual endomorphism and let (G, X) be the associated self-similar action. Then there exist a nilpotent connected and simply connected Lie group L, a co-compact proper* action of G on L by affine transformations and a G-equivariant homeomorphism $\Phi : \mathcal{X}_G \longrightarrow L$. Moreover, the virtual endomorphism ϕ is induced by a contracting automorphism ϕ_L of the Lie group L and $\Phi(\zeta \otimes \phi(g_1)g_2) = \phi_L\left(\Phi(\zeta)g_1\right)g_2$.

COROLLARY 6.1.7. *Let $p : \mathcal{M} \longrightarrow \mathcal{M}$ be an expanding self-covering of a developable orbifold. Then $\mathcal{M}$ is isomorphic to the orbifold of an affine action of $\pi_1(\mathcal{M})$ on a nilpotent connected and simply connected Lie group L. The self-covering p is induced by an expanding automorphism of L, whose inverse induces a virtual endomorphism of $\pi_1(\mathcal{M})$ such that if $\mathcal{X}_{\pi_1(\mathcal{M})}$ is the limit space of the associated self-similar action, then the dynamical systems $\left(\mathcal{X}_{\pi_1(\mathcal{M})}, \pi_1(\mathcal{M})\right)$ and $(L, \pi_1(\mathcal{M}))$ are topologically conjugate.*

PROOF OF COROLLARY 6.1.7. The same arguments as in the proof of Theorem 6.1.3 show that the Julia set of p is equal to $\mathcal{M}$. Then Theorem 6.1.6 and Theorem 5.5.3 (together with the remark after the proof) imply Corollary 6.1.7. □

Let us prove a sequence of auxiliary lemmas before proving Theorem 6.1.6.

LEMMA 6.1.8. *Let ϕ be an injective contracting virtual endomorphism of a finitely generated group G. Suppose that the associated self-similar action of G is faithful. Then the parabolic subgroup $P = \bigcap_{n \geq 0} \mathrm{Dom}\,\phi^n$ is finite.*

PROOF. We have $\phi(P) \leq P$. Let P_0 be the intersection of P with the nucleus of the associated self-similar action. Then $\phi(P_0) \leq P_0$, and since ϕ is injective and P_0 is finite, ϕ is a permutation of P_0.

If $g \in P$, then there exists $n \in \mathbb{N}$ such that $\phi^n(g)$ belongs to the nucleus; i.e., it belongs to P_0. But $\phi^{-n}(P_0) = P_0$; thus $g \in P_0$. Consequently, $P = P_0$ is finite. □

LEMMA 6.1.9. *If conditions of Lemma 6.1.8 are satisfied, then G is a group of polynomial growth and therefore it is virtually nilpotent.*

PROOF. By Proposition 2.13.8, the growth of the action of G on the orbit of every point $w \in \mathsf{X}^\omega$ is polynomial. Take $w = x_0 x_0 x_0 \ldots$, where $x_0 = \phi(1)1 \in \phi(G)G$. Then the parabolic subgroup P is the stabilizer of w in G. The stabilizer of w is finite; hence the growth degree of G is the same as the growth degree of the action of G on $G(w)$. The theorem of M. Gromov (see [**Gro81**]) now implies that G is virtually nilpotent. □

Let G be an arbitrary finitely generated group and let ϕ be a contracting injective and surjective virtual endomorphism of G such that the kernel $\mathcal{K}(\phi)$ of the associated self-similar action is trivial.

We know that G is virtually nilpotent; thus it has a finite-index torsion-free nilpotent subgroup (see [**KM79**]).

LEMMA 6.1.10. *There is a normal nilpotent torsion-free subgroup of finite index $H \trianglelefteq G$ such that*
$$\phi\left(H \cap \mathrm{Dom}\,\phi\right) = H.$$

PROOF. Let $\mathcal{H}$ be the set of the nilpotent torsion-free normal subgroups of G of the least possible index k.

For any $H \in \mathcal{H}$ the group $H \cap \mathrm{Dom}\,\phi$ is a nilpotent torsion-free normal subgroup of $\mathrm{Dom}\,\phi$ of index $k_1 \leq k$. But then $\phi(H \cap \mathrm{Dom}\,\phi)$ is a nilpotent torsion-free normal subgroup of index k_1 in G. Hence, $k = k_1$.

We see that the virtual endomorphism ϕ induces the mapping $H \mapsto \phi(H \cap \mathrm{Dom}\,\phi)$ on the set $\mathcal{H}$. The group G is finitely generated. Therefore the set $\mathcal{H}$ is finite, so that the mapping has a cycle; i.e., there exist subgroups $H_i \in \mathcal{H}$ such that $\phi(H_i \cap \mathrm{Dom}\,\phi) = H_{i+1}$ for $i = 1, \ldots, n-1$ and $\phi(H_n \cap \mathrm{Dom}\,\phi) = H_1$. Then the subgroup $H = \bigcap_{i=1}^{n} H_i$ satisfies the conditions of the lemma. □

PROOF OF THEOREM 6.1.6. Lemma 6.1.9 implies that the group G is virtually nilpotent. There exists a normal nilpotent torsion-free finite-index subgroup H such that $\phi(H \cap \mathrm{Dom}\,\phi) = H$ (see Lemma 6.1.10). Then by Theorem 3.7.1, we have an H-covariant homeomorphism $\mathcal{X}_H \longrightarrow \mathcal{X}_G$.

It follows that every point of $\mathcal{X}_G$ can be written in the form $\ldots y_2 y_1 \cdot h$ for some $y_i \in \mathsf{Y} \subseteq \mathsf{X}$ and $h \in H \leq G$.

By Malcev's theorem [**Mal49**] H is a uniform lattice of a simply connected nilpotent Lie group L. The isomorphism ϕ of the lattices $H \cap \mathrm{Dom}\,\phi$ and H extends to an automorphism of the Lie group L. We will denote this automorphism by ϕ_L. The automorphism $\phi_L : L \longrightarrow L$ is contracting with respect to a right-invariant Riemannian metric on L, since it is contracting on H (see [**Gel95**]).

The Lie group L is a right proper co-compact H-space with a contracting $\phi_H(H)$ H-self-similarity

(6.1) $$\xi \otimes \phi_H(h_1) h_2 = \phi_L(\xi h_1) h_2.$$

The axioms of a self-similarity are directly checked.

Hence, Theorem 3.4.13 defines an H-equivariant homeomorphism $\Phi : \mathcal{X}_H \longrightarrow L$. If $\mathsf{Y} = \{y_1 = \phi_H(r_1)1, \ldots, y_d = \phi_H(r_d)1\}$, then the homeomorphism Φ maps a point $\ldots y_{i_2} y_{i_1} \cdot h$ to the element

$$\ldots \phi_L^3(r_{i_3}) \phi_L^2(r_{i_2}) \phi_L(r_{i_1}) h = \lim_{n \to \infty} \phi_L^n(r_{i_n}) \cdots \phi_L^2(r_{i_2}) \phi_L(r_{i_1}) h$$

of L.

Let us denote $x_0 = \phi_H(1)1 \in \phi_H(H)H \subseteq \phi(G)G$. Then (6.1) implies that

$$\xi \otimes x_0 = \phi_L(\xi).$$

Let us define the action A_g of $g \in G$ on L just conjugating by the homeomorphism $\Phi : \mathcal{X}_H \longrightarrow L$ the action of g on $\mathcal{X}_G = \mathcal{X}_H$ (then equivariance of Φ will be automatically satisfied):

$$\zeta \cdot A_g = \Phi\left(\Phi^{-1}(\zeta) \cdot g\right).$$

If $g \in \mathrm{Dom}\,\phi$, then for every $\zeta \in \mathcal{X}_G$ we have

$$\phi_L(\Phi(\zeta) \cdot A_g) = \Phi(\zeta \cdot g \otimes \phi(1)1) = \Phi(\zeta \otimes \phi(1)1 \cdot \phi(g)) = \phi_L(\Phi(\zeta)) \cdot A_{\phi(g)},$$

since the $\phi_H(H)H$-self-similarity on $\mathcal{X}_H = \mathcal{X}_G$ is the restriction of the $\phi(G)G$-self-similarity. Consequently,

(6.2) $$\phi_L^n(x \cdot A_g) = \phi_L^n(x) \cdot A_{\phi^n(g)}$$

for all $x \in L$ and $g \in \mathrm{Dom}\,\phi^n$.

The group G acts on its normal subgroup H by conjugation. Every automorphism $g : x \mapsto g^{-1}xg$ of H can be extended in a unique way to an automorphism of L. Let us denote this extension by $x \mapsto x^g$.

Let us show that the action of G on L is an action by affine transformations. Consider the restriction of this action onto the lattice $\phi_L^n(H) < L$. Let $\phi_L^n(h)$ be an arbitrary element of $\phi_L^n(H)$. Then, using (6.2), we get for every $g \in G$:

$$\phi_L^n(h) \cdot A_g = \phi_L^n\left(h \cdot A_{\phi^{-n}(g)}\right) = \phi_L^n\left(1 \cdot A_{\phi^{-n}(g)} \cdot A_{\phi^{-n}(g)^{-1}h\phi^{-n}(g)}\right)$$
$$= \phi_L^n\left(1 \cdot A_{\phi^{-n}(g)} \cdot \phi^{-n}(g)^{-1}h\phi^{-n}(g)\right) = (1 \cdot A_g) \cdot \phi_L^n(h)^g.$$

We use here that $x \cdot h = x \cdot A_h$ for all $h \in H \le G$.

Let us denote $1 \cdot A_g = S_g \in L$. Then the action of g on the points of $\phi_L^n(H)$ is given by the formula

$$x \cdot A_g = S_g \cdot x^g.$$

But the set $\bigcup_{n \ge 1} \phi_L^n(H)$ is dense in L, and the action of G on L is continuous; hence it is given by the above formula for all $x \in L$, and thus it is an action by affine transformations. □

6.1.3. Nilpotent self-similar groups. Not every nilpotent group is an iterated monodromy group of an expanding self-covering of an orbifold. For instance, if a torsion-free nilpotent group is co-Hopfian (i.e., if it contains no proper subgroup isomorphic to itself), then it has no recurrent self-similar action. Examples of co-Hopfian torsion-free finitely generated nilpotent groups are discussed in [**Bel03**].

Other interesting restrictions on self-similar actions of nilpotent groups were obtained by S. Sidki and A. Berlatto. They prove the following results. Here an $\mathcal{F}$-group is a finitely generated torsion-free nilpotent group.

THEOREM 6.1.11 (A. Berlatto, S. Sidki). *Let $|\mathsf{X}| = p$ be a prime. Then every nilpotent finitely generated self-similar group acting transitively on the first level of the tree X^* is either free abelian or a finite p-group.*

If $|\mathsf{X}|$ is a product of two primes, then every self-similar $\mathcal{F}$-group acting transitively on the first level of X^* is either free abelian or free abelian by free abelian.

Let $|\mathsf{X}| = p_1^{n_1} \cdots p_k^{n_k}$, where p_i are pairwise different primes. If (G, X) is a recurrent action of an $\mathcal{F}$-group, then $G^{(n_1+n_2+\cdots+n_k)} = \{1\}$.

6.2. Limit spaces of free Abelian groups

6.2.1. Self-coverings of tori and digit tiles. If $p : \mathbb{R}^n/\mathbb{Z}^n \longrightarrow \mathbb{R}^n/\mathbb{Z}^n$ is a d-fold self-covering of a torus, then it induces an injective endomorphism $B : \mathbb{Z}^n \longrightarrow \mathbb{Z}^n$ of the fundamental group. We have $[\mathbb{Z}^n : B(\mathbb{Z}^n)] = d$, i.e., $\det B = \pm d$. The associated virtual endomorphism $A = B^{-1}$ defines the iterated monodromy action of $\mathbb{Z}^n$.

In the other direction, every surjective and injective virtual endomorphism of $\mathbb{Z}^n$ is inverse of an integral matrix B; hence every recurrent self-similar action of $\mathbb{Z}^n$ is a standard iterated monodromy action of the covering $p : \mathbb{R}^n/\mathbb{Z}^n \longrightarrow \mathbb{R}^n/\mathbb{Z}^n$ induced by the linear map $B : \mathbb{R}^n \longrightarrow \mathbb{R}^n$.

The group $\mathbb{Z}^n$ is a uniform lattice in the Lie group $\mathbb{R}^n$, and we can apply Theorem 6.1.6. The extension of the virtual endomorphism $\phi : \mathbb{Z}^n \dashrightarrow \mathbb{Z}^n$ to $\mathbb{R}^n$ is the linear map given by the matrix $A = B^{-1}$.

The covering p is expanding if and only if the matrix B is expanding (i.e., has all the eigenvalues greater than one in absolute value), which is equivalent to the condition that the standard action is finite-state (see Theorem 2.12.1).

By Theorem 6.1.6, if the covering is expanding, then the limit space $\mathcal{X}_{\mathbb{Z}^n}$ is $\mathbb{R}^n$ with the natural action of $\mathbb{Z}^n$, and thus the limit space $\mathcal{J}_{\mathbb{Z}^n}$ is the torus $\mathbb{R}^n/\mathbb{Z}^n$, in accordance with Theorem 5.5.3.

The homeomorphism between $\mathcal{X}_{\mathbb{Z}^n}$ and $\mathbb{R}^n$, by (the proof of) Theorem 6.1.6, is the map

$$\Phi(\ldots x_{i_2} x_{i_1} \cdot g) = \sum_{k=1}^{\infty} \phi^k(r_{i_k}) + g, \qquad (6.3)$$

where $\{x_1 = \phi(r_1)+0, \ldots, x_d = \phi(r_d)+0\} = \mathsf{X}$ is a basis of the bimodule $\phi(\mathbb{Z}^n)+\mathbb{Z}^n$. Recall that this is equivalent to the condition that the *digit set* $R = \{r_i\}$ is a coset transversal of $\operatorname{Dom}\phi = B(\mathbb{Z}^n)$ in $\mathbb{Z}^n$.

Hence the encoding of the limit space $\mathbb{R}^n = \mathcal{X}_{\mathbb{Z}^n}$ by the sequences $\ldots x_2 x_1 \cdot g$ gives the "A-adic" numeration system on $\mathbb{R}^n$.

The respective tile or the *set of fractions* $\mathcal{T}(\phi, R)$ is the set of all possible sums $\sum_{k=1}^{\infty} \phi^k(r_{i_k})$.

The tile $\mathcal{T}(\phi, R)$ is the unique fixed point of the transformation

$$P(C) = \bigcup_{i=1}^{d} \phi(C + r_i)$$

of the space of all non-empty compact subsets of $\mathbb{R}^n$. Moreover, for any non-empty compact set $C \subset \mathbb{R}^n$ the sequence $P^n(C)$ converges in this space to $\mathcal{T}(\phi, R)$ with respect to the Hausdorff metric. This can be used to draw $\mathcal{T}(\phi, R)$.

6.2.2. Examples. In the classical case of the binary numeration system, which corresponds to the virtual endomorphism $\phi(n) = n/2$ of $\mathbb{Z}$ and the digit system $R = \{0, 1\}$, the set of the fractions $\mathcal{T}(\phi, R)$ is the segment $[0, 1]$. Expressions (6.3) are diadic expansions of reals.

Recall that, up to a conjugacy, there exist 6 finite-state self-similar actions of $\mathbb{Z}^2$ on the binary tree (see page 64). Three of them are defined by a virtual endomorphism A with $\det A = 1/2$, and the other three have $\det A = -1/2$.

If $\det A = 1/2$ and A is the matrix of the associated virtual endomorphism, then A is conjugate in $\operatorname{GL}(2, \mathbb{R})$ to one of the matrices

$$\begin{pmatrix} 0 & -\sqrt{2}/2 \\ \sqrt{2}/2 & 0 \end{pmatrix}, \quad \begin{pmatrix} 1/4 & -\sqrt{7}/4 \\ \sqrt{7}/4 & 1/4 \end{pmatrix}, \quad \begin{pmatrix} -1/2 & -1/2 \\ -1/2 & 1/2 \end{pmatrix}.$$

Let us identify $\mathbb{R}^2$ with the complex plane $\mathbb{C}$ identifying the standard basis of $\mathbb{R}^2$ with the $\mathbb{R}$-basis $\{1, i\}$ of $\mathbb{C}$. Then these matrices are the matrices of multiplication by $\alpha = \frac{i\sqrt{2}}{2}, \frac{1+i\sqrt{7}}{4}$ and $\frac{1\pm i}{2}$, respectively.

We can identify $\mathbb{Z}^2$ in each of these cases with the lattice $\Gamma = \mathbb{Z}\left[\alpha^{-1}\right]$. Then multiplication of Γ by α is a virtual endomorphism of index 2 whose matrix is conjugate to the respective 2×2-matrix.

The set $\{0, 1\} \subset \Gamma$ is a coset transversal of the domain of the virtual endomorphism; hence it may be chosen as a digit set. Then the respective encodings of $\mathbb{C} = \mathcal{X}_\Gamma$ by the elements of $\{0, 1\}^{-\omega} + \mathbb{Z}^2$ correspond to the binary numeration systems with the base α^{-1}, i.e., to the expansion of the numbers $z \in \mathbb{C}$ into the series

$$z = a + \sum_{k=1}^{\infty} x_k \alpha^k,$$

where $a \in \Gamma$ and $x_k \in \{0, 1\}$.

See for example a discussion of the numeration system with the base $\alpha^{-1} = (-1 + i)$ in [**Knu69**].

6.2. LIMIT SPACES OF FREE ABELIAN GROUPS

PROPOSITION 6.2.1. *Let A be a contracting virtual endomorphism of $\mathbb{Z}^n$ of index 2. Then the digit tile $\mathcal{T}(A, R)$ depends, up to an affine transformation, only on the conjugacy class of A in $\mathrm{GL}(n, \mathbb{R})$.*

PROOF. It is sufficient to prove that the tiles do not depend on the digit set. Let $R = \{r_0, r_1\}$ and $S = \{s_0, s_1\}$ be two coset transversals of $\mathrm{Dom}\, A$ in $\mathbb{Z}^n$. If we replace R by $R' = \{0, r_1 - r_0\}$, then we replace the tile $\mathcal{T}(A, R)$ by the tile

$$\mathcal{T}(A, R') = \mathcal{T}(A, R) - \sum_{k=1}^{\infty} A^k(r_0).$$

We may assume therefore that $r_0 = s_0 = 0$.

A point belongs to $\mathcal{T}(A, R)$ if and only if it can be represented in the form

$$\sum_{k=1}^{\infty} A^k(r_{x_k}) = \sum_{k=1}^{\infty} x_k A^k(r_1)$$

for some $x_1 x_2 \ldots \in \{0, 1\}^\omega$.

The tile $\mathcal{T}(A, S)$ cannot belong to a proper subspace of $\mathbb{R}^n$, since its $\mathbb{Z}^n$-shifts cover $\mathbb{R}^n$. Hence, there exists a basis $\{e_1, \ldots, e_n\} \subset \mathcal{T}(A, S)$ of $\mathbb{R}^n$. Consequently, we can represent r_1 in the form of a series $\sum_{k=0}^{\infty} a_k A^k(s_1)$, where the sequence $a_k \in \mathbb{R}$ is bounded.

The linear operator $T = \sum_{k=0}^{\infty} a_k A^k$ commutes with A, and we have that $T(s_1) = r_1$. Then

$$\sum_{k=1}^{\infty} x_k A^k(r_1) = \sum_{k=1}^{\infty} x_k A^k(T(s_1)) = T\left(\sum_{k=1}^{\infty} x_k A^k(s_1)\right)$$

for all $x_1 x_2 \ldots \in \{0, 1\}^\omega$; hence $\mathcal{T}(A, R) = T(\mathcal{T}(A, S))$. $\square$

Note that the statement is wrong for actions on the trees of higher degree. See below an example of two different tiles for the matrix $\begin{pmatrix} 1/2 & 0 \\ 0 & 1/2 \end{pmatrix}$.

The tile of the actions defined by the virtual endomorphism $\begin{pmatrix} 0 & -\sqrt{2}/2 \\ \sqrt{2}/2 & 0 \end{pmatrix}$ is a rectangle. For instance, for the conjugate matrix $\begin{pmatrix} 0 & 1 \\ 1/2 & 0 \end{pmatrix}$ and $R = \{(0,0), (1,0)\}$ the set of fractions is the square $[0, 1] \times [0, 1]$.

The tile corresponding to the matrix $\begin{pmatrix} -1/2 & -1/2 \\ 1/2 & -1/2 \end{pmatrix}$ is the "twin dragon" shown on the left-hand side of Figure 6.1.

The tile corresponding to the matrix $\begin{pmatrix} 1/4 & -\sqrt{7}/4 \\ \sqrt{7}/4 & 1/4 \end{pmatrix}$ is the "tame twin dragon" shown on the right-hand side of Figure 6.1.

Some examples of digit tiles are shown in Figure 6.2. They correspond to the following virtual endomorphisms and digit sets:

$\phi = \begin{pmatrix} 1/2 & 0 \\ 0 & 1/2 \end{pmatrix}$	$\phi = \begin{pmatrix} 1/3 & 1/3 \\ -1 & 0 \end{pmatrix}$
$R = \{(0,0), (-1,0), (1,1), (0,-1)\}$	$R = \{(0,0), (1,0), (1,1)\}$
$\phi = \begin{pmatrix} 1/5 & -2/5 \\ 2/5 & 1/5 \end{pmatrix}$	$\phi = \begin{pmatrix} 1/2 & 0 \\ 0 & 1/2 \end{pmatrix}$
$R = \{(0,0), (1,0), (-1,0), (0,1), (0,-1)\}$	$R = \{(0,0), (1,0), (3,3), (0,1)\}$

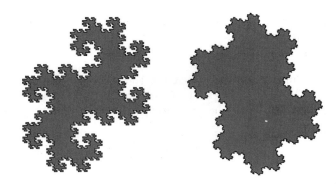

FIGURE 6.1. Twin dragon and tame twin dragon

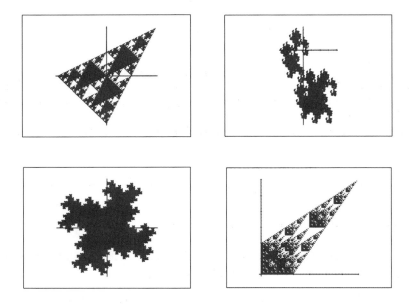

FIGURE 6.2. Digit tiles

6.2.3. Associated tilings of $\mathbb{R}^n$. If the self-similar action of $\mathbb{Z}^n$ satisfies the open set condition (see Definition 3.3.6), then the digit tile $\mathcal{T} = \mathcal{T}(\phi, R)$ is equal to the closure of its interior and the shifts $\mathcal{T} + r$ for $r \in \mathbb{Z}^n$ form a tiling of the space $\mathbb{R}^n$; i.e., they cover $\mathbb{R}^n$ and have disjoint interiors (see Proposition 3.3.7). These tilings are self-affine. Namely, the linear map A^{-1} maps every tile to a union of d tiles.

A part of the tiling by the "twin dragons" is shown in Figure 6.3. The union of two tiles in the center is similar to the original tiles.

See the works [**Ban91, Ken92, Vin95, Vin00**] and their bibliography for the properties of such tilings. They are used in the wavelet theory, computer image processing, toral dynamical systems and other fields. See, for example, the book [**BJ99**] for relations with the C^*-algebras.

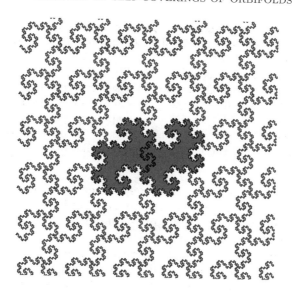

FIGURE 6.3. Plane tiling by twin dragons

Another interesting work is the preprint [**Thu89**] by W. Thurston, describing connections between self-replicating tilings, automata and groups.

6.3. Examples of self-coverings of orbifolds

6.3.1. The dihedral group as IMG($z^2 - 2$). The infinite dihedral group $\mathbb{D}_\infty$ acts on the Lie group $\mathbb{R}$ by the affine transformations of the form $\pm x + n$, where $n \in \mathbb{Z}$. The orbispace of this action is the segment $[0, 1/2]$, whose endpoints are singular with isotropy groups of order 2.

The mapping $x \mapsto dx$, for $d \geq 2$, is an expanding automorphism of $\mathbb{R}$, conjugating the dihedral group with an index d subgroup. It also induces a self-covering of the orbispace.

For example, if $d = 2$, then this self-covering acts on the underlying space by the "tent map", mapping x to $2x$ if $x \in [0, 1/4]$ and to $1 - 2x$, otherwise.

The associated virtual endomorphism ϕ maps a transformation $g : z \mapsto (-1)^k z + n$ to the transformation $\phi(g) : z \mapsto (-1)^k z + n/2$, and $\mathrm{Dom}\,\phi$ is the set of transformations $z \mapsto \pm z + n$ with even n. Let us take the coset transversal $\{r_0(z) = -z + 1,\ r_1(z) = z\}$. Let $a(z) = -z + 1$ and $b(z) = -z$. Then computation of the wreath recursion (see Proposition 2.5.10) gives

(6.4) $$a = \sigma, \qquad b = (a, b).$$

The same self-covering can also be described as the self-covering of the Julia orbispace of the rational function $z^2 - 2$. The critical orbit of $z^2 - 2$ is $0 \mapsto -2 \mapsto 2 \mapsto 2$. Take $t = 0$ as a basepoint and connect it to its preimages $\pm\sqrt{2}$ by straight segments. The fundamental group of the space $\mathcal{M} = \mathbb{C} \setminus \{-2, 2\}$ is generated by the small loops a and b around -2 and 2. We connect these loops to t by straight segments. Computation of the associated self-similar action shows that the respective generators of the iterated monodromy group IMG $(z^2 - 2)$ are defined also by the recursion (6.4).

It follows that $a^2 = 1$ and $b^2 = (a^2, b^2) = (1, b^2) = 1$. Therefore the iterated monodromy groups is isomorphic to the infinite dihedral group $\mathbb{D}_\infty$ (see also [**GNS00**] and [**BGN03**]).

We see that the standard action of IMG $(z^2 - 2)$ coincides with the standard action of the iterated monodromy group of the self-covering $x \mapsto 2x$ of the orbispace $\mathbb{R}/\mathbb{D}_\infty$. It follows that the limit space of this action of $\mathbb{D}_\infty$ is the real segment and that the shift $\mathsf{s} : \mathcal{J}_{\mathbb{D}_\infty} \longrightarrow \mathcal{J}_{\mathbb{D}_\infty}$ is the "tent map".

In general, the self-covering induced by $z \mapsto dz$ on the orbispace $\mathbb{R}/\mathbb{D}_\infty$ is conjugate to the action of the Chebyshev polynomial of degree d on its Julia set (see 6.12.3 below).

6.3.2. Self-coverings of Euclidean spherical orbifolds. Suppose that $\mathcal{M}$ is a 2-dimensional oriented Euclidean orbifold, i.e., that it is an orbispace of a proper action of an orientation-preserving group G of motions of the Euclidean plane $\mathbb{R}^2$. Then $\mathcal{M}$ is developable, $G = \pi_1(\mathcal{M})$ and $\widehat{\mathcal{M}} = \mathbb{R}^2$.

It is known that $\mathcal{M}$ is either the torus without singular points and then $\pi_1(\mathcal{M})$ is a lattice in $\mathbb{R}^2$ or $|\mathcal{M}|$ is a punctured sphere. In the latter case $\mathcal{M}$ has only a finite number of singular points and punctures. Then the *Euler characteristic* of the orbifold $\mathcal{M}$ is the number

$$(6.5) \qquad \chi(\mathcal{M}) = 2 - \sum_{x \in P} \left(1 - \frac{1}{\nu(x)}\right),$$

where P is the set of singular points and punctures and $\nu(x)$ is the order of the isotropy group of x or ∞ if x is a puncture. Since the orbifold is Euclidean, its Euler characteristic is equal to 0.

The following is a well known fact.

PROPOSITION 6.3.1. *If $\mathcal{M}$ is a spherical Euclidean orbifold, then it has at most 4 singular points and punctures and the values of $\nu(x)$ on these points are*

$$(\infty, \infty), \quad (\infty, 2, 2), \quad (3, 3, 3), \quad (6, 3, 2), \quad (4, 4, 2), \quad \text{or} \quad (2, 2, 2, 2).$$

We are interested here in spherical orbifolds with analytic structure on them. Every analytic spherical orbifold is a quotient $\mathbb{C}/G$, where G is the fundamental group acting on $\mathbb{C}$ properly by affine transformations.

The respective actions, up to an affine conjugation, are the following (compare [**DH93**], page 289):

(1) for (∞, ∞): the group of the affine transformations $z \mapsto z + n$, where $n \in \mathbb{Z}$.
(2) for $(\infty, 2, 2)$: the transformations $z \mapsto \pm z + n$, where $n \in \mathbb{Z}$.
(3) for $(3, 3, 3)$: the transformations $z \mapsto e^{2k\pi i/3} z + a$, where $k \in \mathbb{Z}$ and $a \in \mathbb{Z}\left[e^{\pi i/3}\right]$.
(4) for $(6, 3, 2)$: the transformations $z \mapsto e^{k\pi i/3} z + a$, where $k \in \mathbb{Z}$ and $a \in \mathbb{Z}\left[e^{\pi i/3}\right]$.
(5) for $(4, 4, 2)$: the transformations $z \mapsto i^k z + a$, where $k \in \mathbb{Z}$ and $a \in \mathbb{Z}[i]$.
(6) for $(2, 2, 2, 2)$: the transformations $z \mapsto \pm z + a$, where $a \in \Gamma$ for a lattice $\Gamma \subset \mathbb{C}$.

If $f : \mathcal{M} \longrightarrow \mathcal{M}$ is an analytic d-fold self-covering, then it is induced by an affine transformation $z \mapsto \alpha z + \beta$ of $\mathbb{C}$, which conjugates the respective group G to a subgroup of index d. This affine transformation is (see [**DH93**])

6.3. EXAMPLES OF SELF-COVERINGS OF ORBIFOLDS

(1) for (∞, ∞): $z \mapsto nz$;
(2) for $(\infty, 2, 2)$: $z \mapsto nz$ or $z \mapsto nz + \frac{1}{2}$;
(3) for $(3, 3, 3)$: $z \mapsto \alpha z$, $z \mapsto \alpha z + \frac{1}{3}\left(e^{\pi i/3} + 1\right)$, or $z \mapsto \alpha z + \frac{1}{3}i\sqrt{3}$, where $\alpha \in \mathbb{Z}\left[e^{\pi i/3}\right]$;
(4) for $(6, 3, 2)$: $z \mapsto \alpha z$, where $\alpha \in \mathbb{Z}\left[e^{\pi i/3}\right]$;
(5) for $(4, 4, 2)$: $z \mapsto \alpha z$ or $z \mapsto \alpha z + \frac{1}{2}(1 + i)$, where $\alpha \in \mathbb{Z}[i]$;
(6) for $(2, 2, 2, 2)$: $z \mapsto \alpha z + \beta$, where $2\beta \in \Gamma$ and α is an integer in an imaginary quadratic field $\Bbbk$ such that if $\alpha \notin \mathbb{R}$, then Γ is a module over the subring of $\Bbbk$ generated by 1 and α.

The degree of the self-covering $f : \mathcal{M} \longrightarrow \mathcal{M}$ is equal to n for the first two examples and to $|\alpha|^2$ for the rest of them.

The self-covering of the orbifold is defined, in the conditions of Corollary 6.1.7, by an automorphism of the additive Lie group $\mathbb{C}$. This can be achieved in our cases by conjugation of the group G and the affine map $z \mapsto \alpha z + \beta$ by the translation $z \mapsto z + \frac{\beta}{1-\alpha}$ so that the affine map becomes equal to $z \mapsto \alpha z$. The group G will become, however, a less "natural" affine group.

Each of the respective analytic self-coverings is conjugate to an action of a rational function on the (punctured) Riemann sphere.

The covering of (∞, ∞) given by $z \mapsto nz$ is conjugate to the polynomial z^n. The coverings of $(\infty, 2, 2)$ given by $z \mapsto nz$ or $z \mapsto nz + \frac{1}{2}$ are (up to signs) the Chebyshev polynomials of degree n.

Let us consider some of the other cases.

6.3.2.1. *Lattès Examples.* The most well known are the examples of the rational functions considered by S. Lattès [**Lat18**]. They correspond to the case when G is the group of the affine transformations $z \mapsto \pm z + a$, where $a \in \Gamma$ for some lattice $\Gamma \subset \mathbb{C}$. We get in this case the orbispace $(2, 2, 2, 2)$. If α is a multiplicator of the lattice Γ, i.e., if $\alpha \cdot \Gamma \subset \Gamma$, then the branched covering induced on the sphere $\mathbb{C}/G$ by multiplication by α is conjugate to the rational function $f(z)$, which is uniquely defined by the equality

$$\wp(\alpha z) = f(\wp(z)),$$

where $\wp$ is the Weierstrass elliptic function for the lattice Γ given by

(6.6) $$\wp(z) = \frac{1}{z^2} + \sum_{\omega \in \Gamma \setminus \{0\}} \left[\frac{1}{(z+\omega)^2} - \frac{1}{\omega^2}\right].$$

The Weierstrass function is even and induces a two-fold branched covering of the sphere $\widehat{\mathbb{C}}$ by the torus $\mathbb{C}/\Gamma$, identifying the points z and $-z$ of $\mathbb{C}/\Gamma$. Hence, the Weierstrass function realizes the quotient map $\mathbb{C} \mapsto \mathbb{C}/G$.

Corollary 6.1.7 implies that the iterated monodromy group of the rational function f is isomorphic to the group $G = \{\pm z + a : a \in \Gamma\}$. The associated virtual endomorphism is the map $\pm z + a \mapsto \pm z + a/\alpha$.

For example, for $\alpha = 2$ the function f is

$$f(z) = \frac{z^4 + \frac{g_2}{2}z^2 + 2g_3 z + \frac{g_2^2}{16}}{4z^3 - g_2 z - g_3},$$

(see [**Bea91**], p. 74), where $g_2 = 60 s_4$ and $g_3 = 140 s_6$ for $s_m = \sum_{\omega \in \Gamma, \omega \neq 0} \omega^{-m}$.

FIGURE 6.4. Dragon curve

A pair (g_2, g_3) is realized by a lattice Γ if and only if $g_2^3 - 27g_3^2 \neq 0$ (see [**Lan87**], p. 39). In particular, there exists a lattice Γ such that $g_3 = 0$ and $g_2 = 4$, so that

$$(6.7) \qquad f(z) = \frac{(z^2 + 1)^2}{4z(z^2 - 1)}.$$

For the case of the lattice $\Gamma = \mathbb{Z}[i]$ we have $g_3 = 0$; thus $f(z) = \frac{(z^2 + g_2/4)^2}{4z(z^2 - g_2/4)}$, which is also conjugate to (6.7) (the conjugating map is $t(z) = \frac{2z}{\sqrt{g_2}}$).

6.3.2.2. *Heighway dragon.* Consider the group G of the affine transformations of the form $i^k \cdot z + a$ for $k \in \mathbb{Z}$ and $a \in \mathbb{Z}[i]$. The corresponding orbifold $\mathcal{M} = \mathbb{C}/G$ is $(4, 4, 2)$.

Consider the virtual endomorphism

$$\phi : i^k z + a \mapsto i^k z + \frac{1+i}{2} a,$$

i.e., the virtual endomorphism induced by the map $z \mapsto (1 - i)z$. Its domain is the set of the transformations $i^k z + a$ such that $\Re(a) + \Im(a)$ is even.

If we take the coset transversal $D = \{z, z + 1\}$ (or any other coset transversal belonging to $\mathbb{Z}[i]$), then we get the twin dragon as the tile.

But if we take the coset transversal $D = \{z, iz + i\}$, then the tile will be the *dragon curve* (or *Heighway dragon*) shown in Figure 6.4.

If we denote by a the transformation $z \mapsto iz + i$ and by b the transformation $z \mapsto iz$, then the self-similar action defined by the digit set $D = \{1, b\}$ is given by the recursion

$$a = \sigma(1, ba), \quad b = (b, bab^{-1}).$$

The corresponding self-covering of the orbifold $\mathcal{M} = \mathbb{C}/G$ is conjugate to the rational function $f : \widehat{\mathbb{C}} \longrightarrow \widehat{\mathbb{C}}$ such that

$$(6.8) \qquad \wp((1-i)z)^2 = f(\wp(z)^2),$$

where $\wp$ is the Weierstrass function defined by the lattice $\mathbb{Z}[i]$. This follows from the fact that the group of symmetries of $\wp(z)^2$ is the group G of the affine transformations $z \mapsto i^k z + a$, where $a \in \mathbb{Z}[i]$.

It follows from the definition of $\wp$ (see (6.6)) that in our case $\wp(-iz) = -\wp(z)$; hence $\wp'(iz) = -i\wp'(z)$. The classical addition formula

$$\wp(z_1 + z_2) = -\wp(z_1) - \wp(z_2) + \frac{1}{4}\left(\frac{\wp'(z_1) - \wp'(z_2)}{\wp(z_1) - \wp(z_2)}\right)^2$$

implies

$$\wp(z - iz) = -\wp(z) - \wp(-iz) + \frac{1}{4}\left(\frac{\wp'(z) - \wp'(-iz)}{\wp(z) - \wp(-iz)}\right)^2 = \frac{1}{4}\left(\frac{(1+i)\wp'(z)}{2\wp(z)}\right)^2.$$

It is known that

$$(\wp'(z))^2 = 4(\wp(z))^3 - g_2\wp(z) - g_3.$$

We have $g_3 = 0$ in our case; hence

$$\wp((1-i)z) = \frac{i}{8} \cdot \frac{4(\wp(z))^3 - g_2\wp(z)}{(\wp(z))^2} = \frac{i}{2} \cdot \frac{(\wp(z))^2 - g_2/4}{\wp(z)}.$$

Consequently,

$$(\wp((1-i)z))^2 = -\frac{1}{4} \cdot \frac{\left((\wp(z))^2 - g_2/4\right)^2}{(\wp(z))^2};$$

hence the rational function $f(z)$ satisfying (6.8) is $f(t) = -\frac{\left(t - \frac{g_2}{4}\right)^2}{4t}$, which is conjugate to $\left(\frac{2-z}{z}\right)^2$ (the conjugating transformation is $t = \frac{g_2/4}{1-z}$).

6.3.3. Heisenberg group. This example of a self-covering is from [**Shu69**]. Let L be the group of the lower triangular matrices

$$\begin{pmatrix} 1 & 0 & 0 \\ a & 1 & 0 \\ c & b & 1 \end{pmatrix},$$

with $a, b, c \in \mathbb{R}$ and let G be the subgroup of the matrices with $a, b, c \in \mathbb{Z}$. Then for all $p, q \in \mathbb{Z}$, the map

$$f_* : \begin{pmatrix} 1 & 0 & 0 \\ a & 1 & 0 \\ c & b & 1 \end{pmatrix} \mapsto \begin{pmatrix} 1 & 0 & 0 \\ p \cdot a & 1 & 0 \\ pq \cdot c & q \cdot b & 1 \end{pmatrix}$$

is an automorphism of L such that $[G : f_*(G)] = p^2 q^2$. The quotient L/G is a three-dimensional nil-manifold, and the map f_* induces an expanding $p^2 q^2$-fold self-covering.

A modification of this example, due to S. Sidki, gives a self-similar action of the group G over a 4-element alphabet and defines an expanding 4-fold covering of the manifold L/G. (Note that the smallest degree of the covering defined above is $16 = 2^2 \cdot 2^2$.) We have to take the map

$$p_* : \begin{pmatrix} 1 & 0 & 0 \\ a & 1 & 0 \\ c & b & 1 \end{pmatrix} \mapsto \begin{pmatrix} 1 & 0 & 0 \\ 2 \cdot b & 1 & 0 \\ 2 \cdot c & a & 1 \end{pmatrix}$$

It is clearly an injective endomorphism of G and is expanding, since its second iteration is the map f_* for $p = q = 2$.

6.4. Rational functions

6.4.1. Post-critically finite rational functions. Suppose that $f(z) \in \mathbb{C}(z)$ is a non-constant rational function. If $p, q \in \mathbb{C}[z]$ are co-prime and $f(z) = p(z)/q(z)$, then the *degree* of f is $\max(\deg p, \deg q)$ and is denoted $\deg f$.

The function f defines a branched $\deg f$-fold self-covering of the Riemann sphere $\widehat{\mathbb{C}} = \mathbb{C} \cup \{\infty\}$. A point $z \in \widehat{\mathbb{C}}$ is *critical* if f is not a local homeomorphism on any neighborhood of z, i.e., if $f'(z) = 0$.

Let C_f be the set of the critical points of f. We denote by P_f the set of the *post-critical points* of f, i.e., the set

$$P_f = \bigcup_{n \geq 1} f^n(C_f).$$

Here and in the sequel, f^n denotes the nth iteration of f and not the nth degree.

If the closure $\overline{P}_f$ of the post-critical set is such that $\mathcal{M} = \widehat{\mathbb{C}} \setminus \overline{P}_f$ is path-connected, then f defines a d-fold partial self-covering $f : \mathcal{M}_1 \longrightarrow \mathcal{M}$, where $\mathcal{M}_1 = \widehat{\mathbb{C}} \setminus f^{-1}(\overline{P}_f)$, since then $\mathcal{M}_1 \subset \mathcal{M}$.

An important case is when P_f is finite. Such rational functions f are called *post-critically finite*. In this case $\mathcal{M}$ and $\mathcal{M}_1$ are punctured spheres. The fundamental group $\pi_1(\mathcal{M})$ is the free group of rank $|P_f| - 1$.

DEFINITION 6.4.1. Let $f \in \mathbb{C}(z)$ be a post-critically finite rational function. Then the *iterated monodromy group* $\mathrm{IMG}(f)$ is the iterated monodromy group of the partial self-covering $f : \mathcal{M}_1 \longrightarrow \mathcal{M}$, where $\mathcal{M} = \widehat{\mathbb{C}} \setminus P_f$ and $\mathcal{M}_1 = f^{-1}(\mathcal{M})$.

6.4.2. Profinite iterated monodromy groups as Galois groups. The following construction belongs to R. Pink (private communication). It was the origin of the definition of iterated monodromy groups.

Let $f(z) = p(z)/q(z) \in \mathbb{C}(z)$ be a rational function, where $p(z), q(z) \in \mathbb{C}[z]$ are co-prime polynomials. Let $p_n(z), q_n(z) \in \mathbb{C}[z]$ be co-prime polynomials such that $p_n(z)/q_n(z)$ is the nth iteration f^n of f.

Let Ω_n be the field obtained by adjoining all the solutions of the equation $f^n(z) = t$ to the field of rational functions $\mathbb{C}(t)$ in some algebraic closure of $\mathbb{C}(t)$. In other words, Ω_n is the splitting field of the polynomial $F_n(z) = p_n(z) - q_n(z)t \in \mathbb{C}(t)[z]$ over the function field $\mathbb{C}(t)$. It is easy to see that $\Omega_n \subset \Omega_{n+1}$. It is well known that the Galois group $\mathrm{Aut}(\Omega_n/\mathbb{C}(t))$ is isomorphic to the monodromy group of the branched covering $f^n : \widehat{\mathbb{C}} \longrightarrow \widehat{\mathbb{C}}$ (see, for example [**For81**], Theorem 8.12), i.e., to the permutation group of the set $f^{-n}(z_0)$ induced by the action of the fundamental group $\pi_1(\mathbb{C} \setminus P_n, z_0)$, where P_n is the set of the branching points of the function f^n and $z_0 \notin P_n$ is arbitrary.

As a corollary we get the following interpretation of the profinite iterated monodromy group of a rational function.

PROPOSITION 6.4.2. *Let $f \in \mathbb{C}(z)$ be a post-critically finite rational function. Then the profinite iterated monodromy group $\overline{\mathrm{IMG}}(f)$ is isomorphic to the Galois group $\mathrm{Aut}(\Omega/\mathbb{C}(t))$, where $\Omega = \bigcup_{n \geq 1} \Omega_n$.*

6.4.3. Branched coverings and Thurston orbifold. Let S^2 be the real 2-sphere as a topological manifold. A *branched d-fold covering* $f : S^2 \longrightarrow S^2$ is a continuous orientation-preserving map such that there exists a finite set $C \subset S^2$ such that $f : S^2 \setminus C \longrightarrow S^2 \setminus f(C)$ is a d-fold covering. For every $x \in S^2$ the *local degree* $\deg_x(f)$ of f at x is the degree of the map $f : \gamma \longrightarrow f(\gamma)$, where γ is a small simple loop around x. A point x is called *critical* if the local degree $\deg_x(f)$ of f at x is greater than 1. If C_f is the set of the critical points of f, then $f : S^2 \setminus f^{-1}(f(C_f)) \longrightarrow S^2 \setminus f(C_f)$ is a d-fold covering.

There exist local charts $q_x : U_1 \longrightarrow U_x$ and $q_{f(x)} : U_2 \longrightarrow U_{f(x)}$, where U_1, U_2 are neighborhoods of 0 in $\mathbb{C}$ and $U_x, U_{f(x)}$ are neighborhoods of x and $f(x)$ such that f is equal to $z \mapsto z^{\deg_x(f)}$ in these charts; i.e., $q_{f_x}\left(z^{\deg_x(f)}\right) = f\left(q_x(z)\right)$ for all $z \in U_1$.

The number $\deg_x(f) - 1$ is called the *multiplicity* of a critical point x. The Riemann-Hurwitz formula implies that there exist precisely $2d - 2$ critical points of a d-fold branched covering, counting them with multiplicities.

A branched covering $f : S^2 \longrightarrow S^2$ is called *post-critically finite* if the post-critical set $\bigcup_{n \geq 1} f^n(C_f)$ is finite. Post-critically finite branched coverings are also called *Thurston maps*.

The *iterated monodromy group* $\mathrm{IMG}(f)$ of a Thurston map $f : S^2 \longrightarrow S^2$ is the iterated monodromy group of the partial self-covering $f : S^2 \setminus f^{-1}(P_f) \longrightarrow S^2 \setminus P_f$, where P_f is the post-critical set.

Note that if $A \subset S^2$ is any finite set such that $f(A) \subseteq A$ and $f(C_f) \subseteq A$, then $A \supseteq P_f$ and the iterated monodromy group of the partial self-covering $f : S^2 \setminus f^{-1}(A) \longrightarrow S^2 \setminus A$ coincides, by Proposition 5.6.1, with the iterated monodromy group $\mathrm{IMG}(f)$.

In many cases it is not convenient to delete the post-critical set P_f completely. It is more reasonable to define an orbifold with singular points in the post-critical set such that the Thurston map f becomes a partial self-covering of the orbifold.

The construction of such an orbifold is due to W. Thurston (see [**DH93**]) and is called the *Thurston orbifold* of the post-critically finite branched covering.

Let $f : S^2 \longrightarrow S^2$ be a Thurston map with the set of critical points C_f and the post-critical set P_f. Let P' be the union of all the cycles of f which contain a critical point. We obviously have $P' \subseteq P_f$.

Let us find for every $x \in S^2 \setminus P'$ the least common multiple $\nu(x)$ of the local degrees $\deg_z(f^m)$, where $z \in S^2$ and $m \geq 1$ are such that $f^m(z) = x$. It is easy to see that $\nu(x)$ exists (i.e., is finite) for all $x \in S^2 \setminus P'$. It is greater than 1 if and only if $x \in P_f$.

It follows directly from the definition that for any $x \in S^2$ the number $\nu(f(x))$ is divisible by $\deg_x(f) \cdot \nu(x)$. Denote $\nu_0(x) = \frac{\nu(f(x))}{\deg_x(f)}$. Then $\nu(x) | \nu_0(x)$.

Let $\mathcal{M}_\nu$ be the orbispace with the underlying space $|\mathcal{M}| = S^2 \setminus P'$ for which the isotropy group of a point $x \in \mathcal{M}$ is the cyclic group of order $\nu(x)$ acting by rotations of a disc.

Let $\mathcal{M}_{\nu_0}$ be the orbispace with the underlying space $|\mathcal{M}_{\nu_0}| = S^2 \setminus f^{-1}(P')$ defined by the weights $\nu_0(x)$ instead of $\nu(x)$. It follows from the condition $\nu(z) | \nu_0(z)$ that the orbispace $\mathcal{M}_{\nu_0}$ is an open sub-orbispace of the orbispace $\mathcal{M}_\nu$. The embedding acts on the underlying spaces as the identical map.

On the other side, the condition $\deg_x(f) = \frac{\nu_0(x)}{\nu(f(x))}$ implies that the map $f : \mathcal{M}_{\nu_0} \longrightarrow \mathcal{M}_\nu$ is a covering of the orbispaces.

Note that the isotropy groups of the Thurston orbifold $\mathcal{M}$ are represented faithfully in the iterated monodromy group, since we take $\nu(x)$ to be the *least common multiple* of the local degrees $\deg_z(f^n)$ for $f^n(z) = x$ (see Lemma 5.5.2).

6.4.4. Sub-hyperbolic rational functions. Proof of the following result can be found, for example, in [**Mil99**].

THEOREM 6.4.3. *If $f \in \mathbb{C}(z)$ is a post-critically finite rational function, then there exists a Riemannian metric on the Thurston orbifold $\mathcal{M}$ of f such that the partial self-covering $f : \mathcal{M}_1 \longrightarrow \mathcal{M}$ is uniformly expanding on the Julia set.*

Here the *Julia set* of a rational function $f(z)$ is the set of points $z_0 \in \widehat{\mathbb{C}}$ such that the sequence of the iterates $f^n(z)$ is not normal on any neighborhood U of z_0. An equivalent definition is that the Julia set is the closure of the union of repelling cycles of f. For more on Julia sets of rational functions, see, for example, [**Mil99, Bea91**].

A rational function which is expanding with respect to some orbifold metric on a neighborhood of the Julia set is called *sub-hyperbolic*. So, the last theorem says that any post-critically finite rational function is sub-hyperbolic. In fact a rational function is sub-hyperbolic if and only if the orbit of every critical point is either finite or converges to an attracting cycle (see [**Mil99**]).

Thus, in the case when the function is sub-hypberbolic, our definition of the Julia set of an expanding self-covering coincides with the classical definition of the Julia set.

THEOREM 6.4.4. *Let $f \in \mathbb{C}(z)$ be a post-critically finite rational function. Then every standard action of the iterated monodromy group $\mathrm{IMG}(f)$ is contracting and the limit dynamical system $\mathsf{s} : \mathcal{J}_{\mathrm{IMG}(f)} \longrightarrow \mathcal{J}_{\mathrm{IMG}(f)}$ is topologically conjugate with the action of f on the Julia set. Moreover, the partial self-covering of the limit orbispaces $\mathsf{s} : \mathcal{J}^{\circ}_{\mathrm{IMG}(f)} \longrightarrow \mathcal{J}_{\mathrm{IMG}(f)}$ is conjugate with the partial self-covering of the Julia orbispaces.*

Here the Julia orbispace is the restriction of the Thurston orbispace of f onto the Julia set.

PROOF. A direct corollary of Theorems 6.4.3 and 5.5.3. $\square$

Theorem 6.4.4 and the description of the limit space as a quotient of the space of sequences (Theorem 3.6.3 and Proposition 3.6.4) give us nice finite-to-one encodings of the Julia sets by infinite sequences with a complete description of the identifications. These encodings (defined using preimages of paths) were studied before by different authors. See, for example, the papers of M. V. Yacobson [**Yac73, Yac80**].

Different bases of the bimodule associated to a partial self-covering (different standard actions) give encodings of different complexity, i.e., the sizes and the structure of the nuclei (describing the identifications of the sequences) will be different. We will show particularly simple standard actions of the iterated monodromy groups of post-critically finite polynomials in subsequent sections.

6.5. Combinatorial equivalence and Thurston's Theorem

6.5.1. Thurston equivalence. Two Thurston maps $f_1 : S^2 \longrightarrow S^2$ and $f_2 : S^2 \longrightarrow S^2$ with post-critical sets P_{f_1} and P_{f_2} are said to be *combinatorially equivalent* (see [**DH93**]) if there exist orientation-preserving homeomorphisms

$h_0, h_1 : S^2 \longrightarrow S^2$ such that $h_i(P_{f_1}) = P_{f_2}$ for $i = 1, 2$, the diagram

(6.9)
$$\begin{array}{ccc} S^2 & \xrightarrow{f_1} & S^2 \\ \downarrow{h_0} & & \downarrow{h_1} \\ S^2 & \xrightarrow{f_2} & S^2 \end{array}$$

is commutative, and h_0 is isotopic to h_1 through an isotopy constant on P_{f_1}.

DEFINITION 6.5.1. Let $\mathfrak{M}_1, \mathfrak{M}_2$ be permutational bimodules over the groups G_1 and G_2, respectively. We say that an isomorphism $\psi : G_1 \longrightarrow G_2$ *conjugates* the bimodules $\mathfrak{M}_1$ and $\mathfrak{M}_2$ if there exists a bijection $F : \mathfrak{M}_1 \longrightarrow \mathfrak{M}_2$ such that

$$F(g \cdot m \cdot h) = \psi(g) \cdot F(m) \cdot \psi(h)$$

for all $g, h \in G_1$ and $m \in \mathfrak{M}_1$.

In other words, the isomorphism ψ conjugates the bimodules if they become isomorphic after identification of G_1 with G_2 by ψ.

THEOREM 6.5.2. *Let f_1, f_2 be Thurston maps with post-critical sets P_{f_1}, P_{f_2} and let $\mathfrak{M}(f_i)$, $i = 1, 2$, be the respective $\pi_1(S^2 \setminus P_{f_i})$-bimodules.*

Then the maps f_1 and f_2 are combinatorially equivalent if and only if there exists an isomorphism $h_* : \pi_1(S^2 \setminus P_{f_1}) \longrightarrow \pi_1(S^2 \setminus P_{f_2})$ conjugating the bimodules $\mathfrak{M}(f_1)$ and $\mathfrak{M}(f_2)$ and induced by an orientation preserving homeomorphism $h : S^2 \longrightarrow S^2$ such that $h(P_{f_1}) = P_{f_2}$.

COROLLARY 6.5.3. *If two Thurston maps are combinatorially equivalent, then their iterated monodromy groups are isomorphic. Moreover, the respective iterated monodromy actions and the limit dynamical systems are conjugate.*

PROOF OF THEOREM 6.5.2. This theorem easily follows from one of algebraic formulations of the Thurston equivalence relation, found by K. Pilgrim in [**Pil03a, Pil04**] and A. Kameyama in [**Kam01**]. Therefore we give here only a sketch of the proof.

Suppose that the Thurston maps f_1, f_2 are combinatorially equivalent. Let us show that the respective bimodules are isomorphic.

The virtual endomorphism ϕ_i of $\pi_1(S^2 \setminus P_{f_i})$ associated to the partial self-covering f_i is equal to $e_i \circ f_{i*}^{-1}$, where $e_i : \pi_1(S^2 \setminus f_i^{-1}(P_{f_i})) \longrightarrow \pi_1(S^2 \setminus P_{f_i})$ is the homomorphism induced by the embedding $S^2 \setminus f_i^{-1}(P_{f_i}) \hookrightarrow S^2 \setminus P_{f_i}$ and $f_{i*}^{-1} : \pi_1(S^2 \setminus P_{f_i}) \dashrightarrow \pi_1(S^2 \setminus f_i^{-1}(P_{f_i}))$ is the virtual isomorphism induced by the covering $f_i : S^2 \setminus f_i^{-1}(P_{f_i}) \longrightarrow S^2 \setminus P_{f_i}$ (see Lemma 4.7.4). Both are defined uniquely up to conjugations in the fundamental groups.

Let h_0, h_1 be the homeomorphisms as in the definition of a combinatorial equivalence. The isomorphism $h_* : \pi_1(S^2 \setminus P_{f_1}) \longrightarrow \pi_1(S^2 \setminus P_{f_2})$ induced by h_i does not depend, up to conjugation in the fundamental groups, on $i = 0, 1$, since they are isotopic. Commutativity of the diagram (6.9) implies now that the virtual endomorphisms ϕ_1 and $h_*^{-1} \circ \phi_2 \circ h_*$ are conjugate, i.e., that the permutational bimodules $\mathfrak{M}(f_1)$ and $\mathfrak{M}(f_2)$ become isomorphic, if we identify the fundamental groups $\pi_1(S^2 \setminus P_{f_i})$ by the isomorphism h_* (see Corollary 2.5.9).

In the other direction, suppose that there exists a homeomorphism $h_1 : S^2 \longrightarrow S^2$ such that $h_1(P_{f_1}) = P_{f_2}$ and the bimodules $\mathfrak{M}(f_i)$, $i = 1, 2$, become isomorphic after identification of the fundamental groups $\pi_1\left(S^2 \setminus P_{f_i}\right)$ by the induced isomorphism $(h_1)_*$. The isomorphism of the bimodules implies that the monodromy action of $\pi_1\left(S^2 \setminus P_{f_i}\right)$ on the coverings f_i are the same and hence there exists a homeomorphism $h_0 : S^2 \longrightarrow S^2$ making the diagram (6.9) commutative. We also get that the induced isomorphisms $(h_0)_*$ and $(h_1)_*$ of the fundamental groups $\pi_1\left(S^2 \setminus P_{f_i}\right)$ are conjugate. But this implies that the homeomorphisms h_0 and h_1 are isotopic, since a surface homeomorphism is uniquely determined, up to an isotopy, by its action on the fundamental groups. $\square$

6.5.2. Theorem of Thurston. Let $\mathcal{M}$ be an orbifold which has only a finite number of singular points and whose underlying space is a punctured sphere. Let P be the set of the points $x \in S^2$ which are either singular or are deleted from S^2 in $\mathcal{M}$. If $x \in P$, then we denote by $\nu(x)$ the order of the isotropy group of x if $x \in \mathcal{M}$ and ∞ otherwise (i.e., if x is deleted from S^2). Recall that the *Euler characteristic* of the orbifold $\mathcal{M}$ is the number

$$\chi(\mathcal{M}) = 2 - \sum_{x \in P}\left(1 - \frac{1}{\nu(x)}\right).$$

If $\chi(\mathcal{M}) > 0$, then the fundamental group $\pi_1(\mathcal{M})$ is finite. If $\chi(\mathcal{M}) = 0$, then $\pi(\mathcal{M})$ is abelian-by-finite and the orbifold $\mathcal{M}$ is *Euclidean*. Otherwise the fundamental group is Gromov-hyperbolic and the orbifold is *hyperbolic*.

Let $f : S^2 \longrightarrow S^2$ be a Thurston map with the post-critical set P_f. A simple closed curve in $S^2 \setminus P_f$ is said to be *peripheral* if one of the regions that it bounds on the sphere contains less than two points of P_f.

An *f-stable multi-curve* is a finite set Γ of simple, closed, disjoint, non-peripheral, pairwise non-homotopic curves in $S^2 \setminus P_f$ such that for every $\gamma \in \Gamma$ each component of $f^{-1}(\gamma)$ is either peripheral or homotopic in $S^2 \setminus P_f$ to an element of Γ. If Γ is an f-stable multi-curve, then we denote by A_Γ the linear map $A_\Gamma : \mathbb{R}^\Gamma \longrightarrow \mathbb{R}^\Gamma$ given by

$$A_\Gamma(\gamma) = \sum_{\alpha \in f^{-1}(\gamma)} \frac{[\alpha]}{\deg(f : \alpha \longrightarrow \gamma)},$$

where $[\alpha]$ is the element of Γ homotopic to α if α is not peripheral, and 0, otherwise.

The following theorem by Thurston (see its proof in [**DH93**]) gives a criterion when a Thurston map is combinatorially equivalent to a rational function.

THEOREM 6.5.4. *A Thurston map $f : S^2 \longrightarrow S^2$ with hyperbolic orbifold is combinatorially equivalent to a rational function if and only if for any f-stable multi-curve Γ the spectral radius of the operator A_Γ is less than one. In that case the rational function is unique up to a conjugation by a linear fraction.*

Every self-covering of a Euclidean orbifold is equivalent to a unique rational function, except for the orbifold $(2, 2, 2, 2)$. In this case the answer depends on the associated virtual endomorphism ϕ. The self-covering is not equivalent to a rational function if and only if the eigenvalues of ϕ are real and different. If the eigenvalues are real and equal, then there is no uniqueness of the rational function. (Recall that ϕ is an endomorphism of the free abelian subgroup of $\pi_1(\mathcal{M})$ and hence induces a linear transformation of $\mathbb{R}^2$.) The case of self-coverings of Euclidean orbifolds was discussed in Subsection 6.1.7.

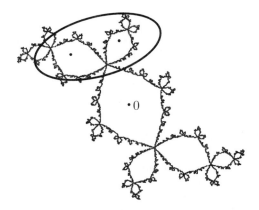

FIGURE 6.5. "Twisted Rabbit"

6.6. "Twisted rabbit" question of J. Hubbard

There are three values of the parameter c such that the critical point 0 of the quadratic polynomial $f(z) = z^2 + c$ belongs to a cycle of length three under the iterations of $f(z)$. These values are the roots of the polynomial $\left((c^2+c)^2+c\right)/c = c^3 + 2c^2 + c + 1$. This polynomial has two complex-conjugate roots $c_r \approx -0.1226 + 0.7449i$ and $\overline{c_r} \approx -0.1226 - 0.7449i$ and one real root $c_a \approx -1.7549$. The polynomial $z^2 + c_r$ (or its Julia set) is called the "Douady rabbit", and the polynomial $z^2 + c_a$ (its Julia set) is called the "airplane". See the corresponding Julia sets in Figure 3.7 on page 116. We shall call the polynomial $z^2 + \overline{c_r}$ the "anti-rabbit".

Let us consider the "rabbit" polynomial $f_r(z) = z^2 + c_r$. Denote $c_1 = c_r$ and $c_2 = c_r^2 + c_r$, so that $0 \mapsto c_1 \mapsto c_2 \mapsto 0$ is the critical cycle. J. Hubbard asked (see, for example, [**Pil03a**]) the following question. Let T be the right Dehn twist around the closed simple curve going around the points c_1 and c_2. See Figure 6.5 where the Julia set of f_r, the curve and the points of the orbit of 0 are shown. The composition $f_r \circ T^m : \widehat{\mathbb{C}} \longrightarrow \widehat{\mathbb{C}}$ is a Thurston map with the post-critical set $P = \{\infty, 0, c_1, c_2\}$. One can prove (see, for example, [**Pil03a**] and [**BFH92**]) that the map $f_r \circ T^m$ has no obstructions; therefore it is equivalent either to the "rabbit" or to the "anti-rabbit" or to the "airplane". The question is, to which of these polynomials is the map $f_r \circ T^m$ equivalent? One can ask this question also for any other homeomorphism T of $\widehat{\mathbb{C}}$, which fixes the set P pointwise.

Let us show how this question can be interpreted and solved using permutational bimodules and iterated monodromy groups. This section is based on joint work [**BN05b**] with L. Bartholdi.

Let $\mathfrak{F}$ be the set of the homotopy classes rel. P of the orientation-preserving branched coverings $f : \widehat{\mathbb{C}} \longrightarrow \widehat{\mathbb{C}}$ such that ∞ and 0 are critical points of f of degree 2, $f(\infty) = \infty$, $f(0) = c_1$, $f(c_1) = c_2$ and $f(c_2) = 0$. So, $P = \{\infty, 0, c_1, c_2\}$ is the post-critical set of every element of $\mathfrak{F}$. Let M be the mapping class group of the punctured sphere $\widehat{\mathbb{C}} \setminus P$, i.e., the set of the homotopy classes rel. P of the homeomorphisms $h : \widehat{\mathbb{C}} \longrightarrow \widehat{\mathbb{C}}$ fixing the set P pointwise.

Then the set $\mathfrak{F}$ is in a natural way an M-bimodule, since the mapping class group acts on $\mathfrak{F}$ by post- and precompositions: $h \cdot f = h \circ f$, $f \cdot h = f \circ h$.

Now let $P \subset S^2$ be any finite subset of the sphere S^2. Then (see [**DH93**]) the *Teichmüller space* $\mathcal{T}_P$ modelled on (S^2, P) is the space of homeomorphisms $\tau : S^2 \longrightarrow \widehat{\mathbb{C}}$, where τ_1 and τ_2 are identified if there exists a biholomorphic isomorphism $h : \widehat{\mathbb{C}} \longrightarrow \widehat{\mathbb{C}}$ such that $h \circ \tau_1 = \tau_2$ on P and $h \circ \tau_1$ is isotopic to τ_2 rel. P.

The *moduli space* $\mathcal{M}_P$ of (S^2, P) is the space of all injective maps $P \hookrightarrow \widehat{\mathbb{C}}$ modulo post-compositions with the elements of the Möbius group (the automorphism group of $\widehat{\mathbb{C}}$).

It is a classical theorem of Teichmüller that the Teichmüller space $\mathcal{T}_P$ coincides with the universal covering of the moduli space $\mathcal{M}_P$. In this case the covering map carries $\tau \in \mathcal{T}_P$ to its restriction $\tau|_P$ onto P.

The mapping class group M of $S^2 \setminus P$ is the fundamental group of the moduli space $\mathcal{M}_P$, and the natural action of M on the Teichmüller space by precompositions coincides with the action of the fundamental group of $\mathcal{M}_P$ by the deck transformations on the universal covering.

Let $f : S^2 \longrightarrow S^2$ be a branched covering with the post-critical set P. Then for every $\tau \in \mathcal{T}_P$ there exists a unique element $\tau' \in \mathcal{T}_P$ such that we have a commutative diagram

(6.10)
$$\begin{array}{ccc} S^2 & \xrightarrow{f} & S^2 \\ \downarrow{\tau'} & & \downarrow{\tau} \\ \widehat{\mathbb{C}} & \xrightarrow{f_\tau} & \widehat{\mathbb{C}} \end{array}$$

where $f_\tau = \tau \circ f \circ (\tau')^{-1} : \widehat{\mathbb{C}} \longrightarrow \widehat{\mathbb{C}}$ is a rational function. Let us denote $\tau' = \sigma_f(\tau)$. The map $\sigma_f : \mathcal{T}_P \longrightarrow \mathcal{T}_P$ is analytic and weakly contracting. For more details see the paper [**DH93**].

Let us compute the defined objects in our situation when $P = \{0, c_1, c_2, \infty\}$. The moduli space $\mathcal{M}_P$ is the set of injective maps $\tau|_P : \{0, c_1, c_2, \infty\} \hookrightarrow \widehat{\mathbb{C}}$ modulo Möbius transformations on $\widehat{\mathbb{C}}$. We may assume, after application of the respective element of the Möbius group, that 0 is mapped by τ to 0, c_1 to 1 and ∞ to ∞. Then the points of $\mathcal{M}_P$ are parametrized by $\tau|_P(c_2) = p$. We have $p \notin \{0, 1, \infty\}$, since the map $\tau|_P$ is injective.

Hence, the moduli space $\mathcal{M}_P$ is homeomorphic to the punctured sphere $\widehat{\mathbb{C}} \setminus \{0, 1, \infty\}$, where a point $p \in \widehat{\mathbb{C}} \setminus \{0, 1, \infty\}$ corresponds to the element $\tau|_P \in \mathcal{M}_P$ such that

$$\tau|_P(\infty) = \infty, \quad \tau|_P(0) = 0, \quad \tau|_P(c_1) = 1, \quad \tau|_P(c_2) = p.$$

Take an arbitrary element $f \in \mathfrak{F}$ of the M-bimodule defined above and consider an arbitrary point τ of the Teichmüller space $\mathcal{T}_P$. Consider its image $\tau' = \sigma_f(\tau)$ under the pull-back map $\sigma_f : \mathcal{T}_P \longrightarrow \mathcal{T}_P$. Suppose that the projection of τ' onto the moduli space is parametrized by a point $p_0 \in \widehat{\mathbb{C}} \setminus \{0, 1, \infty\}$ and the projection of τ is parametrized by p_1.

Then the rational function f_τ in the diagram (6.10) is a degree 2 map with the critical points at 0 and ∞ such that

$$f_\tau(\infty) = \infty, \quad f_\tau(0) = 1, \quad f_\tau(1) = p_1, \quad f_\tau(p_0) = 0,$$

since $f_\tau|_P = \tau|_P \circ f|_P \circ (\tau'|_P)^{-1}$.

We conclude that f_τ is a quadratic polynomial of the form $az^2 + 1$. Consequently

$$a + 1 = p_1, \quad ap_0^2 + 1 = 0;$$

hence $a = -\frac{1}{p_0^2}$, so that
$$p_1 = 1 - \frac{1}{p_0^2}$$
and
(6.11) $$f_\tau(z) = 1 - \frac{1}{p_0^2} z^2.$$

We see that the correspondence $\sigma_f(\tau) \mapsto \tau$ on the Teichmüller space is projected on the moduli space $\widehat{\mathbb{C}} \setminus \{0, 1, \infty\}$ to the rational function
$$z \mapsto 1 - \frac{1}{z^2}.$$

PROPOSITION 6.6.1. *The M-bimodule $\mathfrak{F}$ is isomorphic to the M-bimodule $\mathfrak{M}(F)$ associated to the partial self-covering of $\mathcal{M}_P$ defined by the rational function $F(z) = 1 - \frac{1}{z^2}$.*

The Teichmüller space $\mathcal{T}_P$ is an $\mathfrak{F}$-self-similar right M-space, where the self-similarity is given by
$$\tau \otimes f = \sigma_f(\tau).$$

Here we identify $\mathcal{M}_P$ with $\widehat{\mathbb{C}} \setminus \{0, 1, \infty\}$ and M is identified with the fundamental group of $\mathcal{M}_P$ as described above. For the definition of $\mathfrak{M}(F)$, see 5.1.4, and for the definition of a self-similarity of a G-space, see Definition 3.4.10.

PROOF. Choose a point $\tau_0 \in \mathcal{T}_P$ and let t_0 be its image in $\mathcal{M}_P$. It will be our basepoint for the definitions of the bimodule $\mathfrak{M}(F)$ and of the fundamental group of the moduli space.

For every $f \in \mathfrak{F}$ let $\widetilde{\gamma}_f$ be a path in $\mathcal{T}_P$ starting in τ_0 and ending in $\sigma_f(\tau_0)$. Let γ_f be the image of $\widetilde{\gamma}_f$ in $\mathcal{M}_P$. The paths $\widetilde{\gamma}_f$ and γ_f are defined uniquely up to homotopy, since $\mathcal{T}_P$ is simply connected.

Let us prove that the map $f \mapsto \gamma_f$ is an isomorphism of the M-bimodules. We know that the image of $\sigma_f(\tau_0)$ in $\mathcal{M}_P$ is parametrized by an F-preimage of the point t_0. Consequently, γ_f is an element of the bimodule $\mathfrak{M}(F)$.

Let $g \in M$ and $f \in \mathfrak{F}$ be arbitrary. Consider the following commutative diagram:

(6.12)
$$\begin{array}{ccccccc}
S^2 & \xrightarrow{g} & S^2 & \xrightarrow{f} & S^2 & \xrightarrow{g} & S^2 \\
\downarrow{\sigma_f(\tau)\circ g} & & \downarrow{\sigma_f(\tau)} & & \downarrow{\tau} & & \downarrow{\tau \circ g^{-1}} \\
\widehat{\mathbb{C}} & \xrightarrow{id} & \widehat{\mathbb{C}} & \xrightarrow{f_\tau} & \widehat{\mathbb{C}} & \xrightarrow{id} & \widehat{\mathbb{C}}
\end{array}$$

It implies that
$$\sigma_{f \circ g}(\tau) = \sigma_f(\tau) \circ g, \qquad \sigma_{g \circ f}(\tau \circ g^{-1}) = \sigma_f(\tau)$$
for all $\tau \in \mathcal{T}_P$ and $g \in M$. The second equality is equivalent to
$$\sigma_{g \circ f}(\tau) = \sigma_f(\tau \circ g),$$
which proves in particular that $\tau \otimes f = \sigma_f(\tau)$ is a self-similarity structure on the Teichmüller space.

If $\widetilde{\gamma}$ is a path from τ_0 to $\tau_0 \circ g$ in $\mathcal{T}_P$, then its image γ in $\mathcal{M}_P$ is a loop defining the element corresponding to g in the fundamental group $M = \pi_1(\mathcal{M}_P, t_0)$.

Let $\widetilde{\gamma}_f \circ g$ be the image of the path $\widetilde{\gamma}_f$ under the action of g. The path $\widetilde{\gamma}_f \circ g$ starts in $\tau_0 \circ g$ and ends in $\sigma_f(\tau_0) \circ g = \sigma_{f \circ g}(\tau_0)$. We get therefore the path

$\tilde{\gamma}_{f\circ g} = (\tilde{\gamma}_f \circ g)\tilde{\gamma}$ starting in τ_0 and ending in $\sigma_{f\circ g}(\tau_0)$. The image of this path in $\mathcal{M}_P$ is $\gamma_f \gamma$. Consequently

$$\gamma_{f\circ g} = \gamma_f \cdot \gamma,$$

and the map $f \mapsto \gamma_f$ conjugates the right actions of M on $\mathfrak{F}$ and on $\mathfrak{M}(F)$.

The right M-modules $\mathfrak{F}$ and $\mathfrak{M}(F)$ are free 2-dimensional. This implies that the map $f \mapsto \gamma_f$ is a bijection.

The path $\sigma_f(\tilde{\gamma})$ starts in $\sigma_f(\tau_0)$ and ends in $\sigma_f(\tau_0 \circ g) = \sigma_{g\circ f}(\tau_0)$. We get in this way the path $\tilde{\gamma}_{g\circ f} = \sigma_f(\tilde{\gamma})\tilde{\gamma}_f$ starting in τ_0 and ending in $\sigma_{g\circ f}(\tau_0)$.

But we know that the correspondence $\sigma_f(\tau) \mapsto \tau$ is projected to the rational function $F(z) = 1 - \frac{1}{z^2}$ on the moduli space $\mathcal{M}_P$. Consequently, the image of the path $\sigma_f(\tilde{\gamma})$ in the moduli space is an F-preimage of the path γ. This F-preimage starts at the end of the path γ_f, since $\sigma_f(\tilde{\gamma})$ starts at the end of the path $\tilde{\gamma}_f$. Thus $\gamma_{g\circ f}$ is equal to $F^{-1}(\gamma)[z_f]\gamma_f$, where z_f is the end of the path γ_f. Consequently, the map $f \mapsto \gamma_f$ conjugates the left actions of M on $\mathfrak{F}$ and on $\mathfrak{M}(F)$. □

The action of M on $\mathcal{T}_P$ is not co-compact; therefore the $\mathfrak{F}$-self-similar space $\mathcal{T}_P$ does not satisfy the conditions of Theorem 3.4.13. Note, however, that the rational function $F(z) = 1 - \frac{1}{z^2}$ is hyperbolic; therefore the Julia set of F is a subset of $\mathcal{M}_P = \widehat{\mathbb{C}} \setminus \{\infty, 0, 1\}$ and F is uniformly expanding on a neighborhood of the Julia set. Hence, as a corollary of Theorem 3.4.13 we get that the map

$$\Phi(\ldots f_2 f_1) = \lim_{n\to\infty} \tau \otimes f_n \otimes f_{n-1} \otimes \cdots \otimes f_1 = \lim_{n\to\infty} \sigma_{f_1}(\ldots \sigma_{f_{n-1}}(\sigma_{f_n}(\tau))\ldots)$$

is a homeomorphism between the limit space $\mathfrak{F}^{\otimes-\omega} = \mathcal{X}_M$ and the preimage of the Julia set of F in the Teichmüller space. Moreover, we have

$$\Phi(\zeta \cdot g) = \Phi(\zeta) \cdot g \quad \text{and} \quad \Phi(\zeta \otimes f) = \sigma_f(F(\zeta));$$

i.e., Φ agrees with the actions of M and with the self-similarity structures.

The transformation $\tau \mapsto \sigma_f(\tau)$ of the Teichmüller space has at most one fixed point. Actually, in our case it has precisely one point, since obstructions are not possible. This fixed point is the limit of the sequence $\sigma_f^{\circ n}(\tau_0)$, where τ_0 is arbitrary (see [**DH93**]).

If we start from a point τ_0 in the preimage of the Julia set, then $\tau_0 = \Phi(\zeta)$ for some $\zeta \in \mathfrak{F}^{\otimes-\omega}$. Then

$$\lim_{n\to\infty} \sigma_f^{\circ n}(\tau_0) = \lim_{n\to\infty} \Phi\left(\zeta \otimes f^{\otimes n}\right) = \Phi(\ldots f \otimes f \otimes f).$$

Note that $f^{\otimes-\omega} = \ldots \otimes f \otimes f$ is the unique fixed point of the transformation $\zeta \mapsto \zeta \otimes f$ of the limit space $\mathfrak{F}^{\otimes-\omega}$.

The point $\Phi(f^{\otimes-\omega})$ for every $f \in \mathfrak{F}$ is projected onto a fixed point of F on $\mathcal{M}_P$. The fixed points of F are the solutions of the equation $z = 1 - \frac{1}{z^2}$, i.e., the roots of the polynomial $z^3 - z^2 + 1 = 0$. If the point $\tau = \Phi(f^{\otimes-\omega})$ is projected onto a root z_0, then f is combinatorially equivalent to the polynomial $f_\tau(z) = 1 - \frac{1}{z_0^2}z^2$ (see the diagram (6.10) and formula (6.11), where $\tau' = \sigma_f(\tau)$ will be equal to τ).

The polynomial $f_\tau(z) = 1 - \frac{1}{z_0^2}z^2$ is conjugate to $z^2 - \frac{1}{z_0^2}$ (make the change of variables $z = -z_0^2\tilde{z}$). Note that $-\frac{1}{z_0^2} = z_0 - 1$ and that $z_0 - 1$ is a root of the polynomial $x^3 + 2x^2 + x + 1$.

Consequently, the fixed points of F are equal to $c_r + 1$, $\overline{c_r} + 1$ and $c_a + 1$; and the branched coverings $f \in \mathfrak{F}$ corresponding to them are equivalent to the "rabbit", the "anti-rabbit" and the "airplane" polynomials, respectively.

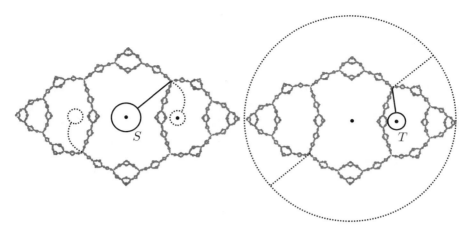

FIGURE 6.6. Computing IMG $\left(1 - \frac{1}{z^2}\right)$

Let us take the fixed point $t_0 = c_r + 1$ as a basepoint of the space $\mathcal{M}_P$. Let S be the loop going in the *negative* direction around the puncture 0 as the solid line shown on the left-hand side of Figure 6.6, and let T be the loop going in the negative direction around the puncture 1 as it is shown on the right-hand side of the same figure. The figure also shows the Julia set of $F(z) = 1 - \frac{1}{z^2}$.

The element T of the fundamental group of $\mathcal{M}_P = \widehat{\mathbb{C}} \setminus \{\infty, 0, 1\}$ corresponds to the Dehn twist of the punctured sphere $\widehat{\mathbb{C}} \setminus P$ along the simple closed curve going around the points c_1 and c_2 (see Figure 6.5). The element S corresponds to the Dehn twist along the analogous curve going around the points c_2 and 0. The mapping class group M (i.e., the fundamental group of the moduli space $\mathcal{M}_P$) is the free group generated by S and T.

PROPOSITION 6.6.2. *The branched covering $f_r \circ T$ is combinatorially equivalent to the "airplane" polynomial, and the branched covering $f_r \circ T^{-1}$ is equivalent to the "anti-rabbit" polynomial.*

SKETCH OF THE PROOF. An algebraic proof, which uses iterated monodromy groups and methods close to Theorem 6.5.2, can be found in [**BN05b**]. The proof from [**BN05b**] gives an algorithmical method to solve similar problems also for bigger post-critical sets and higher degrees of polynomials.

Our situation is rather simple, and we can use Proposition 6.6.1, since we understand now the map σ_f on the Teichmüller space and the moduli space $\mathcal{M}_P$ very well.

We have taken $t_0 = c_r + 1$ as the basepoint of $\mathcal{M}_P$. Choose the basepoint $\tau_0 \in \mathcal{T}_P$ to be the "original" complex structure on $\widehat{\mathbb{C}}$ for the "rabbit" polynomial, i.e., the unique fixed point of σ_{f_r}. Then the element $f_r \in \mathfrak{F}$ corresponds to the trivial path $1_{t_0} \in \mathfrak{M}(F)$ at t_0 (in the isomorphism between $\mathfrak{M}(F)$ and $\mathfrak{F}$, constructed in the proof of Proposition 6.6.1).

It follows from the definition of the self-similarity structure on the universal covering $\mathcal{T}_P$ of $\mathcal{M}_P$ (see Lemma 5.5.4) that the point $\Phi(\ldots \otimes f \otimes f)$ for $f \in \mathfrak{F}$ is given by the path $\ldots \gamma_f^{(2)} \gamma_f^{(1)} \gamma_f$, where each path $\gamma_f^{(n)}$ continues the path $\gamma_f^{(n-1)}$

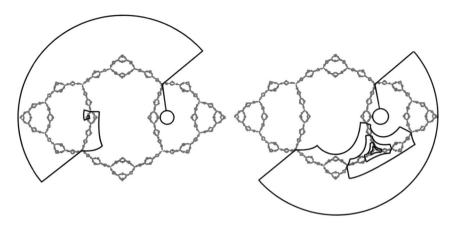

FIGURE 6.7. The paths $f^{\otimes -\omega}$

and is an F-preimage of $\gamma_f^{(n-1)}$. (Here we identify the points of the universal cover with the homotopy classes of paths in $\mathcal{M}_P$ starting in t_0, as usual.)

Figure 6.7 shows the paths of the form $\ldots \gamma_f^{(2)} \gamma_f^{(1)} \gamma_f$, where $f = f_r \circ T$ on the left-hand side of the figure and $f = f_r \circ T^{-1}$ on the right-hand side.

We see that the end of the path $\ldots \gamma_{f_r \circ T}^{(2)} \gamma_{f_r \circ T}^{(1)} \gamma_{f_r \circ T}$ is the point $c_a + 1$; hence $f_r \circ T$ is equivalent to the "airplane" polynomial. In the second case the end is $\overline{c_r} + 1$; hence $f_r \circ T^{-1}$ is equivalent to the "anti-rabbit" polynomial. □

Let us compute IMG (F). The F-preimages of the point $t_0 = c_r + 1$ are t_0 and $-t_0$. Let ℓ_0 be the trivial path at t_0 and let ℓ_1 be the path connecting t_0 with $-t_0$ and going above the puncture 0. Then we get the following wreath recursions for the corresponding standard action (see Figure 6.6):

$$S = (T, 1), \qquad T = \sigma(1, T^{-1}S^{-1}).$$

In particular, the associated virtual endomorphism of M is given by

$$\phi(S) = T, \quad \phi\left(T^2\right) = T^{-1}S^{-1}, \quad \phi\left(T^{-1}ST\right) = 1.$$

A general method of computation of the virtual endomorphism ϕ of the mapping class group (say, for longer critical orbits) is described in [**BN05b**].

Define the map $\varphi : M \longrightarrow M$ by the rule

$$\varphi(g) = \begin{cases} \phi(g) & \text{if } g \in \text{Dom}\,\phi = \langle S, T^2, T^{-1}ST \rangle \\ \phi\left(T^{-1}g\right)T & \text{otherwise.} \end{cases}$$

THEOREM 6.6.3. *The map φ is contracting, and $f_r \circ g$ is combinatorially equivalent to $f_r \circ \varphi(g)$ for every $g \in M$.*

If the 4-adic expansion of the number m has digits 1 or 2, then $f_r \circ T^m$ is combinatorially equivalent to the "airplane" polynomial. Otherwise it is equivalent to the "rabbit" polynomial for non-negative m and to the "anti-rabbit" for negative m.

Here a map $\varphi : M \longrightarrow M$ is called contracting if there exist $n, l_0 \in \mathbb{N}$ such that $l(\varphi^n(g)) < \frac{1}{2}l(g)$ for all $g \in G$, $l(g) > l_0$. Here $l(g)$ is the length of g with respect to some fixed finite generating set.

We use 4-adic expansions without the sign, so that, for instance, -1 is written $\ldots 333$.

PROOF OF THEOREM 6.6.3. We have
$$\ldots \otimes f_r \cdot g \otimes f_r \cdot g \otimes f_r \cdot g = \ldots \otimes g \cdot f_r \otimes g \cdot f_r \otimes g \cdot f_r \cdot g,$$
but
$$\ldots \otimes g \cdot f_r \otimes g \cdot f_r \otimes g \cdot f_r \cdot g = \ldots \otimes f_r \cdot \phi(g) \otimes f_r \cdot \phi(g) \otimes f_r \cdot \phi(g)g$$
if $g \in \text{Dom}\,\phi$; or
$$\ldots \otimes g \cdot f_r \otimes g \cdot f_r \otimes g \cdot f_r \cdot g$$
$$= \ldots \otimes TT^{-1}g \cdot f_r \otimes TT^{-1}g \cdot f_r \otimes TT^{-1}g \cdot f_r \cdot g$$
$$= \ldots \otimes f_r \cdot \phi\left(T^{-1}g\right)T \otimes f_r \cdot \phi\left(T^{-1}g\right)T \otimes f_r \cdot \phi\left(T^{-1}g\right)T \cdot T^{-1}g$$
if $g \in T\,\text{Dom}\,\phi$, i.e., if $g \notin \text{Dom}\,\phi$. This proves that $f_r \cdot g$ and $f_r \circ \varphi(g)$ are combinatorially equivalent.

The map φ is contracting, since the virtual endomorphism ϕ is contracting as a virtual endomorphism associated to an expanding partial self-covering.

Direct computation (see [**BN05b**]) shows that
$$\varphi^3\left(T^{4k}\right) = T^k, \quad \varphi^3\left(T^{4k+1}\right) = T, \quad \varphi^3\left(T^{4k+2}\right) = T, \quad \varphi^3\left(T^{4k+3}\right) = T^k,$$
which, together with Proposition 6.6.2, proves the last statement of the theorem. $\square$

One can prove (see [**BN05b**]) that for every $g \in M$ there exists n such that $\phi^n(g) \in \{1, T, T^{-1}\}$. This gives (together with Theorem 6.6.3) a complete answer to the "twisted rabbit" question.

6.7. Abstract kneading automata

We will describe here the set of iterated monodromy groups of post-critically finite polynomials abstractly as groups generated by automata of a special class.

6.7.1. Kneading automata.

6.7.1.1. *Tree-like sets of permutations.* Let T be a multi-set of permutations of X. Here a *multi-set* is a map $i \mapsto \pi_i$ from a set of indices I to $\mathfrak{S}(\mathsf{X})$. We write $T = \{\pi_i\}_{i \in I}$. Then the *cycle diagram* of T is an oriented 2-dimensional CW-complex whose set of 0-cells is X and where for every cycle $(x_1, x_2, \ldots, x_k)$ of every permutation $\pi_i \in T$ we have a 2-cell equal to a polygon with the vertices $x_1, x_2, \ldots, x_n$ so that their order in the cycle and their order on the boundary of the oriented cell coincide. Two different cells can intersect only along the 0-cells and do not have common 1-cells.

DEFINITION 6.7.1. A multi-set T is said to be *tree-like* if the cycle diagram of T is contractible.

If a multi-set $\{\pi_1, \ldots, \pi_k\}$ is tree-like and $\pi_i = \pi_j$ for $i \neq j$, then π_i and π_j are trivial.

See Figure 6.8, where all possible cycle diagrams of tree-like sets of permutations of X are shown for $|\mathsf{X}| = d$ equal to $2, 3, 4$ and 5. Cycles of length 2 are shown as segments rather than bigons, and cycles of length 1 are not shown.

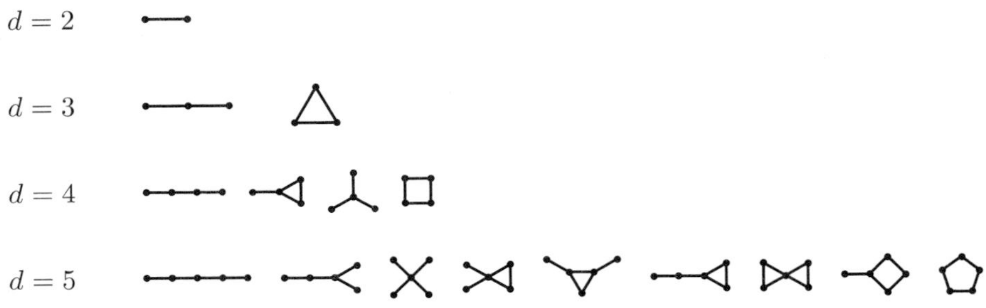

$d = 2$

$d = 3$

$d = 4$

$d = 5$

FIGURE 6.8. Tree-like sets of permutations

We can consider the *cycle graphs* instead of the cycle diagrams, replacing each cell of the cycle diagram by one vertex, which is connected to the vertices of the cell by edges. The cells corresponding to the trivial cycles of the permutations are deleted from the cycle graph. The cycle graph and the cycle diagram are obviously homotopically equivalent; hence a set of permutations is tree-like if and only if its cycle graph is a tree.

PROPOSITION 6.7.2. *Suppose that $T = \{\pi_1, \pi_2, \ldots, \pi_k\}$ is a tree-like set of permutations of* X. *Then the product $\pi = \pi_1 \cdot \pi_2 \cdots \pi_k$ is a transitive cycle on* X.

PROOF. We prove it by induction on $|X|$. The claim is trivial for $|X| = 1$. Suppose that we have proved it for sets of cardinality $d - 1$. Let $|X| = d$. Consider the cycle graph of T. It is a tree; hence there exists a vertex v of degree 1 (a *leaf* of the tree). This vertex is an element of X (i.e., does not correspond to a cycle), since all the vertices corresponding to cycles have degrees greater than one by definition.

Then the point v is fixed under the action of all but one permutation π_i. It is sufficient to prove that some permutation conjugate to π is transitive; therefore we may assume that π_1 is the only permutation moving v. (Otherwise we do a cyclic permutation of the factors $\pi_1 \cdots \pi_k$.)

Consider a new set $X' = X \setminus \{x\}$ and define the permutations π'_i of X' putting $\pi'_i(x) = \pi_i(x)$ for all $x \in X'$ if $i \neq 1$ (using the fact that $\pi_i(v) = v$) and

$$\pi'_1(x) = \begin{cases} \pi_1(x) & \text{if } \pi_1(x) \neq v \\ \pi_1(v) & \text{if } \pi_1(x) = v. \end{cases}$$

In other words, we delete v from the cycle of π_1 to which it belongs: if we had a cycle $(x_1, x_2, \ldots, x_m, v)$ of π_1, then we get the cycle $(x_1, x_2, \ldots, x_m)$ of π'_1.

We get a set of permutations $T' = \{\pi'_1, \pi'_2, \ldots, \pi'_n\}$ of X'. The cycle graph of this set is obtained from the cycle graph of T by deletion of the vertex v together with the unique edge to which it belongs (if v belongs to a cycle (v, y) of length 2 of π_1, then we also have to delete the vertex corresponding to the cycle and the unique edge connecting this vertex with y). Hence, T' is also a tree and, by the inductive hypothesis, the product $\pi' = \pi'_1 \pi'_2 \cdots \pi'_k$ is transitive on X'.

We obviously have $\pi'_i \cdots \pi'_n(x) = \pi_i \cdots \pi_n(x) \neq v$ for all $x \in X'$ and $i = 2, \ldots, n$. Hence:

$$\pi'(x) = \begin{cases} \pi(x) & \text{if } \pi_1(x) \neq v \\ \pi(v) & \text{if } \pi_1(x) = v. \end{cases}$$

Consequently, if $\pi' = (a_1, a_2, \ldots, a_{d-1})$ with $a_{d-1} = \pi_1^{-1}(v)$, then
$$\pi = (a_1, a_2, \ldots, a_{d-1}, v),$$
and π is transitive on X. □

COROLLARY 6.7.3. *Let $T \subset \mathfrak{S}(\mathsf{X})$ be a tree-like set. Then for any partition $\bigsqcup_{i=1}^{k} T_i = T$ the set of permutations $\{\prod T_1, \prod T_2, \ldots, \prod T_k\}$ is tree-like, where $\prod T_i$ is a product of the elements of T_i taken in any order.*

PROOF. The cycles of the permutation $\prod T_i$ are, by Proposition 6.7.2, equal to the connected components of the part of the cycle diagram of T corresponding to the permutations from T_i. This easily implies that the cycle diagram of the set $\{\prod T_i\}_{i=1}^{k}$ is also contractible. □

6.7.1.2. Kneading automata.

DEFINITION 6.7.4. A finite invertible automaton (A, X) is a *kneading automaton* if

(1) every non-trivial state g of A has a unique incoming arrow; i.e., there exist a unique pair $h \in \mathsf{A}$, $x \in \mathsf{X}$ such that $g = h|_x$;
(2) for every cycle $(x_1, x_2, \ldots, x_m)$ of the action of a state $g \in \mathsf{A}$ on X, the state $g|_{x_i}$ is non-trivial for at most one letter x_i;
(3) the multi-set of the permutations defined by the states of A on X is tree-like.

The first condition implies that if we delete the trivial state from the Moore diagram of a kneading automaton together with all the incoming arrows, then the obtained graph will be a disjoint union of cycles with trees attached to them. In particular, every kneading automaton is bounded (see Section 3.9).

Recall that the *dual Moore diagram* of an automaton (A, X) (see 1.3.6 on page 6) is a labeled directed graph with set of vertices identified with X and the set of arrows $\mathsf{A} \times \mathsf{X}$, where the arrow (g, x) starts in x, ends in $g(x)$ and is labeled by the pair $(g, g|_x) \in \mathsf{A} \times \mathsf{A}$.

The dual Moore diagram $\Gamma(\mathsf{A}, \mathsf{X})$ of a kneading automaton is the 1-skeleton of the cycle diagram of the action of A on X, and we can draw $\Gamma(\mathsf{A}, \mathsf{X})$ as

(1) the *cycle diagram* of the action of A on X;
(2) *labeling* of every 2-cell by the state, whose cycle corresponds to the cell;
(3) *labeling* of an arrow from x to $g(x)$ by h, if $g|_x = h \neq 1$.

At most one edge on the boundary of a 2-cell is labeled, due to condition (2) of Definition 6.7.4, and every state $g \in \mathsf{A}$ is a label of exactly one edge, due to condition (1).

Figure 6.9 shows an example of a Moore diagram of a kneading automaton (on the top) and the corresponding dual Moore diagram (on the bottom). The arrows of the Moore diagram, which do not end in the states a, b or c, end in the trivial state. We did not show the arrows ending in the trivial state, which are labeled by pairs of equal letters. We label the cells of the dual Moore diagram by letters inside the cells, and the edges are labeled by letters outside.

PROPOSITION 6.7.5. *If (A, X) is a kneading automaton, then $(\mathsf{A}, \mathsf{X}^n)$ is a kneading automaton for every n.*

188 6. EXAMPLES AND APPLICATIONS

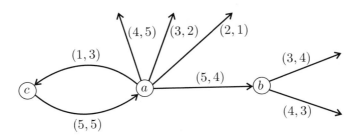

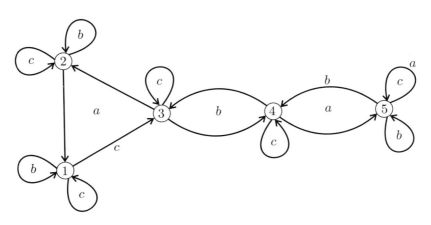

FIGURE 6.9. A kneading automaton

The proof will also be a description of an inductive procedure constructing the dual Moore diagram of the automaton $(\mathsf{A}, \mathsf{X}^n)$. Recall that dual Moore diagrams approximate the limit space of the group generated by the automaton (see Subsection 3.6.3). The dual Moore diagrams of the kneading automata will hence approximate Julia sets of polynomials.

PROOF. Suppose that we have constructed the dual Moore diagram $\Gamma\left(\mathsf{A}, \mathsf{X}^{n-1}\right)$. Let $x_1 \ldots x_{n-1} x_n \in \mathsf{X}^n$ be an arbitrary vertex of the dual Moore diagram $\Gamma\left(\mathsf{A}, \mathsf{X}^n\right)$ and let $g \in \mathsf{A}$ be a state of the kneading automaton. If the edge $(g, x_1 \ldots x_{n-1})$ of $\Gamma\left(\mathsf{A}, \mathsf{X}^{n-1}\right)$ is not labeled, then $g|_{x_1 \ldots x_{n-1}} = 1$ and therefore $g(x_1 \ldots x_{n-1} x_n) = g(x_1 \ldots x_{n-1}) x_n$. In this case the edge $(g, x_1 \ldots x_{n-1} x_n)$ of $\Gamma\left(\mathsf{A}, \mathsf{X}^n\right)$ starts in $x_1 \ldots x_n$, ends in $g(x_1 \ldots x_{n-1}) x_n$ and is not labeled. If the edge $(g, x_1 \ldots x_{n-1})$ is labeled by a state h in $\Gamma\left(\mathsf{A}, \mathsf{X}^{n-1}\right)$, then $g(x_1 \ldots x_{n-1} x_n) = g(x_1 \ldots x_{n-1}) h(x_n)$. In this case the edge $(g, x_1 \ldots x_{n-1} x_n)$ starts in $x_1 \ldots x_n$, ends in $g(x_1 \ldots x_{n-1}) h(x_n)$ and is labeled by $h|_{x_n}$, if $h|_{x_n} \neq 1$.

These arguments show that the dual Moore diagram $\Gamma\left(\mathsf{A}, \mathsf{X}^n\right)$ can be constructed using the following procedure.

Take $|\mathsf{X}|$ copies of $\Gamma\left(\mathsf{A}, \mathsf{X}^{n-1}\right)$. We will denote by $\Gamma\left(\mathsf{A}, \mathsf{X}^{n-1}\right) \cdot x$ the copy corresponding to $x \in \mathsf{X}$. If $v \in \mathsf{X}^{n-1}$ is a vertex of $\Gamma\left(\mathsf{A}, \mathsf{X}^{n-1}\right)$, then the corresponding vertex of the copy $\Gamma\left(\mathsf{A}, \mathsf{X}^{n-1}\right) \cdot x$ will become the vertex vx of the diagram $\Gamma\left(\mathsf{A}, \mathsf{X}^n\right)$.

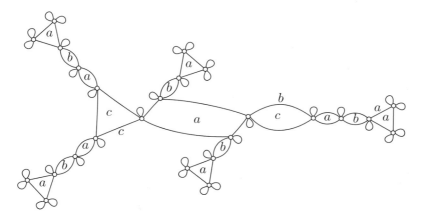

FIGURE 6.10. Dual Moore diagram of $(\mathsf{A},\mathsf{X}^2)$

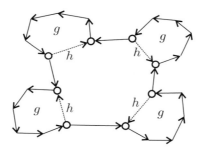

FIGURE 6.11. A cell of $\Gamma(\mathsf{A},\mathsf{X}^n)$

If we have an arrow labeled by h in the copy $\Gamma(\mathsf{A},\mathsf{X}^{n-1})\cdot x$, then we detach it from its end $vx \in \Gamma(\mathsf{A},\mathsf{X}^{n-1})\cdot x$ and attach it to the vertex $vh(x) \in \Gamma(\mathsf{A},\mathsf{X}^{n-1})\cdot h(x)$. If $h|_x \neq 1$, then we label the obtained arrow by $h|_x$. Note that $h|_x$ is the label of the edge (h,x) of $\Gamma(\mathsf{A},\mathsf{X})$. Hence, the copies of $\Gamma(\mathsf{A},\mathsf{X}^{n-1})$ are connected in $\Gamma(\mathsf{A},\mathsf{X}^n)$ in the same way as the vertices of $\Gamma(\mathsf{A},\mathsf{X})$ are. See, for example, in Figure 6.10 the dual Moore diagram $\Gamma(\mathsf{A},\mathsf{X}^2)$, where A is the automaton from Figure 6.9.

It follows immediately that every state $g \in \mathsf{A}$ is a label of exactly one arrow of $\Gamma(\mathsf{A},\mathsf{X}^n)$ (since the labels of $\Gamma(\mathsf{A},\mathsf{X}^n)$ come from the labels of $\Gamma(\mathsf{A},\mathsf{X})$).

The described inductive procedure of constructing the dual Moore diagram can be formulated in the following more geometric way. The diagram $\Gamma(\mathsf{A},\mathsf{X}^n)$ is obtained by gluing discs, corresponding to the cells of $\Gamma(\mathsf{A},\mathsf{X})$, to the copies of $\Gamma(\mathsf{A},\mathsf{X}^{n-1})$ along their labeled edges. More explicitly, if the edge (g,v) is labeled in $\Gamma(\mathsf{A},\mathsf{X}^{n-1})$ by $h = g|_v$ and $x \in \mathsf{X}$ belongs to a cycle $(x, h(x), \ldots, h^{k-1}(x))$ of length k under the action of h, then we have to take a $2k$-sided polygon and glue its every other side to the copies of the edge (g,v) in the diagrams $\Gamma(\mathsf{A},\mathsf{X}^{n-1})\cdot x$, $\Gamma(\mathsf{A},\mathsf{X}^{n-1})\cdot h(x), \ldots, \Gamma(\mathsf{A},\mathsf{X}^{n-1})\cdot h^{k-1}(x)$ in the given cyclic order. We will glue in this way the k copies of a cell of $\Gamma(\mathsf{A},\mathsf{X}^{n-1})$ together and get a cell of $\Gamma(\mathsf{A},\mathsf{X}^n)$. See, for example, Figure 6.11, where the case $k=4$ is shown.

Consequently, we can contract the $|\mathsf{X}|$ copies of $\Gamma(\mathsf{A}, \mathsf{X}^{n-1})$ in $\Gamma(\mathsf{A}, \mathsf{X}^n)$ to points, and get a cellular complex homeomorphic to $\Gamma(\mathsf{A}, \mathsf{X})$, which is contractible. This proves that $\Gamma(\mathsf{A}, \mathsf{X}^n)$ is also contractible.

We also see that every cell of $\Gamma(\mathsf{A}, \mathsf{X}^n)$ has at most 1 labeled side, since the labels come only from the attached $2k$-sided polygons, whose sides are labeled in the same way as the corresponding cell of $\Gamma(\mathsf{A}, \mathsf{X})$. (For every vertex x of $\Gamma(\mathsf{A}, \mathsf{X})$ and every $g \in \mathsf{A}$ there is precisely one cell labeled by g and containing x.) □

COROLLARY 6.7.6. *If a kneading automaton A has only one trivial state, then it is reduced.*

PROOF. Suppose that two non-trivial states $g_1, g_2 \in \mathsf{A}$ define the same permutations on X^*. There exists $n \in \mathbb{N}$ such that g_1 and g_2 define non-trivial permutations of X^n. But then we get that the multi-set of the permutations defined by A on X^n is not tree-like, which contradicts Proposition 6.7.5. □

COROLLARY 6.7.7. *The product of all the states of a kneading automaton (taken in any order) is a level-transitive automorphism of X^*. In particular, the group generated by a kneading automaton is level-transitive.*

PROOF. A direct corollary of Propositions 6.7.2 and 6.7.5. □

6.8. Topological polynomials and critical portraits

6.8.1. Spiders and critical portraits. A *topological polynomial* is a Thurston map $f : S^2 \longrightarrow S^2$ such that $f^{-1}(\infty) = \{\infty\}$. Then after deleting ∞ from S^2 we get a post-critically finite branched covering $f : \mathbb{R}^2 \longrightarrow \mathbb{R}^2$. We denote by $C \subset \mathbb{R}^2$ and P the sets of the critical and post-critical points of f, respectively.

A *spider* (see [**HS94**]) is a collection $\mathcal{S} = \{\gamma_z\}_{z \in P}$ of disjoint simple curves connecting the post-critical points to infinity. The curve connecting $z \in P$ to infinity will be denoted γ_z. The curve γ_z (minus its beginning z) cannot pass through the post-critical points. We identify two spiders if they are isotopic relative to P.

A spider $\mathcal{S}$ is said to be f-*invariant* if $f(\mathcal{S})$ is isotopic (rel. P through spiders) to a subset of $\mathcal{S}$.

Let $z \in C$ be a critical point of local degree d_z. Then $X_z = f^{-1}(\gamma_{f(z)})$ consists of d_z curves connecting z to infinity. The collection $\{X_z : z \in C\}$ is called the *critical portrait* associated to the spider $\mathcal{S}$ and the topological polynomial.

The curves of the critical portrait cut the plane into components, called *sectors*. The polynomial f has no critical points in the interiors of the sectors; therefore it is a homeomorphism of the sector onto an open subset of the plane. This open subset is a complement of a finite collection of disjoint paths belonging to the spider. Consequently, there are $d = \deg f$ sectors and f maps closure of every sector surjectively onto $\mathbb{R}^2$. In other words, partition into sectors gives a choice of d branches of the inverse map f^{-1}. See Figure 6.12 for all (up to an isotopy in $\mathbb{R}^2$) possible critical portraits of topological polynomials of degree $d = 2, \ldots, 5$.

6.8.2. Kneading automaton of a critical portrait. Let $\mathcal{C}$ be a critical portrait of a topological polynomial f associated to an f-invariant spider $\mathcal{S}$, and let $\{S_x : x \in \mathsf{X}\}$ be the set of the corresponding sectors. Here X, $|\mathsf{X}| = d$, is a set of labels of the sectors.

We are going to construct an automaton $\mathsf{K}_{\mathcal{C}, f}$ over the alphabet X which encodes the critical portrait $\mathcal{C}$ and the action of f.

6.8. TOPOLOGICAL POLYNOMIALS AND CRITICAL PORTRAITS

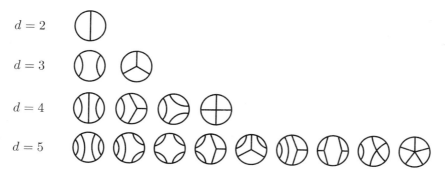

FIGURE 6.12. Critical portraits

Take a point $z \in P$ and a small simple loop α_z going around z in the positive direction. Suppose that $y \in f^{-1}(z)$ is a preimage of z. One of the f-preimages of α_z is a small loop $\alpha_{z,y}$ around y. The degree of the map $f : \alpha_{z,y} \longrightarrow \alpha_z$ is by definition the local degree of f at y.

We will use f-preimages of the loops α_z to define the action of the states g_z of $\mathsf{K}_{\mathcal{C},f}$ on X.

DEFINITION 6.8.1. Let $\mathcal{C}$ be a critical portrait of a topological polynomial f. The corresponding *kneading automaton* $\mathsf{K}_{\mathcal{C},f}$ is the automaton with the set of states $\{g_z\}_{z \in P} \cup \{1\}$ over the alphabet X, $|\mathsf{X}| = \deg f$, where the letters of X label the sectors S_x of $\mathcal{C}$ and the output and the transition functions are defined by the following conditions.

Take an arbitrary post-critical point $z \in P$. If $y \in f^{-1}(z)$ is not critical, i.e., if the local degree $\deg_y(f)$ is equal to one, then y is an internal point of a sector S_x. The loop $\alpha_{z,y}$ also completely belongs to the sector S_x. We encode this by the output and the transition functions of $\mathsf{K}_{\mathcal{C},f}$:

$$g_z \cdot x = x \cdot g_y,$$

where $g_y = 1$ if $y \notin P$.

Suppose now that y is critical of local degree d'. Then y belongs to the boundaries of d' sectors. Let $S_{x_1}, S_{x_2}, \ldots, S_{x_{d'}}$ be these sectors listed according to the circular order in which the curve $\alpha_{z,y}$ meets them (i.e., in the counterclockwise order around y).

If y is not post-critical, then we encode the action of the loop $\alpha_{z,y}$ on the sectors $S_{x_1}, \ldots, S_{x_{d'}}$ setting

$$g_z \cdot x_i = x_{i+1} \cdot 1$$

for $i = 1, \ldots, d'$, where $S_{x_{d'+1}} = S_{x_1}$.

If y is post-critical, then there is a path $\gamma_y \in \mathcal{S}$ connecting y to infinity. We may assume that the sectors $S_{x_1}, \ldots, S_{x_{d'}}$ are labeled in such a way that the curve γ_y is adjacent to S_{x_1} and $S_{x_{d'}}$. Then we set

$$g_z \cdot x_i = x_{i+1} \cdot 1$$

for $i = 1, \ldots, d' - 1$ and

$$g_z \cdot x_{d'} = x_1 \cdot g_y.$$

Thus, in every case the action of g_z on the alphabet is the monodromy action of the small loop α_z around z, and the state transitions in $\mathsf{K}_{\mathcal{C},f}$ show the action

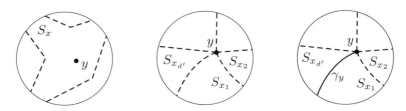

FIGURE 6.13. Cases in Definition 6.8.1

of f on the points of P. See Figure 6.13, where different cases of the definition of $\mathsf{K}_{\mathcal{C},f}$ are shown.

PROPOSITION 6.8.2. *The automaton $\mathsf{K}_{\mathcal{C},f}$ is a kneading automaton.*

PROOF. The conditions (1) and (2) of Definition 6.7.4 follow directly from the construction of the automaton. It is therefore sufficient to prove that the set $\{g_z\}_{z \in P}$ defines a tree-like set of permutations of X.

Consider small loops α_z going around every post-critical point $z \in P$ and connect them to a basepoint t by curves which do not intersect the spider $\mathcal{S}$. Let Γ be the obtained 2-complex. Then its preimage $f^{-1}(\Gamma)$ is precisely the cycle complex of the set of the permutations defined by $\{g_z\}_{z \in P}$ on X.

The complement of Γ in $\mathbb{R}^2$ is homeomorphic to an annulus and does not contain post-critical points of f. Therefore, $f : \mathbb{R}^2 \setminus f^{-1}(\Gamma) \longrightarrow \mathbb{R}^2 \setminus \Gamma$ is a d-fold covering. Consequently, $\mathbb{R}^2 \setminus f^{-1}(\Gamma)$ is also homeomorphic to an annulus; hence $f^{-1}(\Gamma)$ is connected and contractible. □

THEOREM 6.8.3. *Let $f : \mathbb{R}^2 \longrightarrow \mathbb{R}^2$ be a post-critically finite topological polynomial with the critical set C and the post-critical set P. Suppose that there exists an f-invariant spider $\mathcal{S}$ and let $\mathcal{C}$ be the associated critical portrait. Then a standard action of $\operatorname{IMG}(f)$ on X^* is generated by the automaton $\mathsf{K}_{\mathcal{C},f}$.*

PROOF. We will only define the generators g_z of $\operatorname{IMG}(f)$ corresponding to the states g_z, $z \in P$ of the automaton $\mathsf{K}_{\mathcal{C},f}$ and the paths $\ell(x)$ connecting the basepoint to its preimages. The rest, i.e., showing that the formula of the standard action (see Proposition 5.2.2) agrees with the definition of $\mathsf{K}_{\mathcal{C},f}$, will easily follow from the construction.

Let us denote by $\mathcal{M}_{\mathcal{S}}$ the plane $\mathbb{R}^2$ without the curves belonging to the spider $\mathcal{S}$. The set $\mathcal{M}_{\mathcal{S}}$ is simply connected and $f^{-1}(\mathcal{M}_{\mathcal{S}})$ is a subset of $\mathcal{M}_{\mathcal{S}}$ (up to an isotopy rel. P), since $\mathcal{S}$ is f-invariant.

Choose a basepoint $t \in \mathcal{M}_{\mathcal{S}}$. Every sector of the critical portrait contains exactly one point of $f^{-1}(t)$. For every $x \in \mathsf{X}$ let $\ell(x)$ be the path connecting the preimage $t_x \in S_x$ of t to t and going inside $\mathcal{M}_{\mathcal{S}}$ (i.e., not intersecting the spider $\mathcal{S}$). The path $\ell(x)$ is determined by these conditions uniquely up to a homotopy in $\mathcal{M} = \mathbb{C} \setminus P$, since $\mathcal{M}_{\mathcal{S}}$ is simply connected. We compute the standard action of $\operatorname{IMG}(p)$ associated to the obtained set of connecting paths $\{\ell(x)\}_{x \in \mathsf{X}}$.

For every $z \in P$ we take a small simple loop α_z going in the positive direction around z and connect it to t by a path p_z in $\mathcal{M}_{\mathcal{S}}$. Let g_z be the obtained loop $g_z = p_z \alpha_z p_z^{-1}$. The homotopy class of g_z is uniquely determined by the condition that g_z intersects the spider $\mathcal{S}$ only once through the path γ_z and by the direction of the intersection. □

6.9. Iterated monodromy groups of complex polynomials

6.9.1. Critical portraits and invariant spiders. Let us describe how to construct an invariant spider and a critical portrait of a post-critically finite complex polynomial. We will follow the work of A. Poirier [**Poi93**], which extends the paper [**BFH92**] for the general (not only strictly preperiodic) case. Our outline will not contain proofs. The proofs (and references to the proofs) can be found in [**Poi93**]. See also [**DH84, DH85a, Mil99, HS94**].

6.9.1.1. *External and internal rays.* Let $f \in \mathbb{C}[z]$ be a post-critically finite polynomial with the set of (finite) critical points C and the post-critical set P. We assume that f is *monic* and *centered*, i.e., is of the form $f(z) = z^d + a_{d-2}z^{d-2} + a_{d-3}z^{d-3} + \cdots + a_0$.

Let us denote by $\mathcal{J}_f$ and $\mathcal{K}_f$ the Julia set and the filled-in Julia set of f. The *filled-in Julia set* is the set of points $z \in \mathbb{C}$ such that $f^n(z) \to \infty$. Here, as usual, f^n denotes the nth iteration of f.

The Julia set of f is connected, locally connected and coincides with the boundary of the *basin of attraction of infinity* $\widehat{\mathbb{C}} \setminus \mathcal{K}_f$.

The set of the finite critical points C belongs to the filled-in Julia set $\mathcal{K}_f$. If $z_0 \in C$ belongs to the Julia set, then it is strictly preperiodic; i.e., there is no $n \in \mathbb{N}$ such that $f^n(z_0) = z_0$, but there exist $0 < n < m$ such that $f^n(z_0) = f^m(z_0)$. The period of the sequence $z_0, f(z_0), \ldots, f^n(z_0), \ldots$ is a repelling cycle and does not contain critical points.

If $z_0 \in C$ does not belong to the Julia set (i.e., belongs to the *Fatou set*), then it is either periodic or is preperiodic, but then its orbit contains a periodic critical point. In both cases the cycle of the orbit is super-attracting.

Let us denote $\mathbb{D} = \{z \in \mathbb{C} : |z| < 1\}$. There exists a unique bi-holomorphic isomorphism

$$\Phi_\infty : \widehat{\mathbb{C}} \setminus \overline{\mathbb{D}} \longrightarrow \widehat{\mathbb{C}} \setminus \mathcal{K}_f$$

tangent to identity at ∞ and conjugating f with z^d, i.e., such that

$$f(\Phi_\infty(z)) = \Phi_\infty(z^d)$$

for all $z \in \widehat{\mathbb{C}} \setminus \overline{\mathbb{D}}$.

Suppose now that $z_0, z_1 = f(z_0), \ldots, z_{n-1} = f^{n-1}(z_0)$, $f(z_{n-1}) = z_0$ is a cycle containing critical points. Let U_i be the *Fatou component* (i.e., a connected component of the Fatou set) containing z_i. Then $f^n : U_i \longrightarrow U_i$ is a degree d' branched covering, where d' is the product of the local degrees of f at z_i. There exists a uniformizing map $\Phi_{U_i} : \mathbb{D} \longrightarrow U_i$ such that $\Phi_{U_i}(0) = z_i$ and

(6.13) $$f^n(\Phi_{U_i}(z)) = \Phi_{U_i}\left(z^{d'}\right).$$

The functions Φ_{U_i} are determined uniquely up to multiplication of their arguments by roots of unity of degree $d' - 1$.

In general, if U is a Fatou component of f, then its *center* is the point $z_0 \in U$ such that $f^n(z_0)$ belongs to a super-attracting cycle for some n. The center exists and is unique. Let us choose the uniformizing maps $\Phi_U : \mathbb{D} \longrightarrow U$ such that $\Phi_U(0)$ is the center of U.

For $\theta \in \mathbb{R}/\mathbb{Z}$, we denote by $R_{\theta,\infty}$ the curve

$$R_{\theta,\infty}(t) = \Phi_\infty\left(t \cdot e^{\theta 2\pi i}\right), \quad t \in (1, +\infty)$$

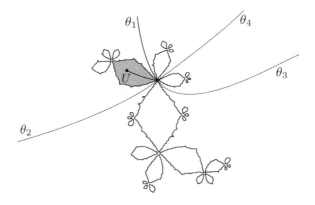

FIGURE 6.14. Supporting ray

and by $R_{\theta,U}$ the curve

$$R_{\theta,U} = \Phi_U\left(t \cdot e^{\theta 2\pi i}\right), \quad t \in [0,1).$$

The rays $R_{\theta,\infty}$ and $R_{\theta,U}$ are called the *external* and the *internal* rays at the *angle* θ. The set of internal rays of a Fatou component does not depend on a particular choice of the map Φ_U.

Since Φ_∞ conjugates f with z^d, the polynomial f acts on the external rays by multiplication of the angles by d:

$$f(R_{\theta,\infty}) = R_{d\theta,\infty}.$$

The Julia set of a post-critically finite polynomial is locally connected; therefore the maps Φ_∞ and Φ_U can be extended to continuous maps of the boundaries. This implies that the rays $R_{\theta,\infty}$ and $R_{\theta,U}$ *land*, i.e., that the limits

$$R_{\theta,\infty}(1) = \lim_{t \searrow 1} R_{\theta,\infty}(t), \quad R_{\theta,U}(1) = \lim_{t \nearrow 1} R_{\theta,U}(t)$$

exist. They obviously belong to $\mathcal{J}_f$. We say that the ray $R_{\theta,*}$ *lands* on the point $R_{\theta,*}(1)$. Every point $z \in \mathcal{J}_f$ is a landing point of at least one external ray. For any given Fatou component U and any point $z \in \partial U$ there exists precisely one internal ray $R_{\theta,U}$ landing on z.

6.9.1.2. *Supporting rays.* Let U be a Fatou component and let $p \in \partial U$ be a point of the boundary. There is only a finite number of external rays

$$R_{\theta_1,\infty}, R_{\theta_2,\infty}, \ldots, R_{\theta_k,\infty}$$

landing on p. Let us order the angles $\theta_1, \ldots, \theta_k$ in the counterclockwise cyclic order (i.e., the natural cyclic order on $\mathbb{R}/\mathbb{Z}$) so that the Fatou component U is between θ_1 and θ_2 (see Figure 6.14).

Then the external ray $R_{\theta_1,\infty}$ is called the *(left) supporting ray* of the Fatou component U. The *extended ray* $R_{U,p}$ is the supporting ray $R_{\theta_1,\infty}$ extended by the internal ray $R_{\theta,U}$ of U landing on p.

The extended ray $R_{U,p}$ is determined uniquely by the Fatou component U and the point $p \in \partial U$. Another important property of the extended rays is that

$$f(R_{U,p}) = R_{f(U),f(p)}.$$

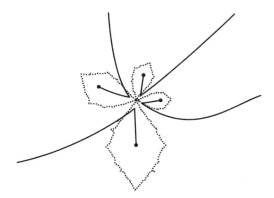

FIGURE 6.15. Construction of invariant spider

6.9.1.3. *Construction of the spiders and the critical portraits.*

THEOREM 6.9.1. *If for every periodic point $z \in P \cap \mathcal{J}_f$ the period of z is equal to the period of an external ray landing on z, then the polynomial has an invariant spider. In particular, an invariant spider exists if the polynomial is hyperbolic.*

There exists for every post-critically finite polynomial f a number $n \in \mathbb{N}$ such that the iteration f^n has an invariant spider.

PROOF. We will only sketch the proof showing how the critical portraits and the invariant spiders are constructed. More details can be found in [**Poi93, HS94**].

Suppose that a critical point $z_0 \in C$ is periodic. Let n be its period, let $z_0, z_1 = f(z_0), \ldots, z_{n-1} = f^{n-1}(z_0)$ be the points of the cycle and let $U_i \ni z_i$ be the corresponding Fatou components. The mapping $f^n : U_0 \longrightarrow U_0$ is conjugate to raising to power d' on $\mathbb{D}$, where d' is the product of the local degrees of f at the points of the cycle. Hence, there exists a point $p \in \partial U_0$ such that $f^n(p) = p$ (we use local connectivity of $\mathcal{J}_f$). One can take $p = \Phi_{U_0}(\zeta)$, where $\Phi_{U_0} : \overline{\mathbb{D}} \longrightarrow \overline{U_0}$ is the uniformizing map satisfying (6.13) and ζ is a root of unity of degree $d' - 1$. The point p is not critical, since it is periodic and belongs to the Julia set.

We get in this way a sequence of extended rays $\gamma_{z_k} = R_{U_k, f^k(p)}$ such that $f(\gamma_{z_k}) = \gamma_{z_{k+1}}$. It is possible that the curves γ_{z_k} are not disjoint (since p may have period less than n), but they become disjoint after a small homotopy: we may slightly move the external rays in the counterclockwise direction (see Figure 6.15). We have to make such moves also when the point p is post-critical.

Suppose that $z_0, z_1 = f(z_0), \ldots, z_{n-1} = f(z_{n-2})$ and $f(z_{n-1}) = z_0$ is a cycle of post-critical points which does not contain critical points. Then the points z_i belong to the Julia set and by the condition of our theorem there exists a ray $R_{\theta,\infty}$ landing on z_0 such that $f^n(R_{\theta,\infty}) = R_{\theta,\infty}$. Then the rays $\gamma_{z_k} = f^k(R_{\theta,\infty}) = R_{k\theta,\infty}$ land on the points z_k.

The curves γ_z are defined now for all cycles of post-critical points. We have $f(\gamma_z) = \gamma_{f(z)}$ for all the defined curves γ_z.

We can define γ_z for the rest of the post-critical points inductively. We choose at each step a point $z \in P$ with minimal n such that γ_z is not defined but $\gamma_{f^n(z)}$ is. Then we set γ_z to be equal to one of the preimages $\gamma_z \in f^{-n}(\gamma_{f^n(z)})$. It is easy to see that the obtained set of curves γ_z is an invariant spider.

If there is no invariant spider, then some of the post-critical cycles are shorter than the lengths of the cycles of the rays landing on them. But then we can pass to an iterate f^n such that all the external rays landing on periodic post-critical points are fixed under f^n. □

6.10. Polynomials from kneading automata

6.10.1. Constructing a topological polynomial.

6.10.1.1. *The adding machines.* Consider the alphabet $\mathsf{X} = \{0, 1, \ldots, d-1\}$. The *d-adic adding machine* is the automorphism a of the tree X^* given by

$$a \cdot i = \begin{cases} (i+1) \cdot id & \text{for } i = 0, 1, \ldots, d-2 \\ 0 \cdot a & \text{for } i = d-1. \end{cases}$$

We denote by id the trivial automorphism of X^* in order to distinguish it from the symbol $1 \in \mathsf{X}$.

The wreath recursion defining the d-adic adding machine is

$$a = \sigma(id, \ldots, id, a),$$

where σ is the cyclic permutation $0 \mapsto 1 \mapsto \cdots \mapsto (d-1) \mapsto 0$.

The d-adic adding machine corresponds to adding 1 to a d-adic integer. We have considered the case $d = 2$ (the *binary adding machine*) in Subsection 1.7.1. The action of the d-adic adding machine is a partial case of the self-similar actions of $\mathbb{Z}^n$, studied in 1.7.2 and in Section 2.9 (it is the case of $n = 1$, the base $A = d$ and the digit system $\{0, 1, \ldots, d-1\}$).

An automorphism g of X^* is conjugate in $\text{Aut}\,\mathsf{X}^*$ with a if and only if it is level-transitive (see [**BORT96, GNS01**]).

6.10.1.2. *Planar kneading automata.* If g is level-transitive and $d = |\mathsf{X}|$, then for every $x \in \mathsf{X}$

$$g^d|_x = g^{d-1}|_{g(x)}g|_x = g^{d-2}|_{g^2(x)}g|_{g(x)}g|_x = \ldots = g|_{g^{d-1}(x)}g|_{g^{d-2}(x)} \cdots g|_{g(x)}g|_x;$$

i.e., $g^d|_x$ is the product of all the states $g|_{x_i}$ for $x_i \in \mathsf{X}$, written in the order opposite to the order of the action of g on X. In particular, if we change x to another letter of X, then the product will be shifted cyclically.

If (A, X) is a kneading automaton, then the product $g = g_1 g_2 \cdots g_n$ of all its states is conjugate to the $|\mathsf{X}|$-ary adding machine by Corollary 6.7.7. Thus $g^d|_x$ for any $x \in \mathsf{X}$ is the product $\prod_{x_i \in \mathsf{X}} (g_1 g_2 \cdots g_n)|_{x_i}$ written in the order opposite to that of the action of g on X.

Condition (1) of Definition 6.7.4 implies that $\prod_{x_i \in \mathsf{X}} (g_1 g_2 \cdots g_n)|_{x_i}$ coincides as a word (if we do not write the trivial states) with the product $g_{i_1} g_{i_2} \cdots g_{i_n}$ for some permutation $i_1, i_2, \ldots, i_n$ of the indices $1, 2, \ldots, n$.

DEFINITION 6.10.1. A kneading automaton (A, X) is said to be *planar* if there exists a circular order $g_1, g_2, \ldots, g_n$ of the set of the non-trivial states of A such that $(g_1 g_2 \cdots g_n)^d|_x$ is a cyclic shift of the word $g_1 g_2 \cdots g_n$ for every letter $x \in \mathsf{X}$, where $d = |\mathsf{X}|$.

It is sufficient to check the condition of the definition only for one letter x.

PROPOSITION 6.10.2. *Let (A, X) be a kneading automaton. Then there exists $n \in \mathbb{N}$ such that the automaton $(\mathsf{A}, \mathsf{X}^n)$ is planar.*

PROOF. Every ordering $g_1, \ldots, g_m$ of the non-trivial states of A uniquely determines the ordering $g_{i_1}, \ldots, g_{i_m}$ such that $(g_1 \cdots g_m)^d|_x = g_{i_1} \cdots g_{i_m}$, and the circular ordering $g_{i_1}, \ldots, g_{i_m}$ depends only on the circular ordering $g_1, \ldots, g_m$ and does not depend on $x \in \mathsf{X}$. Let $R : (g_1, \ldots, g_m) \mapsto (g_{i_1}, \ldots, g_{i_m})$ be the obtained map on the set of circular orderings. Since the number of possible circular orderings is finite, there exists n and a circular ordering $g_1, \ldots, g_m$ fixed under the action of R^n. But this means that $(g_1 \cdots g_m)^{d^n}|_{x^n}$ is equal to a cyclic permutation of the word $g_1 \cdots g_m$, i.e., that the automaton $(\mathsf{A}, \mathsf{X}^n)$ is planar. $\square$

PROPOSITION 6.10.3. *If $\mathcal{C}$ is the critical portrait of a topological polynomial f which is associated to an invariant spider $\mathcal{S}$, then the kneading automaton $\mathsf{K}_{\mathcal{C},f}$ is planar.*

PROOF. We interpret the states of the kneading automaton $\mathsf{K}_{\mathcal{C},f}$ as elements of IMG (f), accordingly to the proof of Theorem 6.8.3. Let $a \in \pi_1(\mathcal{M}, t)$ be the loop going around the post-critical set P in the positive direction. It follows from the definition of the loops g_z that $a = g_{z_1} g_{z_2} \cdots g_{z_n}$ for some ordering $z_1, \ldots, z_n$ of the set P. The loop a is homotopic in $\mathcal{M} = \mathbb{C} \setminus P$ to a simple loop around infinity. Therefore $a^d|_x$ is conjugate to a in $\pi_1(\mathcal{M}, t)$ for every x; i.e., the automaton $\mathsf{K}_{f,\mathcal{S}}$ is planar. $\square$

EXAMPLE. Not every kneading automaton is planar. The following example is described in [**BS02a**]. It is an automaton over the alphabet $\mathsf{X} = \{0, 1\}$ with the set of states $\{id = 1, a_1, a_2, \ldots, a_6\}$ satisfying the following wreath recursions:

$$\begin{aligned} a_1 &= \sigma(a_6, 1) & a_2 &= (1, a_1) \\ a_3 &= (a_2, 1) & a_4 &= (1, a_3) \\ a_5 &= (1, a_4) & a_6 &= (a_5, 1). \end{aligned}$$

Let us show that this automaton cannot be made planar. Suppose that on the contrary, we have some ordering $a_1, a_{i_2}, \ldots, a_{i_6}$ satisfying the conditions of Definition 6.10.1. Looking at the recurrent definitions, we see that

$$a_1 a_{i_2} \cdots a_{i_6} = \sigma\left(a_6 \prod\{a_2, a_5\}, \prod\{a_1, a_3, a_4\}\right),$$

where $\prod S$ denotes the product of the elements of the set S in some order. Hence, in the cyclic ordering $a_1, a_{i_2}, \ldots, a_{i_6}$ we have the elements $\{a_1, a_3, a_4\}$ separated from the elements $\{a_2, a_5, a_6\}$. Then $a_1 a_{i_2} \cdots a_{i_6}$ is equal either to

$$a_1 \prod\{a_3, a_4\} \cdot \prod\{a_2, a_5, a_6\}$$
$$= \sigma(a_6, 1) \cdot (a_2, a_3) \cdot \left(a_5, \prod\{a_1, a_4\}\right)$$
$$= \sigma\left(a_6 a_2 a_5, \ a_3 \prod\{a_1, a_4\}\right),$$

or to

$$a_1 \prod\{a_2, a_5, a_6\} \prod\{a_3, a_4\}$$
$$= \sigma(a_6, 1) \cdot \left(a_5, \prod\{a_1, a_4\}\right) \cdot (a_2, a_3)$$
$$\sigma\left(a_6 a_5 a_2, \ \prod\{a_1, a_4\} a_3\right)$$

or to

$$a_1 a_3 \prod \{a_2, a_5, a_6\} a_4$$
$$= \sigma(a_6, 1)(a_2, 1) \cdot \left(a_5, \prod\{a_1, a_4\}\right)(1, a_3)$$
$$= \sigma\left(a_6 a_2 a_5, \quad \prod\{a_1, a_4\} a_3\right)$$

or to

$$a_1 a_4 \prod \{a_2, a_5, a_6\} a_3$$
$$= \sigma(a_6, 1)(1, a_3) \cdot \left(a_5, \prod\{a_1, a_4\}\right)(a_2, 1)$$
$$= \sigma\left(a_6 a_5 a_2, \quad a_3 \prod\{a_1, a_4\}\right).$$

The first and the second cases do not give us an invariant cyclic order on the generators (the left-hand side and the right-hand side of the equalities give contradictory conditions on the ordering).

In the third case the ordering has to be $(a_1, a_3, a_6, a_2, a_5, a_4)$, but

$$a_1 a_3 a_6 a_2 a_5 a_4 = \sigma(a_6, 1)(a_2, 1)(a_5, 1)(1, a_1)(1, a_4)(1, a_3) = \sigma(a_6 a_2 a_5, a_1 a_4 a_3).$$

In the fourth case the ordering has to be $(a_1, a_4, a_6, a_5, a_2, a_3)$, but

$$a_1 a_4 a_6 a_5 a_2 a_3 = \sigma(a_6, 1)(1, a_3)(a_5, 1)(1, a_4)(1, a_1)(a_2, 1) = \sigma(a_6 a_5 a_2, a_3 a_4 a_1).$$

An efficient criterion for a kneading automaton over $\mathsf{X} = \{0, 1\}$ to be planar is described in [**BS02a**] in terms of Hubbard trees.

THEOREM 6.10.4. *If (A, X) is a planar kneading automaton, then there exists a topological polynomial $f : \mathbb{R}^2 \longrightarrow \mathbb{R}^2$ and an f-invariant spider $\mathcal{S}$ such that (A, X) is isomorphic to the kneading automaton of the corresponding critical portrait.*

PROOF. Let $g_1, g_2, \ldots, g_n$ be a cyclic order on $\mathsf{A} \setminus \{1\}$ satisfying the conditions of Definition 6.10.1.

Let $\Gamma(\mathsf{A}, \mathsf{X})$ be the dual Moore diagram of the kneading automaton A. Recall that we draw it as a 2-dimensional CW-complex with labeled 2-cells and labeled edges.

Let us choose an initial vertex $x_0 \in \mathsf{X}$ of $\Gamma(\mathsf{A}, \mathsf{X})$. There exists a unique arrow e_n of $\Gamma(\mathsf{A}, \mathsf{X})$ starting in x_0 and belonging to a cell labeled by g_n. Let $x_{0,n} = g_n(x_0)$ be its end. Then there exists a unique arrow e_{n-1} starting in $x_{0,n}$ and belonging to the cell labeled by g_{n-1}. Let $x_{0,n-1} = g_{n-1} g_n(x_0)$ be its end. We proceed further and get an oriented path $p_{x_0} = (e_n, e_{n-1}, \ldots e_1)$ such that e_m is adjacent to the cell labeled by g_m and the path goes through the vertices $x_0, g_n(x_0), g_{n-1} g_n(x_0), \ldots, g_1 \cdots g_n(x_0)$. Now we start from $x_1 = g_1 \cdots g_n(x_0)$ and find the path p_{x_1} from x_1 to $x_2 = g_1 \cdots g_n(x_1)$. We continue, and finally we will get a closed oriented path $p = (p_{x_0}, p_{x_1}, \ldots x_{d-1})$, where $x_i = g_1 \cdots g_n(x_{i-1})$.

The word read on the labels of the edges along the path p is equal to $g_n \cdots g_1$, up to a cyclic shift, since $(g_n \cdots g_1)^d$ is the word of the labels of the cells along p and the automaton is planar.

Let us prove that the path p will go through each arrow exactly once. The first arrow of the path p_{x_i} starts at x_i and is adjacent to the cell labeled by g_n. By Proposition 6.7.2, $\{x_0, \ldots, x_{d-1}\} = \mathsf{X}$; hence every edge of the cells labeled by g_n appears exactly once in the path p (since we have exactly one such edge starting

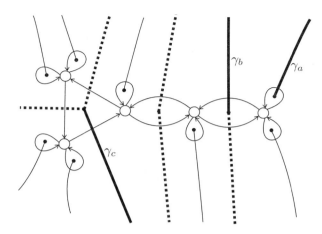

FIGURE 6.16. Rays $R_{z,\theta}$

in each x_i). If we pass to the cyclic shift $g_i, \ldots, g_n, g_1, \ldots, g_{i-1}$, then we will not change the *closed* path p. The same argument will then show that every edge of the cells labeled by g_i appears exactly once in p.

Consequently, there exists an orientation-preserving embedding of the complex $\Gamma(\mathsf{A}, \mathsf{X})$ into the plane $\mathbb{R}^2$ such that p is the path going *around* $\Gamma(\mathsf{A}, \mathsf{X})$ in the positive direction. See, for example, the lower part of Figure 6.9 on page 188, where the dual Moore diagram is embedded in this way into the plane for the ordering a, b, c.

Let us fix the embedding, choose one point in the interior of every 2-cell and one point in the interior of every 1-cell (arrow) of $\Gamma(\mathsf{A}, \mathsf{X})$. We will call the chosen points *midpoints* of the 2-cells and the arrows, respectively.

Connect every midpoint of an arrow with infinity outside $\Gamma(\mathsf{A}, \mathsf{X})$ by disjoint curves (*rays*) and connect the midpoint of every cell by disjoint curves (*rays*) inside the cells with the midpoints of the sides. If z is a midpoint of a cell, then we get, for every midpoint θ of a side of the cell, a ray $R_{z,\theta}$ starting in z, intersecting the boundary of $\Gamma(\mathsf{A}, \mathsf{X})$ only in θ and connecting z with infinity. The rays $R_{z,\theta}$ will intersect only in their endpoints z. We will say that the ray $R_{z,\theta}$ *lands* on z or that z is the *landing point* of the ray.

Denote by C the set of the midpoints of the cells with more than one side (i.e., the set of the starting points of more than one ray $R_{z,\theta}$).

For every $g_i \in \mathsf{A} \setminus \{1\}$ let $\gamma_i = R_{z_i, \alpha_i}$, where α_i is the midpoint of the unique arrow labeled by g_i. Then the *spider* $\mathcal{S}$ will be the set of obtained rays γ_i. Denote by P the set $\{z_1, \ldots, z_n\}$ of their landing points. The counterclockwise cyclic order of the rays γ_i is $\gamma_n, \gamma_{n-1}, \ldots, \gamma_1$, since $g_n, g_{n-1}, \ldots, g_1$ is the order of the labels of the edges along the path p.

See Figure 6.16, where the rays $R_{z,\theta}$ for the automaton from Figure 6.9 are drawn. The dotted rays land on the points of C, and the thick rays γ_g belong to the spider $\mathcal{S}$.

Euler's characteristic of the complex $\Gamma(\mathsf{A}, \mathsf{X})$ is equal to 1, since it is homotopically equivalent to a point. Hence, the number of the edges minus the number of the 2-cells is equal to $d - 1$.

If $z \in C$ is a midpoint of a cell with d' sides, then we add $d'-1$ to the number of the connected components of the plane when we delete the rays $R_{z,\theta}$ from the plane. Consequently, the number of the connected components of the plane with the rays $R_{z,\theta}$ deleted is equal to 1 plus the number of the edges minus the number of the 2-cells, i.e., to d.

Thus the rays $R_{z,\theta}$ divide the plane into d *sectors*. If $R_{z,\theta}$ belongs to the boundary of two sectors, then the arrow to which θ belongs starts in one sector and ends in the other sector. Hence, each sector contains a point of X; therefore each sector contains exactly one point of X. Let us denote by S_x the sector to which $x \in$ X belongs.

Note that here the sectors are the connected components of the plane minus the union of all the rays $R_{z,\theta}$ and not only the rays landing on the points of C. This does not change the number of the sectors, since if $R_{z,\theta}$ lands on a point $z \notin C$, then it is the only ray landing on z, and it does not cut either the plane or the respective sector into disconnected pieces.

Let S_x be a sector. The labels of the cells to which x belongs are pairwise different, and the set of such labels coincides with $\mathsf{A} \setminus \{1\} = \{g_1, \ldots, g_d\}$. Let us denote by U_i the cell labeled by g_i. Exactly one side of U_i is an arrow ending in x, and exactly one side of U_i is an arrow starting in x.

Let e_i be the side of U_i ending in x. The path p goes through e_i exactly once, the next edge of p is adjacent to a cell labeled by g_{i-1} and obviously starts in x. Hence, the edge next to e_i in p belongs to U_{i-1}. The path p is the path going around the diagram $\Gamma(\mathsf{A}, \mathsf{X})$ embedded into the plane. We conclude that the counterclockwise cyclic order of the labels of the cells to which x belongs is $g_n, g_{n-1}, \ldots, g_1$.

Let $y_1, y_2, \ldots, y_n$ be the midpoints of the cells $U_1, U_2, \ldots, U_n$, respectively. Each of the points y_i belongs to the boundary of the sector S_x. If $y_i \in C$, then there are exactly two rays landing on y_i and belonging to the boundary of S_x. Otherwise there is only one ray landing on y_i.

Now we are ready to construct the topological polynomial $f : \mathbb{R}^2 \longrightarrow \mathbb{R}^2$. Let us choose a collection of homeomorphisms

$$\varphi_{(y_1,\theta_1),(y_2,\theta_2)} : R_{y_1,\theta_1} \longrightarrow R_{y_2,\theta_2}$$

such that $\varphi_{(y_2,\theta_2),(y_3,\theta_3)} \circ \varphi_{(y_1,\theta_1),(y_2,\theta_2)} = \varphi_{(y_1,\theta_1),(y_3,\theta_3)}$. For example, we may take rectifiable rays and choose $\varphi_{(y_1,\theta_1),(y_2,\theta_2)}$ to be the isometries between the rays.

We define then f on the closure of S_x so that it is an orientation-preserving homeomorphism between the interior of S_x and $\mathbb{R}^2 \setminus \mathcal{S}$, $f(y_i) = z_i$ and the restriction of f onto a ray R_{y_i,θ_i} belonging to the boundary of S_x coincides with the homeomorphism $\varphi_{(y_i,\theta_i),(z_i,\alpha_i)}$. It is possible, since the cyclic order of the points y_i around x is the same as the cyclic order of the rays R_{z_i,α_i} (see Figure 6.17).

We leave it to the reader to check that f is a well defined topological polynomial with the set of the critical points C, the post-critical set P and the invariant spider $\mathcal{S}$.

The isomorphism $\mathsf{K}_{C,f} \cong (\mathsf{A}, \mathsf{X})$ follows then directly from the construction. $\square$

6.10.2. Finding a complex polynomial.

PROPOSITION 6.10.5. *Let* (A, X) *be an abstract kneading automaton, let* $G = \langle \mathsf{A} \rangle$ *be the group generated by* A *and let* $\mathfrak{M} = \mathsf{X} \cdot G$ *be the associated self-similarity bimodule. Then the following conditions are equivalent.*

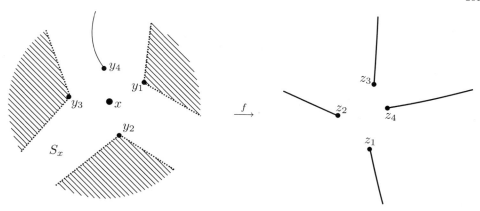

FIGURE 6.17. Construction of a topological polynomial

(1) *There exist non-trivial states $g_1 \ne g_2$ of A, an element $h \in G$ and elements $v_1, v_2 \in \mathfrak{M}^{\otimes n}$ for some $n \ge 1$, such that $g_i \cdot v_i = v_i \cdot g_i$ for $i = 1, 2$, and $h \cdot v_1 = v_2 \cdot h$.*
(2) *There exists a point $\xi \in \mathcal{J}_G$ that is periodic under iterations of the shift $\mathsf{s} : \mathcal{J}_G \longrightarrow \mathcal{J}_G$ and such that there exist its preimages $\xi_1, \xi_2 \in \mathcal{X}_G$ such that $\xi_i \cdot g_i = \xi_i$ for two different non-trivial states g_1, g_2 of A.*

PROOF. Let us prove that (1) implies (2). Let us take ξ_i equal to the image of $v_i^{\otimes -\omega}$ in $\mathcal{X}_G$. The image of the sequence $v_1^{\otimes -\omega}$ in $\mathcal{J}_G$ is equal to the image of $v_2^{\otimes -\omega}$, since $h \cdot v_1 = v_2 \cdot h$. We have $v_i^{\otimes -\omega} = v_i^{\otimes -\omega} \cdot g_i$ in $\mathcal{X}_G$, since $g_i \cdot v_i = v_i \cdot g_i$.

Let us prove that (2) implies (1). Suppose that $\xi_1, \xi_2 \in \mathcal{X}_G$ are such that $\xi_i \cdot g_i = \xi_i$ and the images of ξ_1 and ξ_2 in $\mathcal{J}_G$ are equal and periodic under the action of the shift s. The periodicity implies that there exist elements $u_i \in \mathfrak{M}^{\otimes n}$, for some $n \in \mathbb{N}$, such that $\xi_i \otimes u_i = \xi_i$. Since the images of ξ_1 and ξ_2 in $\mathcal{J}_G$ are equal, there exists a bounded sequence h_k such that $h_k \cdot u_1 = u_2 \cdot h_{k-1}$. We can find an element $h \in \{h_k\}$ such that $h \cdot u_1^{\otimes m} = u_2^{\otimes m} \cdot h$ for some $m \in \mathbb{N}$. Replacing u_i by $u_i^{\otimes m}$ and n by nm, we may assume that $h \cdot u_1 = u_2 \cdot h$.

Equality $\xi_i = \xi_i \otimes u_i$ implies that $u_i^{\otimes -\omega}$ represents the point ξ_i. We have $\xi_i = \xi_i \cdot g_i$; therefore the sequences $u_i^{\otimes -\omega}$ and $u_i^{\otimes -\omega} \cdot g_i$ are asymptotically equivalent. Therefore, there exists a bounded sequence $\{f_{k,i}\}_{k \ge 1}$ such that $f_{k,i} \cdot u_i = u_i \cdot f_{k-1,i}$ and $f_{1,i} \cdot u_i = u_i \cdot g_i$. Again, there exists $r_i \in \mathbb{N}$ and $f_i \in \{f_{k,i}\}$ such that $f_i \cdot u_i^{\otimes r_i} = u_i^{\otimes r_i} \cdot f_i$, and there exists s_i such that $f_i \cdot u_i^{\otimes s_i} = u_i^{\otimes s_i} \cdot g_i$. We may assume that $s_i < r_i$, since we can always replace r_i by its multiple. Moreover, we may assume that $r_1 = r_2 = r$.

Then we have
$$u_i^{\otimes r} \cdot f_i = f_i \cdot u_i^{\otimes r} = f_i \cdot u_i^{\otimes s_i} \otimes u_i^{\otimes (r-s_i)} = u_i^{\otimes s_i} \cdot g_i \cdot u_i^{\otimes (r-s_i)},$$
which implies that $g_i \cdot u_i^{\otimes (r-s_i)} = u_i^{\otimes (r-s_i)} \cdot f_i$; hence
$$g_i \cdot u_i^{\otimes r} = g_i \cdot u_i^{\otimes (r-s_i)} \otimes u_i^{\otimes s_i} = u_i^{\otimes (r-s_i)} \cdot f_i \cdot u_i^{\otimes s_i} = u_i^{\otimes r} \cdot g_i.$$
Thus, we can take $v_i = u_i^{\otimes r}$. □

DEFINITION 6.10.6. We say that an abstract kneading automaton A has *bad isotropy groups* if it satisfies the equivalent conditions of Proposition 6.10.5.

THEOREM 6.10.7. *Let* (A, X) *be an abstract kneading automaton. Then the following conditions are equivalent.*

(1) *There exists a post-critically finite complex polynomial f and an invariant spider $\mathcal{S}$ such that (A, X) is isomorphic to the kneading automaton $\mathsf{K}_{\mathcal{C}, f}$, where $\mathcal{C}$ is the critical portrait defined by $\mathcal{S}$.*
(2) (A, X) *is planar and does not have bad isotropy groups.*

PROOF. Let us prove that (1) implies (2). If (A, X) is isomorphic to a kneading automaton of a complex polynomial f, then it is planar by Proposition 6.10.3. Let $g_1 \ne g_2$ be non-trivial states of A. Then g_1 and g_2 are defined by small loops around two different post-critical points z_1 and z_2, as in the proof of Theorem 6.8.3.

Suppose that there exist $v_1, v_2 \in \mathfrak{M}^{\otimes n}$ such that $g_i \cdot v_i = v_i \cdot g_i$ and suppose that $h \cdot v_1 = v_2 \cdot h$ for some $h \in G$.

Then $g_i \cdot v_i^{\otimes m} = v_i^{\otimes m} \cdot g_i$ and $h \cdot v_1^{\otimes m} = v_2^{\otimes m} \cdot h$ for every $m \in \mathbb{N}$. This implies that there exist f^{mn}-preimages $g_1^{(m)}, g_2^{(m)}$ and $h^{(m)}$ of the loops g_1, g_2 and h, respectively, such that $g_i^{(m)}$ is a loop around z_i starting in the end of $v_i^{\otimes m} \in \mathfrak{M}(f^{mn})$ and $h^{(m)}$ is a path from the end of $v_1^{\otimes m}$ to the end of $v_2^{\otimes m}$. The length of the path $g_2^{(m)} h^{(m)} g_1^{(m)}$ has to go to zero as m goes to infinity, since f is expanding. But, on the other hand, its length is always bigger than the distance between z_1 and z_2. We get a contradiction.

Let us prove now that (2) implies (1). There exists, by Theorem 6.10.4, a topological polynomial f and an invariant spider $\mathcal{S}$ such that $(\mathsf{A}, \mathsf{X}) = \mathsf{K}_{\mathcal{C}, f}$ for the critical portrait $\mathcal{C}$ defined by the spider. It is sufficient to prove that f is combinatorially equivalent to a complex polynomial, i.e., that there is no obstruction.

Suppose that, on the contrary, an obstruction exists. It is known (see [**BFH92**]) that the only possible Thurston obstruction for a topological polynomial is a *Levy cycle*. A Levy cycle is a sequence $\gamma_0, \gamma_1 \ldots, \gamma_k = \gamma_0$ of non-peripheral closed simple curves in $\mathbb{C} \setminus P$ such that γ_i is homotopic rel. P to exactly one component γ'_i of $f^{-1}(\gamma_{i+1})$ and $f : \gamma'_i \longrightarrow \gamma_{i+1}$ has degree 1.

It follows from the definition of a Levy cycle that the curve γ_0 is homotopic rel. P to a component γ' of $f^{-k}(\gamma_0)$ such that $f^k : \gamma' \longrightarrow \gamma_0$ has degree 1. Then f^k is a homeomorphism between the disc bounded by γ' and the disc bounded by γ_0. In particular, f^k is a permutation of the post-critical points which are inside γ_0, and these points are not critical points of f^k. We can find $n = n_1 k$ such that f^n acts trivially on the post-critical points inside γ_0. The post-critical points inside γ_0 will not be critical points of f^n. Let $\widetilde{\gamma} \in f^{-n}(\gamma_0)$ be the curve homotopic rel. P to γ_0.

Consider the standard action of $\mathrm{IMG}(f)$ defined during the proof of Theorem 6.8.3. We may assume that the basepoint t_0 is inside γ_0. Let t_n be the f^n-preimage of t_0 which is inside the curve $\widetilde{\gamma}$. We may also assume that t_n is inside γ_0 (for example, choosing t_0 close to a post-critical point). Let ℓ be a path from t_0 to t_n inside γ_0, seen as an element of the $\pi_1(\mathcal{M})$-bimodule $\mathfrak{M}(f^n)$. Here, as usual, $\mathcal{M} = \mathbb{R}^2 \setminus P_f$.

Let z_1, z_2 be two different post-critical points inside γ_0. If g'_{z_i} is a small loop around z_i connected to t_0 by a path inside γ_0, then

$$g'_{z_i} \cdot \ell = \ell \cdot g''_{z_i},$$

where g''_{z_i} is again a small loop around z_i connected to t_0 by a path (homotopic to a path) inside γ_0. Each of the elements g'_{z_i} and g''_{z_i} is conjugate in $\pi_1(\mathcal{M})$ to the state g_{z_i} of the kneading automaton.

The bimodule $\mathfrak{M}^{\otimes n}$ is hyperbolic (by Theorem 3.9.12); therefore there exist $m \in \mathbb{N}$ and elements $g'_{z_i} \in G$ defined by loops going around z_i inside γ_0 such that
$$g'_{z_i} \cdot \ell^m = \ell^m \cdot g'_{z_i}$$
for $i = 1, 2$. Let $g'_{z_i} = g_i^{-1} g_{z_i} g_i$. Then $g_i^{-1} g_{z_i} g_i \cdot \ell^m = \ell^m \cdot g_i^{-1} g_{z_i} g_i$; hence
$$g_{z_i} \cdot \left(g_i \cdot \ell^m \cdot g_i^{-1} \right) = \left(g_i \cdot \ell^m \cdot g_i^{-1} \right) \cdot g_{z_i}.$$
If we denote $u_i = g_i \cdot \ell^m \cdot g_i^{-1}$, then we get $g_{z_i} \cdot u_i = u_i \cdot g_{z_i}$ and $g_2 g_1^{-1} \cdot u_1 = u_2 \cdot g_2 g_1^{-1}$; i.e., the kneading automaton has bad isotropy groups. □

Theorem 6.10.7 and Proposition 6.10.2 imply the following description of the iterated monodromy groups of the post-critically finite complex polynomials.

THEOREM 6.10.8. *A group is isomorphic to an iterated monodromy group of a post-critically finite polynomial if and only if it is isomorphic to a group generated by a kneading automaton without bad isotropy groups.* □

6.11. Quadratic polynomials

The results of this section come from joint work with Laurent Bartholdi. See [**BN05a**], where algebraic properties of the iterated monodromy groups of quadratic polynomials are discussed.

6.11.1. Kneading sequences of automata. Let (A, X) be a kneading automaton over the binary alphabet $\mathsf{X} = \{0, 1\}$. A tree-like set of permutations of the alphabet $\mathsf{X} = \{0, 1\}$ may contain only one transposition. Therefore A contains only one active state b_1. Every non-trivial state of A has exactly one incoming arrow (by the definition of a kneading automaton), and it is possible to come to b_1 along the arrows of the automaton from every non-trivial state, because b_1 is the only active state of A.

This implies that all the states of A can be ordered into a periodic or preperiodic sequence $b_1, b_2, \ldots$ such that for every $k > 1$ one of the two states $b_k|_0, b_k|_1$ is equal to b_{k-1} and the other one is trivial.

The *kneading sequence* of the automaton A is the sequence of the labels of the arrows along the path $(b_k)_{k=1}^\infty$ (going against the arrows) in the Moore diagram of A. There are four possible labels: $(0,0), (1,1), (0,1)$ and $(1,0)$, and we will denote them by $0, 1, *_0$ and $*_1$, respectively.

If the sequence $(b_k)_{k=1}^\infty$ is periodic with the period
$$a_1 = b_{1+nk}, a_2 = b_{2+nk}, \ldots, a_n = b_{n(k+1)},$$
then the kneading sequence is also periodic with the period of the form $x_1 \ldots x_{n-1} *$, where $x_k \in \{0, 1\}$ and $* \in \{*_0, *_1\}$. The automaton A is defined then by the following wreath recursions:

$$a_1 = \begin{cases} \sigma(a_n, 1) & \text{if } x_n = *_0 \\ \sigma(1, a_n) & \text{if } x_n = *_1 \end{cases}$$
$$a_{i+1} = \begin{cases} (a_i, 1) & \text{if } x_i = 0 \\ (1, a_i) & \text{if } x_i = 1 \end{cases}, \quad i = 1, \ldots, n-1.$$

The corresponding Moore diagram is shown in Figure 6.18. We will denote the respective automaton A by $\mathsf{K}_{x_1 x_2 \ldots x_{n-1}}$.

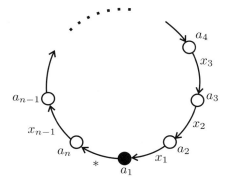

FIGURE 6.18. Automaton with periodic kneading sequence

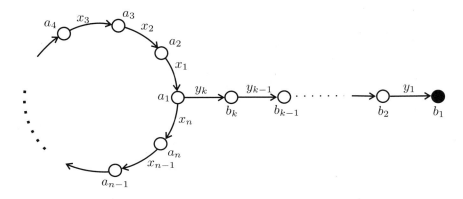

FIGURE 6.19. Automaton with preperiodic kneading sequence

If we replace A by A^{-1}, then the labels $(0,0)$ and $(1,1)$ will not change, while the labels $*_0$ and $*_1$ will be interchanged. Passing to the inverse of an automaton does not change the group it generates; therefore, we may always assume in the case of a periodic kneading sequence that the label $*$ is equal to $*_1$, so that $a_1 = \sigma(1, a_n)$. Therefore we will write just $*$ instead of $*_1$ and $*_0$.

Let us denote by $\mathfrak{K}(v)$ the group generated by the automaton K_v for $v \in \{0,1\}^\omega$. Note that changing the letters of the word $x_1 x_2 \ldots x_{n-1}$ to the opposite ones does not change (the conjugacy class of) the group $\mathfrak{K}(x_1 x_2 \ldots x_{n-1})$, since this corresponds just to renaming the letters of the alphabet.

If the sequence $(b_k)_{k=1}^\infty$ is preperiodic, i.e., is of the form $b_1 \ldots b_k (a_1 \ldots a_n)^\omega$, then the state b_1 appears only once and the kneading sequence is a preperiodic sequence of the form $y_1 y_2 \ldots y_k (x_1 x_2 \ldots x_n)^\omega$, where $y_i, x_i \in \{0,1\}$ and $y_k \ne x_n$ (because y_k and x_n are the labels of different arrows starting in a_1). See the Moore diagram of the automaton A in this case in Figure 6.19.

The wreath recursion defining the automaton A is then

$$b_1 = \sigma,$$

$$b_{i+1} = \begin{cases} (b_i, 1) & \text{if } y_i = 0, \\ (1, b_i) & \text{if } y_i = 1, \end{cases}$$

for $i = 1, \ldots, k-1$,

$$a_1 = \begin{cases} (b_k, a_n) & \text{if } y_k = 0 \text{ and } x_n = 1 \\ (a_n, b_k) & \text{if } y_k = 1 \text{ and } x_n = 0 \end{cases}$$

and

$$a_{i+1} = \begin{cases} (a_i, 1) & \text{if } x_i = 0, \\ (1, a_i) & \text{if } x_i = 1, \end{cases}$$

for $i = 1, \ldots, n-1$.

Let us denote by $\mathsf{K}_{y_1 y_2 \ldots y_k, x_1 x_2 \ldots x_n}$ the respective automaton. The group generated by $\mathsf{K}_{y_1 y_2 \ldots y_k, x_1 x_2 \ldots x_n}$ is denoted $\mathfrak{K}(y_1 y_2 \ldots y_k, x_1 x_2 \ldots x_n)$. Note that here also we may change the letters of the words $y_1 y_2 \ldots y_k, x_1 x_2 \ldots x_n$ to the opposite without changing the group. Remember that the group $\mathfrak{K}(y_1 y_2 \ldots y_k, x_1 x_2 \ldots x_n)$ is defined only if $y_k \neq x_n$.

A binary kneading automaton has bad isotropy groups (see Definition 6.10.6) if and only if it is of the form $\mathsf{K}_{y_1 y_2 \ldots y_k, x_1 x_2 \ldots x_n}$ where the word $x_1 x_2 \ldots x_n$ is a proper power, i.e., if the period of the kneading sequence is less than the period of the respective sequence of states.

6.11.2. Quadratic polynomials.

6.11.2.1. *Review of results in holomorphic dynamics.* Symbolic dynamics of quadratic polynomials is a well studied subject. See, for example, [**BS02a**] for different approaches to it (Hubbard trees, kneading sequences, external rays, etc.) and connections between them.

We will show here only how application of Theorem 6.8.3 to the degree 2 case gives the classical notion of a kneading sequence of a quadratic polynomial.

Let $f(z) = z^2 + c$ be a post-critically finite quadratic polynomial. The critical portrait of f consists of two f-preimages of an external ray $R_\theta = R_{\theta, \infty}$ landing on c (if 0 is preperiodic) or of an extended ray $R_\theta = R_{U,p}$ (if 0 is periodic) consisting of an external ray $R_{\theta, \infty}$ landing on the root p of the Fatou component U of c and an internal ray from the root p to the center c of U (see Section 6.9).

In both cases the curves of the critical portrait are the (external or extended) rays $R_{\theta/2}$ and $R_{(\theta+1)/2}$, since f acts on the external angles by doubling. Both rays land on 0, and they divide the plane into two sectors.

The angle θ shows where the point c is in the *Mandelbrot set*. Suppose that 0 belongs to a cycle of length n for the iteration of $z^2 + c$. Then c belongs to a *hyperbolic component* M_c of the interior of the Mandelbrot set. For any other point c_1 of that component, the quadratic polynomial $z^2 + c_1$ also has a unique attracting cycle of length n. If $\Phi(c_1)$ denotes the multiplier of this cycle, then Φ is a conformal isomorphism of M_c with the open unit disc $\mathbb{D} = \{z \in \mathbb{C} : |z| < 1\}$. We obviously have $\Phi(c) = 0$. The isomorphism $\Phi : M_c \to \mathbb{D}$ extends to a homeomorphism of the boundary of M_c with the unit circle. The preimage of 1 under this homeomorphism is called the *root* of the component M_c. There exist exactly two angles θ such that the *parameter ray* R_θ (i.e., the external ray to the Mandelbrot set) lands on the root of M_c.

In the dynamical plane, the point c belongs to a Fatou component U_c. There is a unique point r on the boundary of U_c, fixed under the map $f^n : U_c \to U_c$ (since $f^n|_{U_c}$ is topologically conjugate via the Boetcher map with the restriction of z^2 onto $\mathbb{D}$). This point is the *root* of the Fatou component U_c.

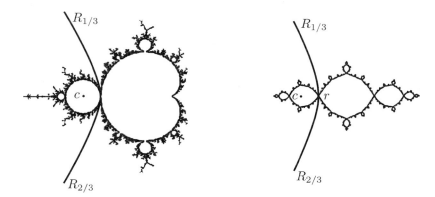

FIGURE 6.20. External rays

See Figure 6.20, where the case $c = -1$ is shown. The root of the component M_c and the root r of the Fatou component U_c are the landing points of the rays $R_{1/3}$ and $R_{2/3}$, which are also shown in Figure 6.20.

A parameter ray R_θ lands on the root of the hyperbolic component M_c if and only if the dynamical ray R_θ (the external ray to the Julia set of $z^2 + c$) lands on the root of the Fatou component U_c. Moreover, the number $\theta \in \mathbb{R}/\mathbb{Z}$ belongs to a cycle of length n under the doubling map $\alpha \mapsto 2\alpha : \mathbb{R}/\mathbb{Z} \to \mathbb{R}/\mathbb{Z}$. In particular the angle θ is equal to $p/(2^n - 1)$ for some integer p and the ray $R_{2^k \theta}$ lands on the root of the Fatou component to which $f^k(c)$ belongs.

The other way around, for every rational number $\theta \in \mathbb{R}/\mathbb{Z}$ with odd denominator, the parameter ray R_θ lands on a root of the hyperbolic component M_c; and if c is the center of the component (i.e., the preimage of 0 under the multiplicator map), then 0 has the same period under iterations of $f(z) = z^2 + c$ as θ has under the doubling map, and the dynamical ray R_θ lands on the root of the Fatou component of $z^2 + c$ containing c.

Suppose now that 0 is preperiodic under iterations of $f(z) = z^2 + c$. Then c belongs to the boundary of the Mandelbrot set (is a *Misiurewicz* point) and there exists a finite set of angles θ such that the parameter rays R_θ land on c. For each of such θ the external ray R_θ in the dynamical plane of $z^2 + c$ lands on c. The preperiod of θ under the doubling map is the same as the preperiod of c under iterations of $z^2 + c$, but the period of θ may be a multiple of the period of c. Here the *preperiod* and the *period* of a point x under a map f are the minimal positive integers k and n such that $f^{k+n}(x) = f^k(x)$.

For proofs and more about the external rays to the Julia and the Mandelbrot sets, see [**DH84**].

6.11.2.2. *Kneading sequences.* Let $f(z) = z^2 + c$ be a post-critically finite quadratic polynomial and let $\mathcal{C} = \{R_{\theta/2}, R_{(\theta+1)/2}\}$ be the critical portrait of f. The rays of $\mathcal{C}$ divide the plane into two sectors. We denote the sector containing c by S_1 and the other sector by S_0. The sector S_0 contains the landing points of the external rays with angles in the interval $\left(\frac{\theta-1}{2}, \frac{\theta}{2}\right) \ni 0$, and S_1 contains the rays with the angles in the interval $\left(\frac{\theta}{2}, \frac{\theta+1}{2}\right) \ni 1/2$. We will denote these intervals also by S_0 and S_1. They are the two semicircles into which the circle $\mathbb{R}/\mathbb{Z}$ is divided by the points $\frac{\theta}{2}$ and $\frac{\theta+1}{2}$.

6.11. QUADRATIC POLYNOMIALS

For every $\alpha \in \mathbb{R}/\mathbb{Z}$ denote by $I_\theta(\alpha)$ the θ-*itinerary* of α, defined as the sequence $a_0 a_1 \ldots$, where
$$a_k = \begin{cases} 0 & \text{if } 2^k \alpha \in S_0 \\ 1 & \text{if } 2^k \alpha \in S_1 \\ * & \text{if } 2^k \alpha \in \{\theta/2, (\theta+1)/2\}. \end{cases}$$

The itinerary $I_\theta(\theta)$ is called the *kneading sequence* of the point $\theta \in \mathbb{R}/\mathbb{Z}$ and is denoted $\widehat{\theta}$.

Not every sequence of the form $(v*)^\omega$ or uv^ω for $v, u \in \{0,1\}^*$ is the kneading sequence $\widehat{\theta}$ of an angle $\theta \in \mathbb{R}/\mathbb{Z}$. A description of the kneading sequence of angles is given in [**BS02a**].

Comparing now the definition of the automaton $\mathsf{K}_{\mathcal{C},f}$ (Definition 6.8.1) with the definitions of the kneading sequence $\widehat{\theta}$ and the kneading sequence of an automaton (Subsection 6.11.1), we see that the kneading sequence of the automaton $\mathsf{K}_{\mathcal{C},f}$ coincides with the kneading sequence $\widehat{\theta}$ of the polynomial f.

If ξ is a periodic sequence of the form $(v*)^\omega$ for $v \in \mathsf{X}^*$, then we denote $\mathfrak{K}(\xi) = \mathfrak{K}(u)$. If $\xi \in \{0,1\}^\omega$ is preperiodic, then we denote $\mathfrak{K}(\xi) = \mathfrak{K}(u,v)$, where v is the shortest period and u is the shortest preperiod of the sequence $\xi = uv^\omega$.

THEOREM 6.11.1. *Let $f(z) = z^2 + c$ be a post-critically finite quadratic polynomial. Suppose that $\theta \in \mathbb{R}/\mathbb{Z}$ is the angle such that the parameter ray R_θ (i.e., the external ray to the Mandelbrot set) lands either on c (if c is a Misiurewicz point) or on the root of the hyperbolic component of c (if f is hyperbolic).*

Then the group $\mathrm{IMG}(f)$ is isomorphic to the group $\mathfrak{K}\left(\widehat{\theta}\right)$ and their actions on the respective rooted trees are conjugate. The dynamical systems $(\mathcal{J}_f, f)$ and $\left(\mathcal{J}_{\mathfrak{K}(\widehat{\theta})}, \mathsf{s}\right)$ are conjugate.

PROOF. If the period of the angle θ under the angle doubling map coincides with the period of c under iterations of f, then the statements of the theorem are partial cases of Theorem 6.8.3 and Theorem 5.5.3. It is sufficient therefore to prove the theorem only for the case when c and θ are preperiodic.

Let P_f be the post-critical set of f. Choose any spider $\mathcal{S}_0$ consisting of external rays landing on the points of P_f. (We choose one external ray $\gamma_{z,0}$ landing on z for every $z \in P_f$.) Since points of P_f are not critical points of f, the preimage $f^{-1}(\mathcal{S}_0)$ is a set of disjoint external rays having pairwise different landing points. Let $\mathcal{S}_1 \subset f^{-1}(\mathcal{S}_0)$ be the subset of the rays landing on the points of P_f. We continue inductively and define the spider $\mathcal{S}_n$ as the set of the rays belonging to $f^{-1}(\mathcal{S}_{n-1})$ and landing on the points of P_f.

We get a sequence of spiders $\mathcal{S}_0, \mathcal{S}_1, \ldots$ such that $\mathcal{S}_n \subset f^{-1}(\mathcal{S}_{n-1})$. Let us denote for $n \geq 0$ and $z \in P_f$ by $\gamma_{z,n}$ the ray belonging to $\mathcal{S}_n$ and landing on z.

The preimage $f^{-1}(\gamma_{c,n})$ of the ray landing on the critical value c consists of two rays landing on 0. They divide the plane into two sectors. Let us denote the sector to which c belongs by $S_{1,n}$ and the other sector by $S_{0,n}$.

The obtained partitions of the plane into $S_{0,n}$ and $S_{1,n}$ are different, but every point $z \in P_f$ belongs either only to the sectors $S_{0,n}$ or only to the sectors $S_{1,n}$ for all n. This follows from the fact that the kneading sequences of all the external angles θ of the point c are equal.

Let $t \in \mathbb{C} \setminus \bigcup_{z \in P_f} \gamma_{z,0}$ be a basepoint and let $T = \bigsqcup_{n \geq 0} f^{-n}(t)$ be the corresponding tree of preimages. We are going to define an isomorphism $I : T \longrightarrow \mathsf{X}^*$

for $X = \{0, 1\}$ using the itineraries with respect to the defined partition of the plane into sectors. Namely, if $y \in f^{-n}(t)$ is a vertex of the tree T, then the corresponding word $I(y) = x_n x_{n-1} \ldots x_1$ is defined by the condition that

$$f^k(y) \in S_{x_k, n-k}$$

for all $k = 1, 2, \ldots, n$.

It follows directly from the definition that if $I(y) = x_n \ldots x_1$, then $I(f(y)) = x_n \ldots x_2$, and it is also easy to see that I is a level-preserving bijection between the trees T and X^*. Thus I is an isomorphism of the rooted trees.

Let us define for $n \geq 0$, $z \in P_f$ and $t_1, t_2 \in \mathbb{C} \setminus \bigcup_{z \in P_f} \gamma_{z,n}$ a path $g_{z,n}(t_1, t_2)$ as the path starting in t_1, ending in t_2 and intersecting the spider $\mathcal{S}_n$ only once through the leg $\gamma_{z,n}$ in such direction that the part of $\gamma_{z,n}$ containing z is on the left of the path $g_{z,n}(t_1, t_2)$. The path $g_{z,n}(t_1, t_2)$ is uniquely defined, up to a homotopy in $\mathcal{M} = \mathbb{C} \setminus P_f$, by these conditions. We also denote by $g_n(t_1, t_2)$ the path starting in t_1, ending in t_2 and disjoint with the legs of the spider $\mathcal{S}_n$. It is also uniquely defined up to a homotopy. If $z \notin P_f$, then we define $g_{z,n}(t_1, t_2) = g_n(t_1, t_2)$.

If $z \neq c$, then it follows from the uniqueness of the defined paths that

$$f^{-1}(g_{z,n}(t_1, t_2)) = \{g_{z_0, n+1}(t_{10}, t_{20}), g_{z_1, n+1}(t_{11}, t_{21})\},$$

where $\{z_0, z_1\} = f^{-1}(z)$, $\{t_{10}, t_{11}\} = f^{-1}(t_1)$, $\{t_{20}, t_{21}\} = f^{-1}(t_2)$, $\{z_0, t_{10}, t_{20}\} \subset S_{0,n}$ and $\{z_1, t_{11}, t_{21}\} \subset S_{1,n}$.

For the case $z = c$, we get

$$f^{-1}(g_{c,n}(t_1, t_2)) = \{g_{n+1}(t_{10}, t_{21}), g_{n+1}(t_{11}, t_{20})\},$$

where t_{ij} satisfy the same conditions as before.

Finally,

$$f^{-1}(g_n(t_1, t_2)) = \{g_{n+1}(t_{10}, t_{20}), g_{n+1}(t_{11}, t_{21})\}.$$

These observations, together with the fact that the itinerary of c with respect to the partitions $S_{0,n}, S_{1,n}$ coincides with the kneading sequence $\widehat{\theta}$, imply that the generators $g_{z,0}(t, t)$ of $\pi_1(\mathcal{M}, t)$ are conjugated by the isomorphism $I : T \longrightarrow X^*$ with the generators a_i, b_i of $\mathfrak{K}\left(\widehat{\theta}\right)$.

The statement about conjugacy of the dynamical systems follows, for instance, from the well known results in symbolic dynamics of quadratic polynomials (see, for example, [**BS02a, Kel00**]). □

6.12. Examples of iterated monodromy groups of polynomials

6.12.1. The iterated monodromy group of $z^2 - 1$ and amenability.
The parameter rays R_θ for $\theta = 1/3$ and $\theta = 2/3$ land on the root of the hyperbolic component with the center $c = -1$ (see Figure 6.20). The orbit of $\theta = 1/3$ under the angle doubling map is $1/3 \mapsto 2/3 \mapsto 1/3$. We have $\{\frac{\theta}{2}, \frac{\theta+1}{2}\} = \{1/6, 2/3\}$. The kneading sequence of θ is $\widehat{\theta} = 1*1*\ldots$. Consequently, by Theorem 6.8.3, a standard action of the iterated monodromy group IMG $(z^2 - 1)$ coincides with $\mathfrak{K}(1)$, which is the group generated by the transformations $a_1 = \sigma(1, a_2)$, $a_2 = (1, a_1)$. Note that this action coincides with the action computed in Subsection 5.2.2.

The group $\mathfrak{K}(1)$ was defined for the first time by R. Grigorchuk and A. Żuk in [**GŻ02a, GŻ02b**] just as an interesting group generated by a three-state automaton. Later R. Pink discovered that $\mathfrak{K}(1)$ is the iterated monodromy group of $z^2 - 1$. More precisely, he defined the profinite iterated monodromy groups as Galois

6.12. EXAMPLES OF ITERATED MONODROMY GROUPS OF POLYNOMIALS

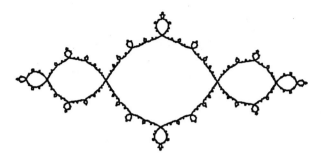

FIGURE 6.21. Julia set of $z^2 - 1$

groups (see Proposition 6.4.2) and computed $\overline{\mathrm{IMG}}(z^2 - 1)$ using only information about the conjugacy classes of a_1, a_2 and $a_1 a_2$ in $\operatorname{Aut} \mathsf{X}^*$.

Theorem 5.5.3 implies that the limit space $\mathcal{J}_{\mathrm{IMG}(z^2-1)}$ is homeomorphic to the Julia set of the polynomial $z^2 - 1$. The Julia set is shown in Figure 6.21. See also its approximation (Figure 3.4, page 112), which is constructed using subdivision rules. It is called sometimes "*basilica*" (it presumably resembles the Basilica San Marco in Venice together with the reflection in the water). This is the reason why the group IMG $(z^2 - 1)$ is also often called "basilica group".

R. Grigorchuk and A. Żuk proved the following properties of IMG $(z^2 - 1)$.

THEOREM 6.12.1. *The group* IMG $(z^2 - 1)$

(1) *is torsion-free;*
(2) *has exponential growth (actually, the semigroup generated by a and b is free);*
(3) *is just non-solvable, i.e., every one of its proper quotients is solvable;*
(4) *has solvable word and conjugacy problems;*
(5) *has no free non-abelian subgroups;*
(6) *is not in the class SG of subexponentially amenable groups.*

The class SG, defined in [**CSGH99**], is a natural generalization of the class EG of the *elementary amenable groups*, which was introduced in [**Day57**]. The class EG of *elementary amenable groups* is the smallest class containing finite and abelian groups and is closed after taking extensions, quotients, subgroups and direct limits. The first example of an amenable, but not elementary amenable, group is the Grigorchuk group. The Grigorchuk group is amenable since it has sub-exponential growth. So, a natural generalization of EG is the class SG, which is the smallest class containing the groups of sub-exponential growth and is closed after taking extensions and direct limits (the other operations are superfluous).

It was proved by L. Bartholdi and B. Virag in [**BV**], using self-similarity of random walks, that the group IMG $(z^2 - 1)$ is amenable. Thus, IMG $(z^2 - 1)$ is the first example of an amenable group not belonging to the class SG. Amenability of the group IMG $(z^2 - 1)$ is a partial case of the following result of L. Bartholdi, V. Kaimanovich, B. Virag and the author. The proof of the theorem also uses self-similar random walks on groups.

THEOREM 6.12.2. *The group of bounded automata $\mathcal{B}_0(\mathsf{X})$ is amenable for every finite alphabet X.*

See Section 3.9 for the definition of the group of bounded automata.

COROLLARY 6.12.3. *Iterated monodromy groups of post-critically finite polynomials are amenable.*

PROOF. Iterated monodromy groups of post-critically finite polynomials are generated by bounded automata due to Theorem 6.10.8. □

It is an open question if any contracting group is amenable.

6.12.2. Belyi polynomials. Probably the first case of computation of an iterated monodromy action is Theorem 4.2 in the paper of K. M. Pilgrim [**Pil00**].

The author considers there the action of the absolute Galois group $\mathrm{Aut}\,(\overline{\mathbb{Q}}/\mathbb{Q})$ on the set of Belyi polynomials.

A *dynamical Belyi polynomial (DBP)* is a complex polynomial f such that its post-critical set P is equal to $\{0, 1\}$, does not contain critical points and $f(1) = f(0) = 0$. (The definition of [**Pil00**] is a bit more restrictive.)

K. M. Pilgrim proves that the action of $\mathrm{Aut}\,(\overline{\mathbb{Q}}/\mathbb{Q})$ on the set of dynamical Belyi polynomials is faithful.

Let us show how Theorem 4.2 of [**Pil00**] follows from Theorem 6.8.3.

PROPOSITION 6.12.4. *The iterated monodromy group of a DBP is generated by two automorphisms* g_0, g_1 *of the tree* X^*, *where* $g_1 = \sigma_1 \in \mathfrak{S}(\mathsf{X})$ *is rooted and*

$$g_0 = \sigma_0 \left(g_0, g_1, 1, 1, \ldots, 1\right),$$

where $\sigma_1, \sigma_0 \in \mathfrak{S}(\mathsf{X})$ *are permutations such that the set* $\{\sigma_0, \sigma_1\}$ *is tree-like and* σ_0 *fixes the letters of* X *corresponding to the first two coordinates of the wreath recursion.*

PROOF. All the critical points of a DBP are strictly preperiodic, and hence the post-critical set $P = \{0, 1\}$ belongs to the Julia set. There exists, up to isotopy, only one spider $\mathcal{S} = \{\gamma_0, \gamma_1\}$ for a two-point post-critical set. Consequently, every DBP has an invariant spider.

Let $\mathcal{C}$ be the critical portrait associated to $\mathcal{S}$ and let $\mathsf{K}_{\mathcal{C},f}$ be the corresponding kneading automaton generating $\mathrm{IMG}\,(f)$.

The kneading automaton consists of three states: g_0, g_1 and 1. The set $f^{-1}(1)$ does not intersect the post-critical set. Therefore, g_1 is a rooted automorphism of X; i.e., it acts on the words by the rule $g_1(x_1 x_2 x_3 \ldots) = \sigma_1(x_1) x_2 x_3 \ldots$, where $\sigma_1 \in \mathfrak{S}(\mathsf{X})$ is a permutation.

We have $P \cap f^{-1}(0) = \{0, 1\}$ and the post-critical points are not critical. Consequently, if 0 and 1 belong to the sectors S_{x_0} and S_{x_1}, then $g_0 \cdot x_0 = x_0 \cdot g_0$ and $g_0 \cdot x_1 = x_1 \cdot g_1$. If $x_i \notin \{x_0, x_1\}$, then $g_0|_{x_i} = 1$. Hence

$$g_0 = \sigma_0 \left(g_0, g_1, 1, 1, \ldots, 1\right),$$

where σ_0 is a permutation fixing x_0 and x_1. □

PROPOSITION 6.12.5. *Let* g_1 *and* g_0 *be automorphisms of* X^* *given by a recursion satisfying the conditions of Proposition 6.12.4. Then there exists a DBP* f *such that the group* $G = \langle g_1, g_0 \rangle$ *coincides with a standard action of* $\mathrm{IMG}\,(f)$.

PROOF. Every kneading automaton containing only two non-trivial states is planar, since there is only one circular order on a two-element set. Consequently, $\mathsf{B} = \{g_0, g_1, 1\}$ is a planar kneading automaton.

The orbispace $\mathcal{J}_G$ of the group generated by B has only two singular points. They are represented by the sequences $x_0^{-\omega}$ and $x_0^{-\omega} x_1$. The second point is not periodic under the action of the shift. The isotropy group of the first point $x_0^{-\omega}$ is the cyclic group generated by g_0. This implies that B does not have bad isotropy groups.

Thus Proposition 6.12.5 implies that there exists a post-critically finite polynomial f and a critical portrait $\mathcal{C}$ such that $\mathsf{K}_{\mathcal{C},f} = \mathsf{B}$. The post-critical set of the polynomial f has two points z_0, z_1 such that $g_0 = g_{z_0}$ and $g_1 = g_{z_1}$. Definitions of B and $\mathsf{K}_{\mathcal{C},f}$ imply that z_0, z_1 are not critical and that $f(z_0) = f(z_1) = z_0$. □

6.12.3. Chebyshev polynomials. Let $T_d(z) = \cos(d \arccos z)$ be the Chebyshev polynomials. They satisfy the recursion

$$T_0(z) = 1, T_1(z) = z, \text{ and } T_d = 2z T_{d-1} - T_{d-2},$$

since

$$\cos(d \arccos z) + \cos((d-2) \arccos z) = 2 \cos((d-1) \arccos z) \cos(\arccos z).$$

PROPOSITION 6.12.6. *The iterated monodromy group* $\mathrm{IMG}(T_d)$ *is isomorphic to the infinite dihedral group* $\mathbb{D}_\infty$.

If d is even, then a standard self-similar action of $\mathrm{IMG}(T_d)$ *is generated by the involutions*

$$a = \sigma_{-1}, \quad b = \sigma_1(b, 1, 1, \ldots, 1, 1, a),$$

where $\sigma_{-1} = (1,2)(3,4) \ldots (d-1, d)$ *and* $\sigma_1 = (2,3)(4,5) \ldots (d-2, d-1)$.

If d is odd, then a standard self-similar action of $\mathrm{IMG}(T_d)$ *is generated by the involutions*

$$a = \sigma_{-1}(1, 1, \ldots, 1, a), \quad b = \sigma_1(b, 1, \ldots, 1, 1),$$

where $\sigma_{-1} = (1,2)(3,4) \ldots (d-2, d-1)$ *and* $\sigma_1 = (2,3)(4,5) \ldots (d-1, d)$.

PROOF. We have $T_d(\cos t) = \cos(dt)$; hence $T_d'(\cos t) = d \cdot \sin(dt)/\sin t$. This implies that the critical points of T_d are $\cos\left(\frac{\pi k}{d}\right)$, where $k \in \mathbb{Z}$ are not divisible by d. We thus get $d - 1$ critical points and every critical point is of local degree 2.

Thus the critical values of T_d are the points of the form $\cos(k\pi)$, where $k \in \mathbb{Z}$ is not divisible by d, i.e., only -1 for $d = 2$ and 1 or -1 for $d \geq 3$.

We have $T_d(-1) = 1$ and $T_d(1) = 1$ for even d and $T_d(-1) = -1$ and $T_d(1) = 1$ for odd $d > 1$.

We see that for even d the Chebyshev polynomial T_d is conjugate to a DBP. Thus, if d is even, then Proposition 6.12.4 implies that $\mathrm{IMG}(T_d)$ is generated by the transformations

$$g_{-1} = \sigma_{-1}, \text{ and } g_1 = \sigma_1(g_1, g_{-1}, 1, 1, \ldots, 1),$$

where σ_{-1} and σ_1 are the monodromy actions of the loops around the critical values -1 and 1, respectively. The permutation σ_1 does not move the letters corresponding to the first two coordinates of the wreath recursion.

If d is odd, then Theorem 6.8.3 implies that $\mathrm{IMG}(T_d)$ is generated by the transformations

$$g_{-1} = \sigma_{-1}(1, g_{-1}, 1, \ldots, 1), \text{ and } g_1 = \sigma_1(g_1, 1, 1, \ldots, 1),$$

where σ_{-1} does not move the letter corresponding to the second coordinate and σ_1 does not move the letter corresponding to the first coordinate.

In both cases $\Sigma = \{\sigma_{-1}, \sigma_1\}$ is a tree-like set of permutations and $\sigma_{-1}^2 = \sigma_1^2 = 1$, since all the critical points are of local degree 2. Consequently, the cyclic diagram of Σ is just a chain of edges. This means that for some indexing $\{1, 2, \ldots, d\}$ of the alphabet we have

$$\sigma_1 = (2,3)(4,5)(5,7)\ldots$$

and

$$\sigma_{-1} = (1,2)(3,4)(5,6)\ldots.$$

If d is even, then the fixed points of σ_1 are 1 and d and σ_{-1} has no fixed points. If d is odd, then σ_1 fixes only 1 and σ_{-1} fixes only d. Consequently, with respect to this indexing the generators g_{-1} and g_1 are defined by the recursions

$$g_{-1} = \sigma_{-1}, \quad g_1 = \sigma_1(g_1, 1, 1, \ldots, 1, 1, g_{-1})$$

when d is even, and by

$$g_{-1} = \sigma_{-1}(1, 1, \ldots, 1, g_{-1}), \quad g_1 = \sigma_1(g_1, 1, \ldots, 1, 1)$$

when d is odd.

The group IMG (T_d) is level-transitive and thus infinite. Consequently, the group IMG (T_d) is isomorphic to the infinite dihedral group $\mathbb{D}_\infty$. □

The same result can be proved using Corollary 6.1.7. We have the following commutative diagram:

$$\begin{array}{ccc} \mathbb{C} & \xrightarrow{d \cdot z} & \mathbb{C} \\ \downarrow \cos z & & \downarrow \cos z \\ \mathbb{C} & \xrightarrow{T_d(z)} & \mathbb{C} \end{array}$$

Its restriction onto the Julia set $[-1, 1]$ of T_d is

$$\begin{array}{ccc} \mathbb{R} & \xrightarrow{d \cdot z} & \mathbb{R} \\ \downarrow \cos z & & \downarrow \cos z \\ [-1, 1] & \xrightarrow{T_d(z)} & [-1, 1] \end{array}$$

The vertical arrows in the diagram are conjugate to the natural quotient map $\mathbb{R} \mapsto \mathbb{R}/\mathbb{D}_\infty$, where $\mathbb{D}_\infty$ acts as the group of the affine transformations $x \mapsto \pm x + a$, $a \in \mathbb{Z}$. The map $x \mapsto d \cdot x$ is an expanding automorphism of the Lie group $\mathbb{R}$. A direct application of Corollary 6.1.7 shows now that the iterated monodromy group of $T_d(z)$ is isomorphic to $\mathbb{D}_\infty$.

6.12.4. Fabrykowski-Gupta group as an iterated monodromy group.

Consider the polynomial $f(z) = z^3(\omega - 1) + 1$, where $\omega = -\frac{1}{2} + \frac{\sqrt{3}}{2}i$ is a root of unity of degree 3. Its unique critical point is $z = 0$. Its orbit is $0 \mapsto 1 \mapsto \omega \mapsto \omega$. Hence, it is affine conjugate to a DBP.

The iterated monodromy group of f is generated by the transformations g_1 and g_ω as it is described in Proposition 6.12.4. The rooted automorphism g_1 acts as a cyclic permutation of the first level of $\mathsf{X}^* = \{0, 1, 2\}^*$, since f has local degree 3 at 0. The point ω is not a critical value; therefore g_ω is not active, and thus it is defined by the recurrent relation

$$g_\omega = (g_\omega, g_1, 1).$$

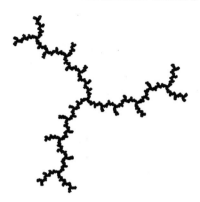

FIGURE 6.22. Julia set of $z^3(-3/2 + i\sqrt{3}/2) + 1$

The group generated by g_1 and g_ω coincides with the group considered by J. Fabrykowski and N. D. Gupta in [**FG91**] as an example of a group of intermediate growth. The Julia set of $z^3(\omega - 1) + 1$ is shown in Figure 6.22.

The Schreier graphs of the action of the Fabrykowski-Gupta group on the levels of the tree X^* were studied by L. Bartholdi and R. Grigorchuk in [**BG00b**]. In particular, they computed their spectra and noticed that the Schreier graphs converge to a fractal set. Their observations served as one of the starting points of the definition of the limit space of a contracting self-similar group.

6.12.5. Groups of intermediate growth and IMG $(z^2 + i)$. A finitely generated group G has *intermediate growth* if the sequence $|B_S(n)|$ grows faster than any polynomial $p(n)$ and slower than any exponential function a^n, $a > 1$. Here $B_S(n) = \{s_1 \cdots s_n \;:\; g_i \in S \cup S^{-1}\}$ for some finite generating set $S \ni 1$.

The first example of a group of intermediate growth is the Grigorchuk group (Section 1.6). The Grigorchuk group grows faster than $\exp\left(n^{0.5157}\right)$ and slower than $\exp\left(n^{0.7675}\right)$ (see [**Gri85, Leo00, Bar98, Bar01**]).

It was also already mentioned that J. Fabrykowski and N. D. Gupta in [**FG91**] considered an example of a group of intermediate growth, which can be defined as IMG $\left(z^3(-3/2 + i\sqrt{3}/2) + 1\right)$.

Another example of a group of intermediate growth is IMG $(z^2 + i)$. This is a result of Kai-Uwe Bux and Rodrigo Pérez (see [**BP04**]). The point i is the landing point of the parameter ray $R_{1/6}$. The orbit of the angle $1/6$ under angle doubling is $1/6 \mapsto 1/3 \mapsto 2/3 \mapsto 1/3$. The orbit of the critical value i under iterations of $f(z) = z^2 + i$ is $i \mapsto i - 1 \mapsto -i \mapsto i - 1$. We see that the period of the dynamical ray landing on i is equal to the period of the critical value and the respective kneading sequence is $1(10)^\omega$. Hence a standard action of IMG $(z^2 + i)$ is generated by the transformations

$$b_1 = \sigma, \quad a_1 = (a_2, b_1), \quad a_2 = (1, a_1).$$

See Figure 6.23, where the Schreier graph of the action of IMG $(z^2 + i)$ on the 6th level of the tree X^* and the Julia set of $z^2 + i$ are shown. The Schreier graph is constructed using the inflation algorithm described in Section 3.10.

An interesting question is to classify all post-critically finite (topological) polynomials whose iterated monodromy groups have intermediate growth.

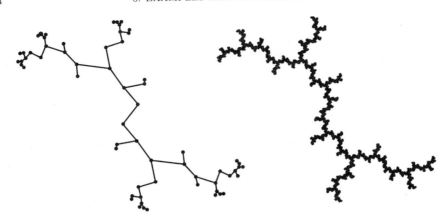

FIGURE 6.23. A Schreier graph of IMG $(z^2 + i)$ and the limit space

The main technique to prove that a self-similar group G has sub-exponential growth is to use the following strong contraction argument.

Let $l(g)$ denote the length of g with respect to some fixed generating set S of G. In many cases *weights* can be used; i.e., we assign a positive number to each generator and then the length of a product of generators is equal to the sum of the lengths of the factors (we assume that g and g^{-1} have the same length).

PROPOSITION 6.12.7. *Let (G, X) be a faithful self-similar action and suppose that there exist numbers $0 \le \eta < 1, 0 < p \le 1$ and $C > 0$ such that for every $r \in \mathbb{N}$ the proportion of the elements in $\{g \in \mathsf{St}_G(1) : l(g) \le r\}$ satisfying*

$$\sum_{x \in \mathsf{X}} l(g|_x) \le \eta r + C$$

is at least p. Then G has sub-exponential growth.

PROOF. This is a classical argument which can be found in [**Gri83, Gri85**]. The present formulation is from [**BP04**]. See also various examples in [**BGŠ03**].

Suppose that, on the contrary, the growth of G is exponential. This means that $\lim_{r \to \infty} \sqrt[r]{|B(r)|} = \lambda > 1$, where $B(r) = \{g \in G : l(g) \le r\}$. The subgroup $H = \mathsf{St}_G(1)$ has finite index in G, hence also $\lim_{r \to \infty} \sqrt[r]{|B(r) \cap H|} = \lambda$. Thus for any $\epsilon > 0$ and all r big enough we have $(\lambda - \epsilon)^r < |B(r) \cap H|$.

The elements $g \in H$ are defined uniquely by the $|\mathsf{X}|$-tuple of restrictions $g|_x$. Let $H_{p,r}$ be the set of the elements in $B(r)$ such that

$$\sum_{x \in \mathsf{X}} l(g|_x) \le \eta r + C.$$

Let us denote $\gamma(n) = |B(n)|$. Then for any $\delta > 0$ there exists $K > 1$ such that $\gamma(n) \le K(\lambda + \delta)^n$ for all n.

We have $|H_{p,r}| \ge p|H \cap B(r)|$ and

$$|H_{p,r}| \le \sum_{\sum_{x \in \mathsf{X}} n_x \le \eta r + C, \, x \in \mathsf{X}} \prod_{x \in \mathsf{X}} \gamma(n_x) \le K^{|\mathsf{X}|} \sum_{\sum_{x \in \mathsf{X}} n_x \le \eta r + C} (\lambda + \delta)^{\eta r + C}.$$

It is well known that the number of natural solutions $(n_x)_{x \in \mathsf{X}}$ of the inequality $\sum_{x \in \mathsf{X}} n_x \le S$ is bounded by a polynomial $F(S)$ of degree $|\mathsf{X}| + 1$.

We get for all r big enough

$$(\lambda - \epsilon)^r < |H \cap B(r)| \leq \frac{1}{p}|H_{p,r}| \leq \frac{K^{|X|}}{p} F(\eta r + C)(\lambda + \delta)^{\eta r + C}.$$

Taking root of degree r and passing to the limit we get $\lambda - \epsilon \leq (\lambda + \delta)^\eta$. But ϵ and δ were arbitrary, so we get a contradictory inequality $\lambda \leq \lambda^\eta$. $\square$

Lower bounds on growth of self-similar groups are obtained using different techniques (see, for instance, [**Bar01, BG00a, Leo00, Ers04**]). A quick argument (not giving a satisfactory lower bound, however) is to use the theorem of M. Gromov on groups of polynomial growth, saying that if a group is not virtually nilpotent, then its growth rate is higher than polynomial (see [**Gro81**]). If the group is additionally residually nilpotent (for example, if it is self-similar and the quotient $G/\mathsf{St}(1)$ is a p-group), then by [**Gri90, LM89**] its growth rate is not less than $\exp\sqrt{r}$.

6.13. Matings

We have proved two results, which are converse to each other in some sense. One is Proposition 5.6.1, which says that if a partial self-covering $p^\circ : \mathcal{M}_1^\circ \longrightarrow \mathcal{M}^\circ$ is a pull-back of a partial self-covering $p : \mathcal{M}_1 \longrightarrow \mathcal{M}$, then IMG (p°) is a self-similar subgroup of IMG (p).

There are many classical cases of pull-backs of partial self-coverings. One of the most important is the notion of *polynomial-like maps*, used in renormalization (see, for example, [**DH85b**]). Proposition 5.6.1 shows therefore that renormalization induces an embedding of the iterated monodromy groups.

The other result is Theorem 3.7.1, which says that every inclusion $H \leq G$ of contracting self-similar groups induces a continuous map $\mathcal{J}_H \longrightarrow \mathcal{J}_G$ of their limit spaces.

Let us show how Theorem 3.7.1 can be used to construct and visualize exotic continuous maps. We are going to construct a continuous surjective map from a dendrite Julia set to the Riemann sphere. See a detailed discussion of such maps by J. Milnor in [**JM04**].

Let f, g be two complex polynomials of equal degree d. Take two copies $\mathbb{C}_f$ and $\mathbb{C}_g$ of the complex plane and let the polynomials f and g act on the corresponding copies. Compactify the planes by the circles at infinity (by the points of the form $+\infty \cdot e^{2\pi i \theta}$, $\theta \in \mathbb{R}/\mathbb{Z}$). The action of each of the polynomials is continuously extended to the action $+\infty \cdot e^{2\pi i \theta} \mapsto +\infty \cdot e^{2\pi i d \theta}$ on the circle at infinity. Therefore, if we glue the compactified planes $\mathbb{C}_f$ and $\mathbb{C}_g$ along the circle at infinity using the identification

$$\mathbb{C}_f \ni +\infty \cdot e^{2\pi i \theta} \rightleftarrows +\infty \cdot e^{-2\pi i \theta} \in \mathbb{C}_g,$$

then we get a branched covering of a sphere whose restrictions onto the hemispheres $\mathbb{C}_f$ and $\mathbb{C}_g$ are equal to f and g, respectively.

The obtained branched covering is called the *(formal) mating* of the polynomials f and g (see [**Tan92, Pil03b**]).

If f and g are post-critically finite polynomials of equal degree, then their formal mating is a post-critically finite Thurston map. The iterated monodromy group of the formal mating is obviously generated by the iterated monodromy groups of f and g. More precisely, if we choose a common basepoint $+\infty$ and connect it to its preimages $+\infty \cdot e^{2\pi i k/d}$, $k = 0, 1, \ldots, d-1$, by paths along the circle at infinity, then

216   6. EXAMPLES AND APPLICATIONS

we can compute the standard actions of IMG (f) and IMG (g) using these paths. The standard action of the iterated monodromy group of the mating will then be the group generated by the standard actions of IMG (f) and IMG (g).

As an example, consider the polynomial $z^2 + i$ and mate it with itself. The corresponding branched covering f of the sphere will have an obstruction. To see this consider the external ray $R_{1/3}$ landing on $i-1$ and the ray $R_{2/3}$ landing on $-i$. Let $\mathbb{C}^{(1)}$ and $\mathbb{C}^{(2)}$ be the two hemispheres on which the polynomial z^2+i acts. Then the ray $R_{1/3}^{(1)}$ in $\mathbb{C}^{(1)}$ has a common point $\mathbb{C}^{(1)} \ni +\infty \cdot e^{\frac{1}{3}2\pi i} \rightleftarrows +\infty \cdot e^{\frac{2}{3}2\pi i} \in \mathbb{C}^{(2)}$ with the ray $R_{2/3}^{(2)}$, and similarly, the ray $R_{2/3}^{(1)}$ in $\mathbb{C}^{(1)}$ will have a common point with the ray $R_{1/3}^{(2)}$ in $\mathbb{C}^{(2)}$. Consider the closed simple curves γ_1 and γ_2 going around the obtained curves $R_{1/3}^{(1)} \cup R_{2/3}^{(2)}$ and $R_{2/3}^{(1)} \cup R_{1/3}^{(2)}$, respectively. Then $f(\gamma_1) = \gamma_2$, $f(\gamma_2) = \gamma_1$ and the degrees of the respective mappings of closed curves are equal to one. Consequently, $\{\gamma_1, \gamma_2\}$ is a Levy cycle. Levy cycles are essentially the only obstructions which a mating of two quadratic polynomials may have (see [**Tan92**]).

It is possible, however, to remove this obstruction. We just have to contract the curves $R_{1/3}^{(1)} \cup R_{2/3}^{(2)}$ and $R_{2/3}^{(1)} \cup R_{1/3}^{(2)}$ to points z_1 and z_2. We obtain then a degree 2 Thurston map $\widehat{f}$ with two critical points (copies $0^{(1)}$ and $0^{(2)}$ of 0 in each of the hemispheres) and four post-critical points: the images $i^{(1)}$, $i^{(2)}$ of i in both hemispheres and the points z_1 and z_2. We have $\widehat{f}(0^{(1)}) = i^{(1)}$, $\widehat{f}(0^{(2)}) = i^{(2)}$, $\widehat{f}(i^{(1)}) = z_1$, $\widehat{f}(i^{(2)}) = z_2$, $\widehat{f}(z_1) = z_2$ and $\widehat{f}(z_2) = z_1$.

Let us compute the iterated monodromy groups of the Thurston maps f and $\widehat{f}$. We take $0^{(1)}$ as the basepoint and connect it to its preimages $\left(\frac{i-1}{2}\right)^{(1)}$ and $\left(\frac{-i+1}{2}\right)^{(1)}$ by the lines not intersecting the external rays $R_{1/6}^{(1)}, R_{1/3}^{(1)}$ and $R_{2/3}^{(1)}$. Let a_1, b_1, c_1 be small loops around the points $i^{(1)}$, $(-1+i)^{(1)}$ and $(-i)^{(1)}$, respectively, connected to the basepoint by the lines not intersecting the external rays. By a_2, b_2, c_2 we denote the small loops around the points i_2, $(-1+i)_2$ and $(-i)_2$, respectively, connected to the basepoint as shown in Figure 6.24. Here the connecting paths go near the punctures of the plane $\mathbb{C}^{(1)}$, leaving them on their right-hand side, then go along the respective external rays $R_{5/6}^{(1)}$, $R_{2/3}^{(1)}$, $R_{1/3}^{(1)}$ to the circle at infinity, and after that go along the external rays $R_{1/6}^{(2)}$, $R_{1/3}^{(2)}$ and $R_{2/3}^{(2)}$ in the plane $\mathbb{C}^{(2)}$ to the small loops around the respective punctures.

The standard action of the loops a_1, b_1, c_1 is given (see Subsection 6.12.5) by

$$a_1 = \sigma, \quad b_1 = (c_1, a_1), \quad c_1 = (1, b_1),$$

where $\sigma \in \mathfrak{S}(\mathsf{X})$ is the transposition.

The preimages of the loops a_2, b_2, c_2 under the action of the map f are shown in Figure 6.25 (the paths connecting the basepoint to its preimages are shown by dotted lines). It shows that the standard action of these loops is given by

$$a_2 = \sigma(c_2 b_1 a_1, a_1 b_1 c_2), \quad b_2 = (a_2, c_2), \quad c_2 = (b_2, 1).$$

The iterated monodromy group IMG (f) is the group generated by the set of transformations $\{a_1, b_1, c_1, a_2, b_2, c_2\}$. Note that each of these generators is of order 2.

The iterated monodromy group IMG $\left(\widehat{f}\right)$ is the subgroup of IMG (f) generated by the loops, which do not intersect the lines $R_{1/3}^{(1)} \cup R_{2/3}^{(2)}$ and $R_{2/3}^{(1)} \cup R_{1/3}^{(2)}$. Therefore,

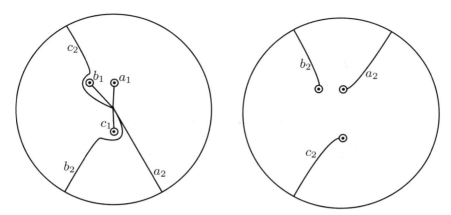

FIGURE 6.24. Mating two copies of $z^2 + i$

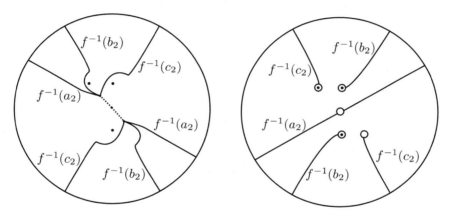

FIGURE 6.25. Preimages of the loops a_2, b_2, c_2

$\mathrm{IMG}\left(\widehat{f}\right) = \langle a_1, a_2, c_1b_2, b_1c_2 \rangle$. Let us denote $c = c_1b_2 = (a_2, b_1c_2)$ and $b = b_1c_2 = (c_1b_2, a_1)$. Then we get recursions

$$a_1 = \sigma, \qquad a_2 = \sigma(b^{-1}a_1, a_1b) = \sigma(ba_1, a_1b)$$
$$c = (a_2, b), \qquad b = (c, a_1).$$

LEMMA 6.13.1. *If $b_1(v) \neq v$ for $v \in X^*$, then $b_1(v) = b(v)$. If $c_1(v) \neq v$, then $c_1(v) = c(v)$. If $b_2(v) \neq v$, then $b_2(v) = c(v)$. If $c_2(v) \neq v$, then $c_2(v) = b(v)$.*

The simplicial Schreier graph of the action of the group $\langle a_1, b_1, c_1 \rangle$ on X^n is a subgraph of the simplicial Schreier graph of the action of $\mathrm{IMG}\left(\widehat{f}\right) = \langle a_1, a_2, b, c \rangle$ on X^n.

The simplicial Schreier graphs of the actions on X^n of the groups $\mathrm{IMG}(f) = \langle a_1, b_1, c_1, a_2, b_2, c_2 \rangle$ and $\mathrm{IMG}\left(\widehat{f}\right) = \langle a_1, a_2, b, c \rangle$ coincide.

PROOF. It follows from the recursion for a_1, b_1, c_1 that if $b_1(v) \neq v$, then v is of the form $(01)^n 1u$ for some $n \geq 0$. Then $b_1((01)^n 1u) = (01)^n 1a_1(u) = b((01)^n 1u)$. Similar arguments work for the other cases. □

Corollary 3.6.7 and Lemma 6.13.1 (see also Theorem 3.7.1) imply that the identical map $\mathsf{X}^{-\omega} \longrightarrow \mathsf{X}^{-\omega}$ induces a homeomorphism $\mathcal{J}_{\mathrm{IMG}(\widehat{f})} \longrightarrow \mathcal{J}_{\mathrm{IMG}(f)}$ and a surjective continuous map $\mathcal{J}_{z^2+i} = \mathcal{J}_{\mathrm{IMG}(z^2+i)} \longrightarrow \mathcal{J}_{\mathrm{IMG}(\widehat{f})}$.

We see from the dynamics of the Thurston map $\widehat{f}$ on its post-critical set that the corresponding Thurston orbifold has four singular points $i^{(1)}$, $i^{(2)}$, z_1 and z_2, all of them having isotropy groups of order two. Hence it is the Euclidean orbifold $(2, 2, 2, 2)$. Consequently, the fundamental group of the orbifold is isomorphic to the group of affine transformations $z \mapsto \pm z + r$ of $\mathbb{Z}^2$ (see Subsection 6.3.2, page 170).

The generators a_1, a_2, b, c correspond to simple loops around singular points; therefore they correspond to elements of order two in the fundamental group, i.e., to affine transformations of the form $z \mapsto -z + r$.

More detailed information is contained in the following description of the partial self-covering $\widehat{f}$.

PROPOSITION 6.13.2. *Let G be the group of affine transformations of $\mathbb{C}$ generated by $A_1 : z \mapsto -z - \lambda/2$, $B : z \mapsto -z - \lambda/2 - 1$ and $C : -z + \lambda/2 - 1$, where $\lambda = \frac{1}{2} + \frac{i\sqrt{7}}{2}$. Let ϕ be the virtual endomorphism of G mapping an affine transformation $z \mapsto (-1)^k z + \beta$ to the transformation $z \mapsto (-1)^k z + \lambda^{-1} \beta$. Then the map $a_1 \mapsto A_1$, $b \mapsto B$, $c \mapsto C$ extends to an isomorphism $\psi : \mathrm{IMG}\left(\widehat{f}\right) \longrightarrow G$. The isomorphism ψ agrees with the standard action of $\mathrm{IMG}\left(\widehat{f}\right)$ on X^* and the self-similar action of G defined by the virtual endomorphism ϕ and the digit set $\{id, A_1\}$. The self-covering $\widehat{f}$ is Thurston equivalent to the self-covering of $\mathbb{C}/G$ induced by the expanding automorphism $z \mapsto \lambda z$ of $\mathbb{C}$.*

PROOF. We have $\lambda^2 - \lambda + 2 = 0$ and $\lambda^{-1} = \frac{1-\lambda}{2}$.

Note that the affine transformations A_1, B and C are of order 2 and hence the subgroup G_1 of G generated by $X = A_1 C$ and $Y = A_1 B$ has index 2 in G. The transformation X is equal to

$$z \mapsto -(-z + \lambda/2 - 1) - \lambda/2 = z - \lambda + 1,$$

and Y is equal to

$$z \mapsto -(-z - \lambda/2 - 1) - \lambda/2 = z + 1.$$

Consequently, G_1 is the group of translations $z \mapsto z + r$, where $r \in \Gamma = \mathbb{Z}[\lambda]$. Note that $\lambda \mathbb{Z}[\lambda]$ is a subgroup of index 2 in $\mathbb{Z}[\lambda]$, since $\lambda^2 - \lambda + 2 = 0$. The complement of G_1 in G is equal to $G_1 A_1$; therefore it is equal to the set of affine transformations of the form $z \mapsto -z - \lambda/2 + r$, where $r \in \mathbb{Z}[\lambda]$ is arbitrary.

We have $cba_1 a_2 = 1$ (since the loop $cba_1 a_2$ is contractible on the sphere minus the post-critical set). The corresponding equality can also be seen from the wreath recursion for the iterated monodromy group, since

$$cba_1 a_2 = (a_2, b)(c, a_1)\sigma\sigma(ba_1, a_1 b) = (a_2 cba_1, 1).$$

It follows that $\mathrm{IMG}\left(\widehat{f}\right)$ is generated by a_1, b and c.

It is sufficient to check that the recursions defining the transformations a_1, b and c agree with their interpretation as affine transformations.

Let ϕ' be the virtual endomorphism associated to the self-similar action of $\mathrm{IMG}\left(\widehat{f}\right)$ and the first coordinate of the wreath recursion (i.e., $\phi'(g) = g_0$ if $g =$

(g_0, g_1)). We have $a_1 = \sigma$; therefore $g = (\phi'(g), \phi'(a_1ga_1))$ if g is inactive and $g = \sigma(\phi'(a_1g), \phi'(ga_1))$ otherwise.

Let ϕ be the virtual endomorphism of G induced by the automorphism $z \mapsto \lambda^{-1}z$ of $\mathbb{C}$. If $g \in G$ is a transformation $z \mapsto (-1)^k z + r$, then $\phi(g)$ is the transformation $z \mapsto (-1)^k z + \lambda^{-1}r$. This implies that $A_1 \notin \mathrm{Dom}\,\phi$, since $\phi(A_1)$ is the transformation $z \mapsto -z - 1/2$, which does not belong to G. Let us take then $\{id, A_1\}$ as a digit set.

Let us compute the recursions defining the self-similar action of G on X^* associated to ϕ and the chosen digit set. Recall that this recursion is given by $g = (\phi(g), \phi(A_1gA_1))$ if $g \in \mathrm{Dom}\,\phi$ and $g = \sigma(\phi(A_1g), \phi(gA_1))$ if $g \notin \mathrm{Dom}\,\phi$ (see Proposition 2.5.10).

We get immediately that $A_1 = \sigma(1,1) = \sigma$. The transformations B and C belong to $\mathrm{Dom}\,\phi$. The affine transformation $\phi(B)$ is equal to

$$z \mapsto -z + \lambda^{-1}\left(-\frac{\lambda}{2} - 1\right) = -z - \frac{1}{2} - \lambda^{-1} = -z + \frac{\lambda}{2} - 1,$$

i.e., to C, while $\phi(A_1BA_1)$ is equal to

$$z \mapsto -z + \lambda^{-1}\left(-\lambda + \frac{\lambda}{2} + 1\right) = -z - \frac{\lambda}{2},$$

i.e., to A_1. Consequently,

$$B = (C, A_1),$$

which agrees with the recursion for b in $\mathrm{IMG}\left(\widehat{f}\right)$.

The element $\phi(C)$ is equal to the affine transformation

$$z \mapsto -z + \lambda^{-1}\left(\frac{\lambda}{2} - 1\right) = -z + \frac{\lambda}{2}.$$

Let us denote this transformation by A_2. We have that the transformation CBA_1A_2 is equal to

$$z \mapsto -\left(-\left(-\left(-z + \frac{\lambda}{2}\right) - \frac{\lambda}{2}\right) - \frac{\lambda}{2} - 1\right) + \frac{\lambda}{2} - 1$$

$$= z - \frac{\lambda}{2} - \frac{\lambda}{2} + \frac{\lambda}{2} + 1 + \frac{\lambda}{2} - 1 = z;$$

therefore $A_2 = A_1BC$.

The element $\phi(A_1CA_1)$ is equal to the affine transformation

$$z \mapsto -z + \lambda^{-1}\left(-\lambda - \frac{\lambda}{2} + 1\right) = -z - \frac{\lambda}{2} - 1,$$

i.e., to B. We have therefore

$$C = (A_1BC, B),$$

which also agrees with the recursion for c in $\mathrm{IMG}\left(\widehat{f}\right)$. $\square$

Proposition 6.13.2 makes it possible to describe the Schreier graphs of the action of the group $G = \mathrm{IMG}\left(\widehat{f}\right)$ on the levels of the tree X^*, i.e., the adjacency graphs of the tiles of the nth level of $\mathcal{J}_G$. These graphs converge as n goes to infinity to the limit space of G, i.e., to the Julia set of the rational function equivalent to $\widehat{f}$, which is homeomorphic to the sphere.

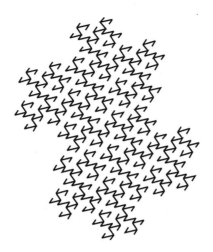

FIGURE 6.26. An approximation of the map $\mathcal{J}_{z^2+i} \longrightarrow \widehat{\mathbb{C}}$

Let us denote now (in view of Proposition 6.13.2) by $\widehat{f}$ also the self-covering of the orbifold $\mathbb{C}/G$ induced by the map $z \mapsto \lambda z$ on $\mathbb{C}$. The limit space $\mathbb{C}/G$ of G is homeomorphic to the quotient of the space $\mathsf{X}^{-\omega}$ by the asymptotic equivalence relation. The quotient map $\mathsf{X}^{-\omega} \longrightarrow \mathbb{C}/G$ maps a sequence $\ldots x_2 x_1$ to the image of the point $\lim_{n \to \infty} \zeta \otimes x_n \ldots x_1 \in \mathcal{X}_G$ in $\mathcal{J}_G$. Let $\mathsf{X} = \{0, 1\}$, where 0 corresponds to the coset representative id and 1 corresponds to A_1 (i.e., $0 = \phi(1)1$ and $1 = \phi(A_1)1$ in $\phi(G)G$).

The space $\mathcal{X}_G$ is homeomorphic, by Theorem 6.1.6, to the complex plane $\mathbb{C}$ with the original action of G on it. We have $\zeta \otimes 0 = \lambda^{-1} \zeta$ and $\zeta \otimes 1 = \lambda^{-1}(-\zeta - \lambda/2) = -\lambda^{-1} \zeta - 1/2$.

The self-similar action of (G, X) is recurrent (i.e., the associated virtual endomorphism is surjective), so (by Corollary 2.8.5) every element of the bimodule $\mathfrak{M}^{\otimes n}$ can be written in the form $\phi^n(g_1) g_2$. Here, as usual, $\mathfrak{M} = \mathsf{X} \cdot G$ is the self-similarity bimodule. We know that $\zeta \otimes \phi^n(g_1) g_2 = \phi^n(\zeta \cdot g_1) g_2$ (see Theorem 6.1.7); therefore the images in $\mathcal{J}_G = \mathbb{C}/G$ of the points $\zeta \otimes v$, $v \in \mathfrak{M}^{\otimes n}$, are of the form $\lambda^{-n}(\zeta + r)$ and $\lambda^{-n}(-\zeta - \lambda/2 + r)$, where $r \in \mathbb{Z}[\lambda]$.

Let us choose a point ζ which belongs to the digit tile $\mathcal{T} \subset \mathcal{X}_G = \mathbb{C}$ (for example, $\zeta = 0$). Then the point $\zeta \otimes v$, where $v \in \mathfrak{M}^{\otimes n}$, belongs to the tile $\mathcal{T} \otimes v$, and the image of $\zeta \otimes v$ in $\mathcal{J}_G$ belongs to the tile $\mathcal{T}_v$ of $\mathcal{J}_G$. Let us draw the Schreier graphs $\Gamma(G, \mathsf{X}^n)$ of the action of G on the nth level of the tree X^* on the sphere $\mathbb{C}/G$. The vertex $v \in \mathsf{X}^n$ of the graph will be identified with the image of $\zeta \otimes v$. This will make our graphs agree with the adjacency of the tiles of $\mathcal{J}_G$. (Actually, the adjacency graph is the Schreier graph defined by the generating set equal to the nucleus, but the graphs are not changed "too much" when we use a different generating set.)

A vertex $v \in \mathsf{X}^n$ of the Schreier graph $\Gamma(G, \mathsf{X}^n)$ is connected to the vertices of the form $g(v)$, where g belongs to the generating set of G. The images in $\mathcal{J}_G$ of the points $\zeta \otimes g(v)$ and $\zeta \cdot g \otimes v = \zeta \otimes g \cdot v$ are equal. Consequently, we have the following edges of the Schreier graph $\Gamma(G, \mathsf{X}^n)$:

- The vertex $\lambda^{-n}(\zeta + r)$ is connected to
$\lambda^{-n}(-\zeta - \lambda/2 + r), \lambda^{-n}(-\zeta - \lambda/2 - 1 + r)$, and to $\lambda^{-n}(-\zeta + \lambda/2 - 1 + r)$.
- The vertex $\lambda^{-n}(-\zeta - \lambda/2 + r)$ is connected to
$\lambda^{-n}(\zeta + r), \lambda^{-n}(\zeta + 1 + r)$, and to $\lambda^{-n}(\zeta - \lambda + 1 + r)$.

If we take $\zeta = 0$, then the Schreier graph $\Gamma(G, \mathsf{X}^n)$ is the image under the composition of $z \mapsto \lambda^{-n}z$ and the quotient map $\mathbb{C} \longrightarrow \mathbb{C}/G$ of the graph with the set of vertices $\mathbb{Z}[\lambda] \cup (\mathbb{Z}[\lambda] - \lambda/2) \subset \mathbb{C}$ and the set of edges

$$\{r, r - \lambda/2\}, \quad \{r, r - \lambda/2 - 1\}, \quad \{r, r + \lambda/2 - 1\}.$$

The Schreier graphs $\Gamma(\langle a_1, b_1, c_1 \rangle, \mathsf{X}^n)$ are, by Lemma 6.13.1, subgraphs of the Schreier graphs $\Gamma(G, \mathsf{X}^n)$. The graphs $\Gamma(\langle a_1, b_1, c_1 \rangle, \mathsf{X}^n)$ describe the adjacency of the tiles of the nth level of the limit space $\mathcal{J}_{\mathrm{IMG}(z^2+i)}$. Therefore they show how the Julia set $\mathcal{J}_{z^2+i} = \mathcal{J}_{\mathrm{IMG}(z^2+i)}$ is mapped onto the sphere $\mathbb{C}/G$ in the same way as, for example, the classical approximations of the Peano curve show how the segment is mapped onto the square. See Figure 6.26 for the graph $\Gamma(\langle a_1, b_1, c_1 \rangle, \mathsf{X}^9)$ drawn as a subgraph of the graph $\Gamma(G, \mathsf{X}^9)$ on the plane $\mathbb{C}$.

Bibliography

[Ale83] S. V. Aleshin, *A free group of finite automata*, Moscow University Mathematics Bulletin **38** (1983), 10–13.

[Ban91] Christoph Bandt, *Self-similar sets. V: Integer matrices and fractal tilings of $\mathbb{R}^n$*, Proc. Am. Math. Soc. **112** (1991), no. 2, 549–562.

[Bar98] Laurent Bartholdi, *The growth of Grigorchuk's torsion group*, Internat. Math. Res. Notices **20** (1998), 1049–1054.

[Bar01] _____, *Lower bounds on the growth of Grigorchuk's torsion group*, Internat. J. Algebra Comput. **11** (2001), no. 1, 73–88.

[Bar03a] _____, *Endomorphic presentations of branch groups*, Journal of Algebra **268** (2003), no. 2, 419–443.

[Bar03b] _____, *A Wilson group of non-uniformly exponential growth*, C. R. Acad. Sci. Paris. Sér. I Math. **336** (2003), no. 7, 549–554.

[Bau93] Gilbert Baumslag, *Topics in combinatorial group theory*, Lectures in Mathematics, ETH Zürich, Birkhäuser Verlag, Basel, 1993.

[Bea91] Alan F. Beardon, *Iteration of rational functions. Complex analytic dynamical systems*, Graduate Texts in Mathematics, vol. 132, Springer-Verlag. New York, 1991.

[Bel03] Igor Belegradek, *On co-Hopfian nilpotent groups*, Bull. London Math. Soc. **35** (2003), 805–811.

[BFH92] Ben Bielefeld, Yuval Fisher, and John H. Hubbard, *The classification of critically preperiodic polynomials as dynamical systems*, Journal of the A.M.S. **5** (1992), no. 4, 721–762.

[BG00a] Laurent Bartholdi and Rostislav I. Grigorchuk, *Lie methods in growth of groups and groups of finite width*, Computational and Geometric Aspects of Modern Algebra (Michael Atkinson et al., eds.), London Math. Soc. Lect. Note Ser., vol. 275, Cambridge Univ. Press, Cambridge, 2000, pp. 1–27.

[BG00b] _____, *On the spectrum of Hecke type operators related to some fractal groups*, Proceedings of the Steklov Institute of Mathematics **231** (2000), 5–45.

[BG02] _____, *On parabolic subgroups and Hecke algebras of some fractal groups*, Serdica Math. J. **28** (2002), 47–90.

[BGN03] Laurent Bartholdi, Rostislav Grigorchuk, and Volodymyr Nekrashevych, *From fractal groups to fractal sets*, Fractals in Graz 2001. Analysis – Dynamics – Geometry – Stochastics (Peter Grabner and Wolfgang Woess, eds.), Birkhäuser Verlag, Basel, Boston, Berlin, 2003, pp. 25–118.

[BGŠ03] Laurent Bartholdi, Rostislav I. Grigorchuk, and Zoran Šuniḱ, *Branch groups*, Handbook of Algebra, Vol. 3, North-Holland, Amsterdam, 2003, pp. 989–1112.

[BH99] Martin R. Bridson and André Haefliger, *Metric spaces of non-positive curvature*, Grundlehren der Mathematischen Wissenschaften, vol. 319, Springer, Berlin, 1999.

[Bha95] Meenaxi Bhattacharjee, *The ubiquity of free subgroups in certain inverse limits of groups*, J. Algebra **172** (1995), 134–146.

[BJ99] Ola Bratelli and Palle E. T. Jorgensen, *Iterated function systems and permutation representations of the Cuntz algebra*, vol. 139, Memoirs of the American Mathematical Society, no. 663, A. M. S., Providence, Rhode Island, 1999.

[BL01] Hyman Bass and Alexander Lubotzky, *Tree lattices*, Progress in Mathematics, vol. 176, Birkhäuser Boston Inc., Boston, MA, 2001, With appendices by Bass, L. Carbone, Lubotzky, G. Rosenberg and J. Tits.

[BN03] Evgen Bondarenko and Volodymyr Nekrashevych, *Post-critically finite self-similar groups*, Algebra and Discrete Mathematics **2** (2003), no. 4, 21–32.

[BN05a] Laurent Bartholdi and Volodymyr V. Nekrashevych, *Iterated monodromy groups of quadratic polynomials* (preprint), 2005.

[BN05b] Laurent Bartholdi and Volodymyr V. Nekrashevych, *Thurston equivalence of topological polynomials* (preprint), 2005.

[BORT96] Hyman Bass, Maria Victoria Otero-Espinar, Daniel Rockmore, and Charles Tresser, *Cyclic renormalization and automorphism groups of rooted trees*, Lecture Notes in Mathematics, vol. 1621, Springer-Verlag, Berlin, 1996.

[Bou71] Nicolas Bourbaki, *Éléments de mathématique. Topologie générale. Chapitres 1 à 4*, Hermann, Paris, 1971.

[BP04] Kai-Uwe Bux and Rodrigo Pérez, *On the growth of iterated monodromy groups* (preprint), 2004.

[BS97] Andrew M. Brunner and Said N. Sidki, *On the automorphism group of the one-rooted binary tree*, J. Algebra **195** (1997), 465–486.

[BS98] _____, *The generation of $GL(n,Z)$ by finite state automata*, Internat. J. Algebra Comput. **8** (1998), no. 1, 127–139.

[BS02a] Henk Bruin and Dierk Schleicher, *Symbolic dynamics of quadratic polynomials*, Institut Mittag-Leffler, Report No. 7, 2001/2002.

[BS02b] Andrew M. Brunner and Said N. Sidki, *Wreath operations in the group of automorphisms of the binary tree*, J. Algebra **257** (2002), 51–64.

[BV] Laurent Bartholdi and Bálint Virág, *Amenability via random walks*, to appear in Duke Math Journal.

[CDP90] Michel Coornaert, Thomas Delzant, and Athanase Papadopoulos, *Geometrie et theorie des groupes: Les groupes hyperboliques de Gromov*, Lectures Notes in Mathematics, vol. 1441, Springer Verlag, 1990.

[CJ85] Alain Connes and Vaughan Jones, *Property T for von Neumann algebras*, Bull. London Math. Soc. **17** (1985), 57–62.

[Con94] Alain Connes, *Noncommutative geometry*, San Diego, CA: Academic Press, 1994.

[CSGH99] Tullio Ceccherini-Silberstein, Rostislav I. Grigorchuk, and Pierre de la Harpe, *Amenability and paradoxical decompositions for pseudogroups and discrete metric spaces*, Trudy Mat. Inst. Steklov. **224** (1999), no. Algebra. Topol. Differ. Uravn. i ikh Prilozh., 68–111, Dedicated to Academician Lev Semenovich Pontryagin on the occasion of his 90th birthday (Russian).

[Day57] Mahlon M. Day, *Amenable semigroups*, Illinois J. Math. **1** (1957), 509–544.

[DH84] Adrien Douady and John H. Hubbard, *Étude dynamiques des polynômes complex. (Première partie)*, Publications Mathematiques d'Orsay, vol. 02, Université de Paris-Sud, 1984.

[DH85a] _____, *Étude dynamiques des polynômes complex. (Deuxième partie)*, Publications Mathematiques d'Orsay, vol. 04, Université de Paris-Sud, 1985.

[DH85b] Adrien Douady and John H. Hubbard, *On the dynamics of polynomial-like mappings*, Ann. Sci. Éc. Norm. Supér. IV. Sér. **18** (1985), 287–343.

[DH93] _____, *A proof of Thurston's topological characterization of rational functions*, Acta Math. **171** (1993), no. 2, 263–297.

[Dra00] Alexander N. Dranishnikov, *Asymptotic topology*, Uspekhi Mat. Nauk **55** (2000), no. 6(336), 71–116.

[Eil74] Samuel Eilenberg, *Automata, languages and machines*, vol. A, Academic Press, New York, London, 1974.

[Eng68] Ryszard Engelking, *Outline of general topology*, Amsterdam: North-Holland Publishing Company, 1968.

[Eng77] _____, *General topology*, Monografie Matematyczne, vol. 60, Państwowe Wydawnictwo Naukove, Warszawa, 1977.

[Ers04] Anna Erschler, *Boundary behaviour for groups of subexponential growth*, Annals of Mathematics **160** (2004), 1183–1210.

[FG91] Jacek Fabrykowski and Narain D. Gupta, *On groups with sub-exponential growth functions. II*, J. Indian Math. Soc. (N.S.) **56** (1991), no. 1-4, 217–228.

[For81] Otto Forster, *Lectures on Riemann surfaces*, Graduate Texts in Mathematics, vol. 81, New York – Heidelberg – Berlin: Springer-Verlag, 1981.

[Fra70] John M. Franks, *Anosov diffeomorphisms*, Global Analysis, Berkeley, 1968, Proc. Symp. Pure Math., vol. 14, Amer. Math. Soc., 1970, pp. 61–93.

BIBLIOGRAPHY

[Gan59] F. R. Gantmacher, *The theory of matrices*. Vols. 1, 2, translated by K. A. Hirsch, Chelsea Publishing Co., New York, 1959.

[Gel95] Götz Gelbrich, *Self-similar tilings and expanding homomorphisms of groups*, Arch. Math. **65** (1995), no. 6, 481–491.

[GH90] Étienne Ghys and Pierre de la Harpe, *Sur les groupes hyperboliques d'après Mikhael Gromov*, Progress in Mathematics, vol. 83, Birkhäuser Boston Inc., Boston, MA, 1990, Papers from the Swiss Seminar on Hyperbolic Group held in Bern, 1988.

[GLSŻ00] Rostislav I. Grigorchuk, Peter Linnell, Thomas Schick, and Andrzej Żuk, *On a question of Atiyah*, C. R. Acad. Sci. Paris Sér. I Math. **331** (2000), no. 9, 663–668.

[GM03] Yair Glasner and Shahar Mozes, *Automata and square complexes*, to appear in Geom. Dedicata, 2005.

[GNS00] Rostislav I. Grigorchuk, Volodymyr V. Nekrashevich, and Vitaliĭ I. Sushchanskii, *Automata, dynamical systems and groups*, Proceedings of the Steklov Institute of Mathematics **231** (2000), 128–203.

[GNS01] Piotr W. Gawron, Volodymyr V. Nekrashevych, and Vitaly I. Sushchansky, *Conjugation in tree automorphism groups*, Int. J. of Algebra and Computation **11** (2001), no. 5, 529–547.

[Gri80] Rostislav I. Grigorchuk, *On Burnside's problem on periodic groups*, Functional Anal. Appl. **14** (1980), no. 1, 41–43.

[Gri83] _____, *On the Milnor problem of group growth*, Dokl. Akad. Nauk SSSR **271** (1983), no. 1, 30–33.

[Gri85] _____, *Degrees of growth of finitely generated groups and the theory of invariant means*, Math. USSR Izv. **25** (1985), no. 2, 259–300.

[Gri88] _____, *Semigroups with cancellations of polynomial growth*, Mat. Zametki **43** (1988), no. 3, 305–319, 428.

[Gri90] _____, *On the Hilbert-Poincaré series of graded algebras that are associated with groups*, Math. USSR-Sb. **66** (1990), no. 1, 211–229.

[Gri98] _____, *An example of a finitely presented amenable group that does not belong to the class EG*, Mat. Sb. **189** (1998), no. 1, 79–100.

[Gri99] _____, *On the system of defining relations and the Schur multiplier of periodic groups generated by finite automata*, Groups St. Andrews 1997 in Bath, I, Cambridge Univ. Press, Cambridge, 1999, pp. 290–317.

[Gri00] _____, *Just infinite branch groups*, New Horizons in pro-p Groups (Aner Shalev, Marcus P. F. du Sautoy, and Dan Segal, eds.), Progress in Mathematics, vol. 184, Birkhäuser Verlag, Basel, 2000, pp. 121–179.

[Gro81] Mikhael Gromov, *Groups of polynomial growth and expanding maps*, Publ. Math. I. H. E. S. **53** (1981), 53–73.

[Gro87] _____, *Hyperbolic groups*, Essays in Group Theory (S. M. Gersten, ed.), M.S.R.I. Pub., no. 8, Springer, 1987, pp. 75–263.

[Gro93] _____, *Asymptotic invariants of infinite groups*, Geometric Group Theory, Vol. 2 (Sussex, 1991), London Math. Soc. Lecture Note Ser., vol. 182, Cambridge Univ. Press, Cambridge, 1993, pp. 1–295.

[GS83] Narain D. Gupta and Said N. Sidki, *On the Burnside problem for periodic groups*, Math. Z. **182** (1983), 385–388.

[GW03] Rostislav I. Grigorchuk and John S. Wilson, *The uniqueness of the actions of certain branch groups on rooted trees*, Geom. Dedicata **100** (2003), 103–116.

[GŻ01] Rostislav I. Grigorchuk and Andrzej Żuk, *The lamplighter group as a group generated by a 2-state automaton and its spectrum*, Geom. Dedicata **87** (2001), no. 1–3, 209–244.

[GŻ02a] _____, *On a torsion-free weakly branch group defined by a three state automaton*, Internat. J. Algebra Comput. **12** (2002), no. 1, 223–246.

[GŻ02b] Rostislav I. Grigorchuk. and Andrzej Żuk, *Spectral properties of a torsion-free weakly branch group defined by a three state automaton*, Computational and Statistical Group Theory (Las Vegas, NV/Hoboken, NJ, 2001), Contemp. Math., vol. 298, Amer. Math. Soc., Providence, RI, 2002, pp. 57–82.

[Hae01] André Haefliger, *Groupoids and foliations*, Groupoids in Analysis, Geometry, and Physics. AMS-IMS-SIAM joint summer research conference, University of Colorado, Boulder, CO, USA, June 20-24, 1999 (Arlan Ramsay et al., eds.), Contemp. Math, vol. 282, Providence, RI: A.M.S., 2001, pp. 83–100.

BIBLIOGRAPHY

[Har00] Pierre de la Harpe, *Topics in geometric group theory*, University of Chicago Press, 2000.

[Hir70] Morris W. Hirsch, *Expanding maps and transformation groups*, Global Analysis, Proc. Sympos. Pure Math., vol. 14, American Math. Soc., Providence, Rhode Island, 1970, pp. 125–131.

[HR32] H. Hopf and W. Rinow, *Über den Begriff der vollständigen differentialgeometrischen Fläche*, Comment. Math. Helv **3** (1932), 209–225.

[HS94] John H. Hubbard and Dierk Schleicher, *The spider algorithm*, Complex Dynamical Systems. The Mathematics Behind the Mandelbrot and Julia Sets (Robert L. Devaney, ed.), Proceedings of Symposia in Applied Mathematics, vol. 49, 1994, pp. 155–180.

[JM04] John J. Milnor, *Pasting together Julia sets: a worked out example of mating*, Experiment. Math. **13** (2004), no. 1, 55–92.

[JS97] Vaughan Jones and V.S. Sunder, *Introduction to subfactors*, London Mathematical Society Lecture Note Series, vol. 234, Cambridge University Press, 1997.

[Kai03] Vadim A. Kaimanovich, *Random walks on Sierpiński graphs: hyperbolicity and stochastic homogenization*, Fractals in Gratz 2001 (W. Woess, ed.), Trends Math., Birkhäuser, Basel, 2003, pp. 145–183.

[Kam01] Atsushi Kameyama, *The Thurston equivalence for postcritically finite branched coverings*, Osaka J. Math. **38** (2001), no. 3, 565–610.

[Kel00] Karsten Keller, *Invariant factors, Julia equivalences and the (abstract) Mandelbrot set*, Lecture Notes in Mathematics, vol. 1732, Springer, 2000.

[Ken92] Richard Kenyon, *Self-replicating tilings*, Symbolic Dynamics and Its Applications (P. Walters, ed.), Contemp. Math., vol. 135, Amer. Math. Soc., Providence, RI, 1992, pp. 239–264.

[Kig92] Jun Kigami, *Laplacians on self-similar sets — analysis on fractals*, Transl., Ser. 2, Amer. Math. Soc. 161, 75-93 (1994); translation from Sugaku 44, No.1, 13-28 (1992) (1992).

[Kig01] ———, *Analysis on fractals*, Cambridge Tracts in Mathematics, vol. 143, Cambridge University Press, 2001.

[KL05] Vadim A. Kaimanovich and Mikhail Lyubich, *Conformal and harmonic measures on laminations associated with rational maps*, vol. 173, Memoirs of the A.M.S., no. 820, A.M.S., Providence, Rhode Island, 2005.

[KM79] M. I. Kargapolov and Ju. I. Merzljakov, *Fundamentals of the theory of groups*, Graduate Texts in Mathematics, vol. 62, Springer-Verlag, New York, Heidelberg, Berlin, 1979.

[Knu69] Donald E. Knuth, *The art of computer programming, Vol. 2, Seminumerical algorithms*, Addison-Wesley Publishing Company, 1969.

[Kur61] Kazimierz Kuratowski, *Topologie*, vol. II, Warszawa, 1961.

[Lan87] Serge Lang, *Elliptic functions. Second edition*, Graduate Texts in Mathematics, vol. 112, Springer-Verlag, New York, 1987.

[Lat18] S. Lattès, *Sur l'itération des substitutions rationelles et les fonctions de Poincaré*, C. R. Acad. Sci. Paris **166** (1918), 26–28.

[Lav99] Yaroslav Lavreniuk, *Automorphisms of wreath branch groups*, Visnyk Kyivskogo Universytetu (1999), no. 1, 50–57 (in Ukrainian).

[Leo00] Yuriĭ G. Leonov, *On a lower bound for the growth function of the Grigorchuk group*, Mat. Zametki **67** (2000), no. 3, 475–477.

[Lin90] Tom Lindstrøm, *Brownian motion on nested fractals*, Mem. Am. Math. Soc. **420** (1990), 128 pp.

[LM89] Alexander Lubotzky and Avinoam Mann, *Residually finite groups of finite rank*, Math. Proc. Cambridge Philos. Soc. **106** (1989), no. 3, 385–388.

[LM97] Mikhail Lyubich and Yair Minsky, *Laminations in holomorphic dynamics*, J. Differ. Geom. **47** (1997), no. 1, 17–94.

[LMZ94] Alexander Lubotzky, Shahar Mozes, and Robert J. Zimmer, *Superrigidity for the commensurability group of tree lattices*, Comment. Math. Helvetici **69** (1994), 523–548.

[LN02] Yaroslav V. Lavreniuk and Volodymyr V. Nekrashevych, *Rigidity of branch groups acting on rooted trees*, Geom. Dedicata **89** (2002), no. 1, 155–175.

[LPS88] Alexander Lubotzky, Ralph Philips, and Peter Sarnak, *Ramanujan graphs*, Combinatorica **8** (1988), no. 3, 261–277.

[Lys85]   Igor G. Lysionok, *A system of defining relations for the Grigorchuk group*, Mat. Zametki **38** (1985), 503–511.
[Mal49]   A. I. Malcev, *On a class of homogeneous spaces*, Izv. Akad. Nauk SSSR Ser. Mat. **13** (1949), 9–32.
[Mer83]   Yuriĭ I. Merzlyakov, *Infinite finitely generated periodic groups*, Dokl. Akad. Nauk SSSR **268** (1983), no. 4, 803–805.
[Mil99]   John W. Milnor, *Dynamics in one complex variable. Introductory lectures*, Wiesbaden: Vieweg, 1999.
[MNS00]   Olga Macedońska, Volodymyr V. Nekrashevych, and Vitaliĭ I. Sushchansky, *Commensurators of groups and reversible automata*, Dopov. Nats. Akad. Nauk Ukr., Mat. Pryr. Tekh. Nauky (2000), no. 12, 36–39.
[Nek99]   Volodymyr V. Nekrashevych, *Uniformly bounded spaces*, Voprosy Algebry **14** (1999), 47–97.
[Nek00]   _____, *Stabilizers of transitive actions on locally finite graphs*, Int. J. of Algebra and Computation **10** (2000), no. 5, 591–602.
[Nek02]   _____, *Virtual endomorphisms of groups*, Algebra and Discrete Mathematics **1** (2002), no. 1, 96–136.
[Nek04]   _____, *Cuntz-Pimsner algebras of group actions*, Journal of Operator Theory **52** (2004), no. 2, 223–249.
[Neu86]   Peter M. Neumann, *Some questions of Edjvet and Pride about infinite groups*, Illinois J. Math. **30** (1986), no. 2, 301–316.
[NS04]   Volodymyr Nekrashevych and Said Sidki, *Automorphisms of the binary tree: state-closed subgroups and dynamics of 1/2-endomorphisms*, Groups: Topological, Combinatorial and Arithmetic Aspects (T. W. Müller, ed.), LMS Lecture Notes Series, vol. 311, 2004, pp. 375–404.
[Oli98]   Andrij S. Oliĭnyk, *Free groups of automatic permutations*, Dop. NAS Ukraine (1998), no. 7, 40–44 (in Ukrainian).
[Oli99]   _____, *Free products of C_2 as groups of finitely automatic permutations*, Voprosy Algebry (Gomel) **14** (1999), 158–165.
[Pil00]   Kevin M. Pilgrim, *Dessins d'enfants and Hubbard trees*, Ann. Sci. École Norm. Sup. (4) **33** (2000), no. 5, 671–693.
[Pil03a] _____, *An algebraic formulation of Thurston's combinatorial equivalence*, Proc. Amer. Math. Soc. **131** (2003), no. 11, 3527–3534.
[Pil03b] _____, *Combinations of complex dynamical systems*, Lecture Notes in Mathematics, vol. 1827, Springer, 2003.
[Pil04]   _____, *A Hurwitz-like classification of Thurston combinatorial classes*, Osaka J. Math. **41** (2004), 131–143.
[Poi93]   Alfredo Poirier, *On Post Critically Finite Polynomials. Part One: Critical Portraits*, arXiv:math.DS/9305207 v1, 1993.
[Pri80]   Stephen J. Pride, *The concept of "largeness" in group theory*, Word Problems II (S. I. Adian, W. W. Boone, and G. Higman, eds.), Studies in Logic and Foundations of Math., 95, North-Holland Publishing Company, 1980, pp. 299–335.
[Roe03]   John Roe, *Lectures on coarse geometry*, University Lecture Series, vol. 31, American Mathematical Society, Providence, Rhode Island, 2003.
[Röv02]   Claas E. Röver, *Commensurators of groups acting on rooted trees*, Geom. Dedicata **94** (2002), 45–61.
[Roz96]   A. V. Rozhkov, *Finiteness conditions in automorphism groups of trees*, Cheliabinsk, 1996, Habilitation thesis.
[Rub89]   Matatyahu Rubin, *On the reconstruction of topological spaces from their groups of homeomorphisms*, Trans. Amer. Math. Soc. **312** (1989), no. 2, 487–538.
[Sab97]   Christophe Sabot, *Existence and uniqueness of diffusions of finitely ramified self-similar fractals*, Ann. Sci. Éc. Norm. Supér., IV. Sér. **30** (1997), no. 5, 605–673.
[Sat56]   Ichiro Satake, *On a generalization of the notion of a manifold*, Proc. Nat. Acad. Sci. U.S.A. **42** (1956), 359–363.
[Sco83]   Peter Scott, *The geometries of 3-manifolds*, Bull. London Math. Soc. **15** (1983), 401–487.
[Ser80]   Jean-Pierre Serre, *Trees*, New York: Springer-Verlag, 1980.

[Shu69]   Michael Shub, *Endomorphisms of compact differentiable manifolds*, Am. J. Math. **91** (1969), 175–199.

[Shu70]   _____, *Expanding maps*, Global Analysis, Proc. Sympos. Pure Math., vol. 14, American Math. Soc., Providence, Rhode Island, 1970, pp. 273–276.

[Sid87a]   Said N. Sidki, *On a 2-generated infinite 3-group: subgroups and automorphisms*, J. Algebra **110** (1987), no. 1, 24–55.

[Sid87b]   _____, *On a 2-generated infinite 3-group: the presentation problem*, J. Algebra **110** (1987), no. 1, 13–23.

[Sid97]   _____, *A primitive ring associated to a Burnside 3-group*, J. London Math. Soc. (2) **55** (1997), 55–64.

[Sid98]   _____, *Regular trees and their automorphisms*, Monografias de Matematica, vol. 56, IMPA, Rio de Janeiro, 1998.

[Sid00]   _____, *Automorphisms of one-rooted trees: growth, circuit structure and acyclicity*, J. of Mathematical Sciences (New York) **100** (2000), no. 1, 1925–1943.

[Sid04a]   _____, *Finite automata of polynomial growth do not generate a free group*, Geom. Dedicata **108** (2004), 193–204.

[Sid04b]   _____, *Tree-wreathing applied to generation of groups by finite automata* (preprint), 2004.

[SS]   Pedro V. Silva and Benjamin Steinberg, *On a class of automata groups generalizing lamplighter groups*, to appear in Internat. J. Algebra Comput.

[Sus79]   Vitaliĭ I. Sushchansky, *Periodic permutation p-groups and the unrestricted Burnside problem*, DAN SSSR. **247** (1979), no. 3, 557–562 (in Russian).

[Sus98]   _____, *Groups of automatic permutations*, Dop. NAN Ukrainy (1998), no. 6, 47–51 (in Ukrainian).

[Sus99]   _____, *Groups of finitely automatic permutations*, Dop. NAN Ukrainy (1999), no. 2, 29–32 (in Ukrainian).

[SW03]   Said N. Sidki and John S. Wilson, *Free subgroups of branch groups*, Arch. Math. **80** (2003), 458–463.

[Tan92]   Lei Tan, *Matings of quadratic polynomials*, Ergodic Theory Dynam. Systems **12** (1992), no. 3, 589–620.

[Thu89]   William P. Thurston, *Groups, tilings and finite state automata* (AMS Colloqium Lecture Notes), 1989.

[Thu90]   _____, *Three-dimensional geometry and topology*, Univ. of Minnesota Geometry Center preprint, 1990.

[Vin95]   Andrew Vince, *Rep-tiling Euclidean space*, Aequationes Mathematicae **50** (1995), 191–213.

[Vin00]   _____, *Digit tiling of Euclidean space*, Directions in Mathematical Quasicrystals, Amer. Math. Soc., Providence, RI, 2000, pp. 329–370.

[Wil71]   John S. Wilson, *Groups with every proper quotient finite*, Math. Proc. Cambridge Philos. Soc. **69** (1971), 373–391.

[Wil00]   _____, *On just infinite abstract and profinite groups*, New Horizons in pro-p Groups (Aner Shalev, Marcus P. F. du Sautoy, and Dan Segal, eds.), Progress in Mathematics, vol. 184, Birkhäuser Verlag, Basel, 2000, pp. 181–203.

[Wil04a]   _____, *Further groups that do not have uniformly exponential growth*, Journal of Algebra **279** (2004), 292–301.

[Wil04b]   _____, *On exponential growth and uniform exponential growth for groups*, Inventiones Mathematicae **155** (2004), 287–303.

[Yac73]   M. V. Yacobson, *On the question of topological classification of rational mappings of the Riemann sphere*, Uspekhi Mat. Nauk **28** (1973), no. 2, 247–248.

[Yac80]   _____, *Markov partitions for rational endomorphisms of the Riemann sphere*, Multicomponent Random Systems, Dekker, New York, 1980, pp. 381–396.

Index

action
   contracting, 57
   defined by a bimodule and a basis, 35
   finite-state, 11
   fractal, 45
   level-transitive, 2
   monodromy, 136
   recurrent, 45
   self-similar, 10
   standard, 142, 143
adding machine, 16, 196
airplane, 116
asymptotic equivalence, 73
atlas, 121
automaton, 4
   bi-reversible, 23
   complete, 11
   dual, 6
   invertible, 8
   kneading, 187
   planar, 196
   reduced, 8

bad isotropy groups, 201
basilica, 209
basis of a bimodule, 32
Belyi polynomial, 210
Bernoulli measure, 50
bimodule
   $\phi(G)G$, 39
   associated to a self-covering, 140
   associated to an action, 32
   covering, 31
   d-fold, 31
   hyperbolic, 59
   irreducible, 38
   over algebras, 40
   permutational, 31
boundary
   of a hyperbolic space, 101
   of a rooted tree, 1, 50
bounded automatic transformation, 106
branched covering, 174

Chebyshev polynomials, 211
cocycle, 125
covering defined by a cocycle, 126
covering of orbispaces, 124
critical portrait, 190
cycle diagram, 185
cycle graph, 186

DBP, 210
depth of a finitary automorphism, 104
digit system, 45
digit tile, 78
Douady rabbit, 116
dragon curve, 172

embedding of orbispaces, 123
equivalence of groupoids, 120
Euclidean orbifold, 170
Euler characteristic, 170, 178
expanding self-covering, 149
extended ray, 194
external ray, 194

Fabrykowski-Gupta group, 213
Fatou component, 193
Fatou set, 193
finitary automorphism, 104

$\mathcal{G}$-path, 131
$\mathcal{G}$-set, 118
graded covering, 125
Grigorchuk group, 13
Grigorchuk groups G_w, 55
Gromov-hyperbolic space, 99
group
   of A-adic vectors, 49
   basilica, 209
   of bounded automata, 104
   branch, 3
   contracting, 57
   Fabrykowski-Gupta, 213
   of finitary automorphisms, 104
   of finite automata, 8
   finite-state, 11

of functionally recursive automorphisms, 12
generated by an automaton, 12
Grigorchuk, 13
Gupta-Sidki, 18
G_w, 55
Heisenberg, 173
of intermediate growth, 213
isotropy, 118
iterated monodromy, 137
just-infinite, 17
lamplighter, 22
level-transitive, 2
profinite iterated monodromy, 137
recurrent, 45
regular branch, 66
self-similar, 10
weakly branch, 3
groupoid, 117
of action, 118
of changes of charts, 120, 121
étale, 117
free, 120
of germs, 118
proper, 119
Gupta-Sidki group, 18

Heighway dragon, 172
Heisenerg group, 173
hyperbolic space, 99

index of a virtual homomorphism, 37
internal ray, 194
isotropy group, 118
iterated monodromy action, 138, 140
iterated monodromy group, 137, 138

Julia set, 149

kneading automaton, 187, 191
kneading sequence, 203, 207

L-presentation, 66
Lattès examples, 171
length structure, 148
level-transitive
automorphism, 25
group, 2
tree, 50
limit dynamical system, 93
limit G-space, 73
limit solenoid, 156
limit space
$\mathcal{J}_G$, 92
$\mathcal{X}_G$, 73
linear recursion, 41
localization of a groupoid, 121

mating, 215
monodromy action, 136

Moore diagram, 5
dual, 7

nucleus, 57

odometer, 16
open map of orbispaces, 122
open set condition, 80
orbifold, 122
orbispace, 121
orbit of a groupoid, 118
output function, 4

parameter ray, 205
partial self-covering, 127
path in an orbispace, 131
path-connected
groupoid, 132
orbispace, 132
portrait
of an automorphism, 4
critical, 190
post-critical point, 174
post-critically finite, 174, 175
pseudogroup, 117
proper, 119
pseudogroup of changes of charts, 121
pull back of a partial self-covering, 127
pull-back, 126

quasi-isometry, 97

rabbit, 116
ray
extended, 194
external, 194
internal, 194
supporting, 194
restriction, 4, 35
of a groupoid, 120, 121
of a partial self-covering, 127
rigid orbispace, 122
rigid stabilizer, 2
rooted automorphism, 10

saturated isomorphism, 54
Schreier graph, 94
sectors (of a critical portrait), 190
self-covering, 127
self-similarity graph, 97
shift, 93
Sierpinski gasket, 112
skew product, 125
spider, 190
stabilizer
of a level, 2
rigid, 2
of a vertex, 2
standard action, 142, 143
sub-hyperbolic rational function, 176

subgroup
    ϕ-invariant, 41
    ϕ-semi-invariant, 41
    self-similar, 42, 96
supporting ray, 194

tame twin dragon, 167
tensor power of an action, 35
tensor product of bimodules, 33
Thurston map, 175
Thurston orbifold, 175
tile, 78
tile diagram, 110
topological polynomial, 190
transition function, 4
tree-like set of permutations, 185
twin dragon, 167

underlying space, 121
uniformizing map, 121
union of atlases, 122
universal covering, 133

virtual endomorphism, 37
    associated to an action, 38
    associated to a bimodule, 38
    associated to a self-covering, 141
virtual homomorphism, 37

wreath product, 9
wreath recursion, 10, 33

Titles in This Series

117 **Volodymyr Nekrashevych**, Self-similar groups, 2005

116 **Alexander Koldobsky**, Fourier analysis in convex geometry, 2005

115 **Carlos Julio Moreno**, Advanced analytic number theory: L-functions, 2005

114 **Gregory F. Lawler**, Conformally invariant processes in the plane, 2005

113 **William G. Dwyer, Philip S. Hirschhorn, Daniel M. Kan, and Jeffrey H. Smith**, Homotopy limit functors on model categories and homotopical categories, 2004

112 **Michael Aschbacher and Stephen D. Smith**, The classification of quasithin groups II. Main theorems: The classification of simple QTKE-groups, 2004

111 **Michael Aschbacher and Stephen D. Smith**, The classification of quasithin groups I. Structure of strongly quasithin K-groups, 2004

110 **Bennett Chow and Dan Knopf**, The Ricci flow: An introduction, 2004

109 **Goro Shimura**, Arithmetic and analytic theories of quadratic forms and Clifford groups, 2004

108 **Michael Farber**, Topology of closed one-forms, 2004

107 **Jens Carsten Jantzen**, Representations of algebraic groups, 2003

106 **Hiroyuki Yoshida**, Absolute CM-periods, 2003

105 **Charalambos D. Aliprantis and Owen Burkinshaw**, Locally solid Riesz spaces with applications to economics, second edition, 2003

104 **Graham Everest, Alf van der Poorten, Igor Shparlinski, and Thomas Ward**, Recurrence sequences, 2003

103 **Octav Cornea, Gregory Lupton, John Oprea, and Daniel Tanré**, Lusternik-Schnirelmann category, 2003

102 **Linda Rass and John Radcliffe**, Spatial deterministic epidemics, 2003

101 **Eli Glasner**, Ergodic theory via joinings, 2003

100 **Peter Duren and Alexander Schuster**, Bergman spaces, 2004

99 **Philip S. Hirschhorn**, Model categories and their localizations, 2003

98 **Victor Guillemin, Viktor Ginzburg, and Yael Karshon**, Moment maps, cobordisms, and Hamiltonian group actions, 2002

97 **V. A. Vassiliev**, Applied Picard-Lefschetz theory, 2002

96 **Martin Markl, Steve Shnider, and Jim Stasheff**, Operads in algebra, topology and physics, 2002

95 **Seiichi Kamada**, Braid and knot theory in dimension four, 2002

94 **Mara D. Neusel and Larry Smith**, Invariant theory of finite groups, 2002

93 **Nikolai K. Nikolski**, Operators, functions, and systems: An easy reading. Volume 2: Model operators and systems, 2002

92 **Nikolai K. Nikolski**, Operators, functions, and systems: An easy reading. Volume 1: Hardy, Hankel, and Toeplitz, 2002

91 **Richard Montgomery**, A tour of subriemannian geometries, their geodesics and applications, 2002

90 **Christian Gérard and Izabella Łaba**, Multiparticle quantum scattering in constant magnetic fields, 2002

89 **Michel Ledoux**, The concentration of measure phenomenon, 2001

88 **Edward Frenkel and David Ben-Zvi**, Vertex algebras and algebraic curves, second edition, 2004

87 **Bruno Poizat**, Stable groups, 2001

86 **Stanley N. Burris**, Number theoretic density and logical limit laws, 2001

85 **V. A. Kozlov, V. G. Maz'ya, and J. Rossmann**, Spectral problems associated with corner singularities of solutions to elliptic equations, 2001

TITLES IN THIS SERIES

84 **László Fuchs and Luigi Salce,** Modules over non-Noetherian domains, 2001
83 **Sigurdur Helgason,** Groups and geometric analysis: Integral geometry, invariant differential operators, and spherical functions, 2000
82 **Goro Shimura,** Arithmeticity in the theory of automorphic forms, 2000
81 **Michael E. Taylor,** Tools for PDE: Pseudodifferential operators, paradifferential operators, and layer potentials, 2000
80 **Lindsay N. Childs,** Taming wild extensions: Hopf algebras and local Galois module theory, 2000
79 **Joseph A. Cima and William T. Ross,** The backward shift on the Hardy space, 2000
78 **Boris A. Kupershmidt,** KP or mKP: Noncommutative mathematics of Lagrangian, Hamiltonian, and integrable systems, 2000
77 **Fumio Hiai and Dénes Petz,** The semicircle law, free random variables and entropy, 2000
76 **Frederick P. Gardiner and Nikola Lakic,** Quasiconformal Teichmüller theory, 2000
75 **Greg Hjorth,** Classification and orbit equivalence relations, 2000
74 **Daniel W. Stroock,** An introduction to the analysis of paths on a Riemannian manifold, 2000
73 **John Locker,** Spectral theory of non-self-adjoint two-point differential operators, 2000
72 **Gerald Teschl,** Jacobi operators and completely integrable nonlinear lattices, 1999
71 **Lajos Pukánszky,** Characters of connected Lie groups, 1999
70 **Carmen Chicone and Yuri Latushkin,** Evolution semigroups in dynamical systems and differential equations, 1999
69 **C. T. C. Wall (A. A. Ranicki, Editor),** Surgery on compact manifolds, second edition, 1999
68 **David A. Cox and Sheldon Katz,** Mirror symmetry and algebraic geometry, 1999
67 **A. Borel and N. Wallach,** Continuous cohomology, discrete subgroups, and representations of reductive groups, second edition, 2000
66 **Yu. Ilyashenko and Weigu Li,** Nonlocal bifurcations, 1999
65 **Carl Faith,** Rings and things and a fine array of twentieth century associative algebra, 1999
64 **Rene A. Carmona and Boris Rozovskii, Editors,** Stochastic partial differential equations: Six perspectives, 1999
63 **Mark Hovey,** Model categories, 1999
62 **Vladimir I. Bogachev,** Gaussian measures, 1998
61 **W. Norrie Everitt and Lawrence Markus,** Boundary value problems and symplectic algebra for ordinary differential and quasi-differential operators, 1999
60 **Iain Raeburn and Dana P. Williams,** Morita equivalence and continuous-trace C^*-algebras, 1998
59 **Paul Howard and Jean E. Rubin,** Consequences of the axiom of choice, 1998
58 **Pavel I. Etingof, Igor B. Frenkel, and Alexander A. Kirillov, Jr.,** Lectures on representation theory and Knizhnik-Zamolodchikov equations, 1998
57 **Marc Levine,** Mixed motives, 1998
56 **Leonid I. Korogodski and Yan S. Soibelman,** Algebras of functions on quantum groups: Part I, 1998
55 **J. Scott Carter and Masahico Saito,** Knotted surfaces and their diagrams, 1998

For a complete list of titles in this series, visit the
AMS Bookstore at **www.ams.org/bookstore/**.